Lecture Notes in Computer Science   10445

Commenced Publication in 1973
Founding and Former Series Editors:
Gerhard Goos, Juris Hartmanis, and Jan van Leeuwen

Editorial Board

More information about this series at http://www.springer.com/series/7410

Josep Carmona · Gregor Engels
Akhil Kumar (Eds.)

Business Process Management

15th International Conference, BPM 2017
Barcelona, Spain, September 10–15, 2017
Proceedings

 Springer

Editors
Josep Carmona (iD)
Department of Computer Science
Universitat Politècnica de Catalunya
Barcelona
Spain

Gregor Engels
Department of Computer Science
University of Paderborn
Paderborn
Germany

Akhil Kumar (iD)
Department of Supply Chain
 and Information Systems
Pennsylvania State University
University Park, PA
USA

ISSN 0302-9743   ISSN 1611-3349 (electronic)
Lecture Notes in Computer Science
ISBN 978-3-319-64999-3   ISBN 978-3-319-65000-5 (eBook)
DOI 10.1007/978-3-319-65000-5

Library of Congress Control Number: 2017948188

LNCS Sublibrary: SL4 – Security and Cryptology

Printed on acid-free paper

This Springer imprint is published by Springer Nature
The registered company is Springer International Publishing AG
The registered company address is: Gewerbestrasse 11, 6330 Cham, Switzerland

Preface

We are pleased to present to you the proceedings of the 15th edition of BPM, which was held in Barcelona during September 10–15, 2017. That this conference is in its 15th year is a clear sign of the maturity of the business process management (BPM) area. During this time the conference has clearly established itself as the most important academic event in BPM. It is the premium forum for researchers, practitioners, and developers in this area.

The Program Committee (PC) comprised 21 senior PC members and 103 regular PC members. We received 116 full paper submissions. Out of all submissions, 19 papers were accepted or conditionally accepted, with an acceptance rate of 16%. This conference has very rigorous reviewing criteria. Each paper is reviewed by a team comprising a senior PC and four regular PC members who engage in a discussion phase after the initial reviews are prepared. The authors receive four review reports, and a meta-review that summarizes the reviews and the discussion. In the end, a conference program is only as good as the PC members. The program chairs are, for the most part, coordinators of the review process and messengers who relay the recommendations of the review team to the authors. We were very fortunate to have an excellent set of PC members who were very diligent and conscientious. We cannot thank them enough for their support and cooperation. The review process was conducted entirely on the EasyChair conference management system that was an invaluable resource for us.

There were six main sessions into which the papers were organized, spanning process modeling, process mining or discovery, process knowledge, and decisions and understanding. There were also papers that relate to the novel Blockchain paradigm and business process as a service. The conference program was enriched by talks from three distinguished keynote speakers: Alan Brown (University of Surrey), Miguel Valdés (Bonitasoft) and Mathias Weske (HPI Potsdam); along with panel and tutorial sessions. In conjunction with the main program of BPM, there was an industry track and a BPM Forum, a sub-track of the conference to host innovative research which has potential to stimulate discussions. The papers included in the industry track and the forum for presentation at the BPM Conference will be published in a separate volume.

We are grateful for the generous support of our sponsors: Signavio, Celonis, IBM, Diputacio de Tarragona, MyInvenio, DCR, Bizagi, CA Technologies, Mysphera and Springer. We very much hope you will enjoy reading the research papers in this volume.

September 2017

Josep Carmona
Gregor Engels
Akhil Kumar

Organization

BPM 2017 was organized by the Universitat Politècnica de Catalunya, and took place in Barcelona, Spain.

Steering Committee

Wil van der Aalst (Chair)	Eindhoven University of Technology, The Netherlands
Boualem Benatallah	University of New South Wales, Australia
Jörg Desel	University of Hagen, Germany
Schahram Dustdar	Vienna University of Technology, Austria
Marlon Dumas	University of Tartu, Estonia
Manfred Reichert	University of Ulm, Germany
Stefanie Rinderle-Ma	University of Vienna, Austria
Barbara Weber	Technical University of Denmark, Denmark
Mathias Weske	HPI, University of Potsdam, Germany
Michael zur Muehlen	Stevens Institute of Technology, USA

Executive Committee

Conference Chair

Josep Carmona	Universitat Politècnica de Catalunya, Spain

Program Chairs

Josep Carmona	Universitat Politècnica de Catalunya, Spain
Gregor Engels	Paderborn University, Germany
Akhil Kumar	Penn State University, USA

Industry Chairs

Marco Brambilla	Politecnico Milano, Italy
Thomas Hildebrandt	IT University of Copenhagen, Denmark
Victor Muntès	CA Technologies, Spain
Darius Silingas	No Magic Europe and ISM UME, Lithuania

Workshops

Matthias Weidlich	Humboldt-Universität zu Berlin, Germany
Ernest Teniente	Universitat Politècnica de Catalunya, Spain

Tutorial and Panel Chairs

Joaquin Ezpeleta   University of Zaragoza, Spain
Dirk Fahland   Eindhoven University of Technology, The Netherlands
Barbara Weber   Technical University of Denmark, Denmark

Demo Chairs

Robert Clarisó   Universitat Oberta de Catalunya, Spain
Henrik Leopold   VU University Amsterdam, The Netherlands

Doctoral Consortium Chairs

Antonio Ruiz Cortés   University of Seville, Spain
Mathias Weske   HPI, University of Potsdam, Germany

Publicity Chairs

Jordi Cabot   Open University of Catalonia, Spain
Marcos Sepúlveda   Pontificia Universidad Católica de Chile, Chile
Marco Montali   Free University of Bozen-Bolzano, Italy

Sponsorship Chairs

Carlos Fernandez-Llatas   Universidad Politecnica de Valencia, Spain
Pedro Álvarez   University of Zaragoza, Zaragoza
Rubén Mondéjar   Universitat Rovira i Virgili, Spain

Co-located Events Chairs

Manuel Lama   University of Santiago de Compostela, Spain
Alberto Manuel   Microsoft, Lisbon
Antonio Valle   G2, Spain

Web and Social Media Chairs

Jorge Munoz-Gama   Pontificia Universidad Católica de Chile, Chile
Andrea Burattin   University of Innsbruck, Austria

Proceedings Chair

Alexander Teetz   Paderborn University, Germany

Senior Program Committee

Marlon Dumas   University of Tartu, Estonia
Schahram Dustdar   TU Wien, Austria
Avigdor Gal   Technion, Israel
Richard Hull   IBM T.J. Watson Research Center, USA
Fabrizio Maria Maggi   University of Tartu, Estonia
Massimo Mecella   Sapienza Università di Roma, Italy

Jan Mendling	Wirtschaftsuniversität Wien, Austria
Marco Montali	Free University of Bozen-Bolzano, Italy
Artem Polyvyanyy	Queensland University of Technology, Australia
Manfred Reichert	University of Ulm, Germany
Hajo A. Reijers	Vrije Universiteit Amsterdam, The Netherlands
Stefanie Rinderle-Ma	University of Vienna, Austria
Michael Rosemann	Queensland University of Technology, Australia
Antonio Ruiz-Cortés	University of Seville, Spain
Pnina Soffer	University of Haifa, Israel
Jianwen Su	University of California at Santa Barbara, USA
Boudewijn Van Dongen	Eindhoven University of Technology, The Netherlands
Irene Vanderfeesten	Eindhoven University of Technology, The Netherlands
Barbara Weber	Technical University of Denmark, Denmark
Matthias Weidlich	Humboldt-Universität zu Berlin, Germany
Mathias Weske	HPI, University of Potsdam, Germany

Program Committee

Mari Abe	IBM Research, Japan
Shivali Agarwal	IBM, India Research Lab, India
Ahmed Awad	Cairo University, Egypt
Hyerim Bae	Pusan National University, South Korea
Bart Baesens	KU Leuven, Belgium
Seyed-Mehdi-Reza Beheshti	University of New South Wales, Australia
Boualem Benatallah	University of New South Wales, Australia
Giorgio Bruno	Politecnico di Torino, Italy
Joos Buijs	Eindhoven University of Technology, The Netherlands
Andrea Burattin	University of Innsbruck, Austria
Jorge Cardoso	University of Coimbra, Portugal
Fabio Casati	University of Trento, Italy
Jan Claes	Ghent University, Belgium
Florian Daniel	Politecnico di Milano, Italy
Massimiliano de Leoni	Eindhoven University of Technology, The Netherlands
Jochen De Weerdt	KU Leuven, Belgium
Patrick Delfmann	European Research Center for Information Systems (ERCIS), Germany
Jörg Desel	University of Hagen, Germany
Alin Deutsch	University of California San Diego, USA
Chiara Di Francescomarino	Fondazione Bruno Kessler-IRST, Italy
Remco Dijkman	Eindhoven University of Technology, The Netherlands
Dirk Draheim	Tallinn University of Technology, Estonia
Johann Eder	Alpen Adria Universität Klagenfurt, Austria
Rik Eshuis	Eindhoven University of Technology, The Netherlands
Joerg Evermann	Memorial University of Newfoundland, Canada
Dirk Fahland	Technische Universiteit Eindhoven, The Netherlands
Marcelo Fantinato	University of São Paulo, Brazil

Peter Fettke   DFKI/Saarland University, Germany
Hans-Georg Fill   University of Vienna, Austria
Walid Gaaloul   Télécom SudParis, France
Luciano García-Bañuelos   University of Tartu, Estonia
Christian Gerth   Osnabrück University of Applied Sciences, Germany
Chiara Ghidini   FBK-irst, Italy
María Teresa Gómez-López   University of Seville, Spain
Guido Governatori   Data61, Australia
Sven Graupner   Hewlett-Packard Laboratories, USA
Paul Grefen   Eindhoven University of Technology, The Netherlands
Daniela Grigori   University of Paris-Dauphine, France
Thomas Hildebrandt   IT University of Copenhagen, Denmark
Mieke Jans   Hasselt University, Belgium
Anup Kalia   IBM T.J. Watson Research Center, USA
Dimka Karastoyanova   Kühne Logistics University, Germany
Ekkart Kindler   Technical University of Denmark, Denmark
Agnes Koschmider   Karlsruhe Institute of Technology, Germany
John Krogstie   Norwegian University of Science and Technology,
                                  Norway
Jochen Kuester   Bielefeld University of Applied Sciences, Bielefeld
Marcello La Rosa   Queensland University of Technology, Australia
Geetika Lakshmanan   IBM T.J. Watson Research Center, USA
Manuel Lama Penin   University of Santiago de Compostela, Spain
Alexei Lapouchnian   University of Toronto, Canada
Ralf Laue   University of Applied Sciences Zwickau, Germany
Henrik Leopold   VU University Amsterdam, The Netherlands
Rong Liu   IBM Research, USA
Irina Lomazova   National Research University Higher School
                                  of Economics, Russia
Peter Loos   DFKI/Saarland University, Germany
Heiko Ludwig   IBM Research, USA
Hamid Motahari   IBM Research, USA
Juergen Muench   Reutlingen University, Germany
John Mylopoulos   University of Toronto, Canada
Nanjangud Narendra   Ericsson Research Bangalore, India
Selmin Nurcan   Université Paris 1 Panthéon-Sorbonne, France
Hye-Young Paik   University of New South Wales, Australia
Oscar Pastor Lopez   Universitat Politecnica de Valencia, Spain
Dietmar Pfahl   University of Tartu, Estonia
Geert Poels   Ghent University, Belgium
Frank Puhlmann   Bosch Software Innovations, Germany
Mu Qiao   IBM Almaden Research Center, USA
Jan Recker   Queensland University of Technology, Australia
Manuel Resinas   University of Seville, Spain
Maximilian Roeglinger   FIM Research Center, Germany
Shazia Sadiq   The University of Queensland, Australia

Flavia Santoro   Federal University of the State of Rio de Janeiro, Brazil
Rainer Schmidt   Munich University of Applied Sciences, Germany
Heiko Schuldt   University of Basel, Switzerland
Marcos Sepúlveda   Pontificia Universidad Católica de Chile, Chile
Quan Z. Sheng   Macquarie University, Australia
Renuka Sindhgatta   IBM Research, India
Sergey Smirnov   SAP Research, Germany
Marc Sole   CA Strategic Research Labs, CA Technologies, Spain
Minseok Song   Pohang University of Science and Technology,
                             South Korea
Harald Störrle   Danmarks Tekniske Universitet, Denmark
Heiner Stuckenschmidt   University of Mannheim, Germany
Keith Swenson   Fujitsu, USA
Samir Tata   IBM Research, USA
Pankaj Telang   SAS Institute Inc., USA
Ernest Teniente   Universitat Politècnica de Catalunya, Spain
Arthur Ter Hofstede   Queensland University of Technology, Australia
Lucinéia Heloisa Thom   Federal University of Rio Grande do Sul, Brazil
Farouk Toumani   LIMOS/Blaise Pascal University, France
Peter Trkman   University of Ljubljana, Slovenia
Roman Vaculín   IBM T.J. Watson Research Center, USA
Wil van der Aalst   Eindhoven University of Technology, The Netherlands
Amy Van Looy   Ghent University, Belgium
Jan Vanthienen   KU Leuven, Belgium
Hagen Voelzer   IBM Research Zurich, Switzerland
Jianmin Wang   Tsinghua University, China
Ingo Weber   Data61 CSIRO, Australia
Lijie Wen   Tsinghua University, China
Karsten Wolf   Universität Rostock, Germany
Moe Wynn   Queensland University of Technology, Australia
Liang Zhang   Fudan University, China

Additional Reviewers

Alexander Norta   Javier de San Pedro   Riccardo De Masellis
Bernardo Nugroho Yahya   Johannes De Smedt   Rick Gilsing
Carlos Rodriguez   Julius Köpke   Seyed-Mehdi-Reza
Chun Ouyang   Marigianna Skouradaki   Beheshti
David Sanchez-Charles   Mauro Dragoni   Toon Jouck
Erik Proper   Mirela Madalina Botezatu   Sander Peters
Gert Janssenswillen   Montserrat Estañol   Wasana Bandara
Jaehun Park   Pavlos Delias

Sponsors

Platinum Sponsor

Gold Sponsor

Gold Sponsor

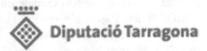

Silver Sponsor

Bronze Sponsor

Bronze Sponsor

Bronze Sponsor

Demo Sponsor

Keynotes

A Leaders Guide to Understanding New Business Models in the Digital Economy

Alan W. Brown

Surrey Centre for the Digital Economy, University of Surrey, Guildford, UK
alan.w.brown@surrey.ac.uk

Many organizations have been preoccupied in recent years with their repeated efforts to upgrade to digital technologies, digital media, and digital delivery channels. However, digital transformation has also opened the opportunity for organizations to question major assumptions about their business model – the users being served, the experiences offered to them, and the most efficient ways to deliver those experiences in a coordinated, consistent and cost-effective way. Consequently, real world thinking and experiences in business model innovation have changed significantly over the past few years. How can business model innovation can keep pace with changing business needs? This session explores business model innovation in digital transformation and presents a range of simple frameworks and models that can be used to help explain core concepts of business model innovation, and techniques that help accelerate business model experimentation

Intelligent Continuous Improvement, When BPM Meets AI

Miguel Valdés

Bonitasoft
miguel.valdes@bonitasoft.com

Artificial Intelligence (AI) technologies are evolving faster than ever thanks to the maturity of cloud computing, BigData and the accessibility of predictive and machine learning algorithms and frameworks. But, is BPM software ready to embrace AI? Through continued modernisation BPM platforms goes beyond traditional process automation and optimisation use cases to play a key role in digital transformation in organisations of all sizes. Modern BPM applications requirements include advanced end user interfaces (UIs), access to big volumes of business data and real time updates of those processes, UIs and data. AI will be the next major wave of innovation in BPM. In this session we will discuss the challenges and opportunities involved in the shift towards the use of AI technologies in BPM. We will particularly cover uses cases in which AI enables intelligent continuous improvement of business processes and BPM applications. We will also discuss about pros and cons of different AI technologies when it relates to BPM.

BPM: Reflections on a Broad Discipline

Mathias Weske

University of Potsdam, Germany
weske@hpi.de

BPM is a broad discipline. Topics addressed in business process management range from formal methods in computer science to behavioral science methods in management. These distant points in the spectrum are linked by information systems engineering methods. Computer science, information systems engineering, and management share business processes as a common interest, as a joint research area. There are few disciplines that share this breadth. Under a BPM umbrella, not only different research topics are addressed, but also different research methods are employed that ask for different evaluation criteria.

In this talk, the breadth of the BPM discipline is illustrated by highlighting research results from its sub-fields and discussing their respective research objectives. In the second part of the talk, the implications of these observations on the BPM conference are discussed. The goal of our conference series has always been to provide a forum for all aspects of BPM research. Despite this claim the center of gravity has been initially in formal aspects of business process models and, more recently, in business process intelligence. While these research areas will continue to be cornerstones of our conference, an important area of business process management is not well represented: management aspects that focus on the interplay between process technology, persons, and organizations. These topics clearly deserve more attention in a conference on business process – management.

To further develop the conference and to match in the conference structure the breadth of the field, the Steering Committee proposes a novel structure of BPM conferences. This structure is based on different tracks, each of which has a track chair, a dedicated program committee, and specific evaluation criteria.

By this new structure we hope to broaden the BPM community and, ultimately, to be a forum for all aspects of the broad business process management discipline.

Contents

Decisions and Understanding

Process Knowledge

Process Mining 2

Process Modeling

Temporal Network Representation of Event Logs for Improved Performance Modelling in Business Processes

Arik Senderovich[1]([⊠]), Matthias Weidlich[2], and Avigdor Gal[1]

[1] Technion – Israel Institute of Technology, Haifa, Israel
sariks@technion.ac.il, avigal@ie.technion.ac.il
[2] Humboldt-Universität zu Berlin, Berlin, Germany
matthias.weidlich@hu-berlin.de

Abstract. Analysing performance of business processes is an important vehicle to improve their operation. Specifically, an accurate assessment of sojourn times and remaining times enables bottleneck analysis and resource planning. Recently, methods to create respective performance models from event logs have been proposed. These works are severely limited, though: They either consider control-flow and performance information separately, or rely on an ad-hoc selection of temporal relations between events. In this paper, we introduce the Temporal Network Representation (TNR) of a log, based on Allen's interval algebra, as a complete temporal representation of a log, which enables simultaneous discovery of control-flow and performance information. We demonstrate the usefulness of the TNR for detecting (unrecorded) delays and for probabilistic mining of variants when modelling the performance of a process. In order to compare different models from the performance perspective, we develop a framework for measuring performance fitness. Under this framework, we provide guarantees that TNR-based process discovery dominates existing techniques in measuring performance characteristics of a process. To illustrate the practical value of the TNR, we evaluate the approach against three real-life datasets. Our experiments show that the TNR yields an improvement in performance fitness over state-of-the-art algorithms.

1 Introduction

Modern process-aware information systems (PAIS) support the design, enactment, and analysis of business processes in various domains [1]. Based on a formalisation of the supported business process in terms of a process model, they control how the execution of a set of activities is coordinated to reach a certain outcome for an instance of the process. The operation of business processes can be improved by modelling their performance. Specifically, an accurate assessment of key performance measures, such as sojourn times and remaining times, enables bottleneck analysis and optimised resource planning [2].

Recently, to enable performance analysis of business processes, methods that construct process models from event logs that contain transactional data have

© Springer International Publishing AG 2017
J. Carmona et al. (Eds.): BPM 2017, LNCS 10445, pp. 3–21, 2017.
DOI: 10.1007/978-3-319-65000-5_1

been proposed [3,4]. Yet, these methods consider control-flow and performance information separately [5–7]. They first create a process model that captures causal dependencies between activities (commonly referred to as *discovery*), which is later annotated with performance details (referred to as *enhancement*). Hence, any bias introduced in control-flow discovery carries over to the performance analysis.

To illustrate the problem implied by this 2-step approach, we consider a claim handling example, where discovery may yield the BPMN model in Fig. 1a. Annotating the model with activity durations, however, does not capture delays between actual activity executions. This potentially yields inaccuracies when conducting performance analysis. That is, once a claim is received (A), a system may automatically fetch previous claims (C). Yet, the plausibility check (B), supposed to be done in parallel, involves a knowledge worker, who is not available immediately. Hence, the start of the activity is delayed (see Fig. 1b). Similarly, after the automatic status update (D), another staff member needs to enter the decision (E), which also introduces a delay.

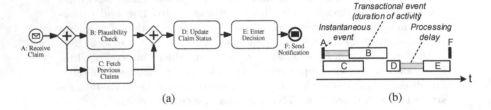

(a)   (b)

Fig. 1. Claim handling process (a); common actual execution of activities (b).

In transactional event logs that record the start *and* end of activity execution, delays are directly visible for individual process instances, as shown in Fig. 1b. However, such an instance may represent noise in the event log, which raises the question of how to consider delays on the model-level. When constructing a process model for performance analysis, the observed delays of individual instances need to be generalised.

The challenge of incorporating delays in the construction of performance models has been recognised in the literature. Specifically, Tsinghua-α [8] and variants of the Inductive Miner [4] incorporate performance details by considering temporal relations between the start and end of activity executions. Yet, these approaches are limited in two ways: (i) They take an ad-hoc decision on the type of temporal relation to consider in model discovery (e.g., to distinguish interleaved and concurrent execution of activities [4]); and (ii) they require a model to represent a single temporal relation per pair of activities (e.g., two activities are *always* interleaved or concurrent [4]).

In this paper, to overcome the above limitations, we introduce the Temporal Network Representation (TNR) of an event log as a formalism that is grounded in Allen's interval algebra [9]. The TNR is a compact representation of all (pairwise) temporal relations between activity executions as observed in the event log. As such, it generalises different notions of dependency graphs commonly used in process model discovery and enables us to incorporate performance information in terms of processing delays in model discovery. Our contributions and the structure of the paper are summarised as follows:

- *The Temporal Network Representation (TNR) of an event log:* Following an introduction of preliminaries (Sect. 2), in Sect. 3, we present the TNR of transactional event logs. The TNR generalises common representations of event logs.
- *Inductive Mining with the TNR:* In Sect. 4, we propose an algorithm to integrate *delay unfolding* in inductive mining, exploiting the TNR to include processing delays explicitly. We then show how the TNR enables *probabilistic variant mining*, which handles noisy event logs, but preserves performance details in the discovered model.
- *Measuring Performance Fitness:* Sect. 5 introduces a framework for measuring performance fitness between an event log and a model. We also show that under this framework, TNR-based inductive mining is guaranteed to discover unbiased models.

To demonstrate the practicality of the TNR, we evaluated our approach with three real-world healthcare datasets. As detailed in Sect. 6, TNR based reasoning yields up-to 40% improvement in performance fitness with respect to existing approaches. Finally, we discuss related work in Sect. 7, before concluding in Sect. 8.

2   Preliminaries

This section reviews preliminaries for our work in terms of event logs, process trees as a formalism for process modelling, and Allen's algebra to reason on temporal intervals.

Event Logs. We adopt a notion of a transactional event log that relates events to their activity labels (activities, for short), start times, and completion times. Let $\mathcal{E}$ be the universe of events produced by an information system and let $\mathcal{A}$ be the set of supported activities. Then, by $e.a \in \mathcal{A}$, $e.s \in \mathbb{R}_0^+$, and $e.c \in \mathbb{R}_0^+$, we denote the activity that corresponds to the event, its start time, and completion time, respectively.

A case $\xi \in 2^{\mathcal{E}} \setminus \emptyset$ is a finite set of events, assuming that no event may occur in more than one case and that a case comprises at least one event. An event log $L \subseteq 2^{\mathcal{E}}$ is a set of cases. Table 1 presents an example event log for the claim handling process in Fig. 1. Note that some of the events are instantaneous (i.e., have a duration of 0). We denote by $A_\xi \subseteq \mathcal{A} \times \mathbb{N}^+$ the multi-set of activities that appear in ξ, namely $A_\xi = \{(e.a, k) | e \in \xi\}$ with k being the frequency of $e.a$ in ξ.

Process Trees. To represent the process executed by an information system, we adopt the notion of a process tree [10] that is enriched with time information.

Table 1. Example event log for the claim handling process.

Case	Activity	Start	Complete	Case	Activity	Start	Complete
1	A: Receive Claim	9:05	9:05	3	B: Plausibility Check	10:25	10:28
1	C: Fetch Previous Claim	9:05	9:10	3	C: Fetch Previous Claim	10:25	10:30
1	B: Plausibility Check	9:08	9:20	2	B: Plausibility Check	10:30	10:55
1	D: Update Claim Status	9:20	9:22	3	D: Update Claim Status	10:30	10:30
1	E: Enter Decision	9:40	12:05	2	D: Update Claim Status	10:55	10:55
2	A: Receive Claim	10:23	10:23	2	E: Enter Decision	11:10	11:28
2	C: Fetch Previous Claim	10:23	10:34	2	F: Send Notification	11:28	11:28
3	A: Receive Claim	10:25	10:25	1	F: Send Notification	12:05	12:05

Traditionally, a process tree encodes the control-flow of a process in terms of its possible traces, i.e., sequences of activity executions. We recall the intuition behind process trees and refer the reader to [10] for a complete formalisation of their syntax and semantics.

An untimed process tree is a rooted tree, in which the leaf nodes are activities in $\mathcal{A}_\tau = \mathcal{A} \cup \{\tau_1, \ldots, \tau_n\}$ with τ_i, $1 \le i \le n$, denoting *silent activities* that cannot be observed during the execution of the process (but which may have different durations, so that they need to be distinguished from one another). All non-leaf nodes are control-flow operators, denoted by $\mathcal{O}$. Common control-flow operators are sequence ($\rightarrow$), exclusive choice ($\times$), concurrency ($\wedge$), interleaving ($\|$) and structured loops ($\circlearrowleft$). Figure 2 shows the process tree for the BPMN model in Fig. 1a. Semantics of a process tree is defined by recursively constructing a set of traces: For a leaf node labelled with $a \in \mathcal{A}$, the set of traces contains a single trace, $\{\langle a \rangle\}$, whereas it contains the empty trace $\{\langle \rangle\}$ for a silent activity. Semantics of a non-leaf node is formalised by a language function that joins the traces of the subtrees of the node. For instance, the set of traces of the exclusive choice operator is given by the union of the trace sets of its children.

We extend process tress by adding durations to leaf nodes. Each activity $a \in \mathcal{A}_\tau$ is assigned a duration of the (potentially silent) execution of a, which comes from a cumulative distribution function (CDF) D_a. This induces a timed semantics of the process tree in terms of sequences of events. From a trace of the untimed process tree, a set of events is constructed by drawing a duration from D_a for each activity $a \in \mathcal{A}_\tau$ of the trace and constructing the start time and

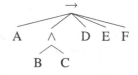

Fig. 2. Process tree of the claim handling process.

completion time as follows: the start time is the completion time of the event for the previous activity in the trace (or 0, if the activity is the first one) and the completion time is the start time plus the duration. This way we model instantaneous activities (with a constant duration of 0) and processing delays (silent activities of a certain duration).

As another extension to the common model of process trees, we consider its enrichment with branching probabilities. For our purposes, it suffices to assign a probability distribution P_o to each n-ary exclusive choice operator $o \in \mathcal{O}$, so that P_o models the occurrence probabilities of the n children of the operator.

Allen's Interval Algebra. To reason about temporal relations of events, Allen presented an interval algebra [9] that defines 13 relations between two intervals. Each of them formalises a different partial order of the start and completion times of interval events, see Fig. 3. Instantiating these relations for the above notion of events, for instance, a pair of events $x, y \in \mathcal{E}$ is in the *overlaps* relation, if and only if, $x.s < y.s < x.c < y.c$.

The interval relations are mutually exclusive and partition the Cartesian product of events. As shown in Fig. 3, each relation between two events, except *equals*, has a counterpart that holds for the reversed pair of events. To avoid this kind of redundancy, in the remainder, we consider the following 7 out of the 13 relations: *precedes, meets, overlaps, is finished by, contains, starts,* and *equals.* The set of these relations is denoted by $\mathcal{R}$.

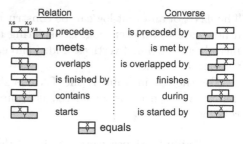

Fig. 3. Allen's interval relations, see [11].

3 The Temporal Network Representation of an Event Log

This section introduces our notion of the temporal network representation (TNR) of an event log. It is based on Allen's interval relations to capture the temporal information in the log. We further discuss how the TNR can be collapsed to obtain commonly used models of event logs, namely the directly-follows graph and the concurrency graph.

3.1 Definition

The TNR is grounded in the notion of *temporal evidence*, which is needed when lifting the interval relations from events to activities. Since in an event log, there may be many pairs of events related to the same pair of activities, the temporal evidence captures the frequency of a particular interval relation being observed among the respective events.

Definition 1 (Temporal Evidence). *Temporal evidence is a tuple* $(R, f) \in \mathcal{R} \times \mathbb{N}^+$ *with R being an interval relation and f being its frequency.*

Given an event log, its TNR is a directed graph where nodes are activities and edge labels assign temporal evidence to the respective pairs of activities.

Definition 2 (Temporal Network Representation (TNR)). *Let L be an event log. Its Temporal Network Representation is a directed, edge-labelled graph* $G = (V, E, \lambda)$, *such that*

- $V = \bigcup_{\xi \in L} \bigcup_{e \in \xi} e.a$, the nodes are all activities of events of cases in the log;
- $E = \{(v_1, v_2) \in V \times V \mid \exists\, \xi \in L : \exists\, e_1, e_2 \in \xi : e_1.a = v_1 \wedge e_2.a = v_2\}$, edges are defined between all pairs of activities that occur jointly in cases;
- $\lambda : E \rightarrow 2^{\mathcal{R} \times \mathbb{N}^+}$, with $\lambda(d) \mapsto (R, f)$, $R \in \mathcal{R}$, and $f = |\{(e_1, e_2) \in \mathcal{E} \times \mathcal{E} \mid \exists\, \xi \in L : e_1, e_2 \in \xi : (e_1.a, e_2.a) = d \wedge (e_1, e_2) \in R\}|$, the edge labelling maps temporal evidence as observed in the log to edges.

The TNR of an event log can be constructed incrementally upon the addition of a new case to an event log. Considering the new information from the log may introduce additional vertices, additional edges, or increase the frequency of some temporal evidence.

For the log in Table 1, the TNR is shown in Table 2. Here, the time of an instantaneous event is considered as a completion time when deriving the interval relations.

Table 2. Matrix representation of the TNR of the event log in Table 1.

	A	B	C	D	E	F
A		{(precedes, 2), (meets, 1)}	{(meets, 3)}	{(precedes, 3)}	{(precedes, 2)}	{(precedes, 2)}
B			{(starts, 1)}	{(precedes, 1), (meets, 2)}	{(precedes, 2)}	{(precedes, 2)}
C		{(overlaps, 2)}		{(precedes, 2), (meets, 1)}	{(precedes, 2)}	{(precedes, 2)}
D					{(precedes, 2)}	{(precedes, 2)}
E						{(is finished by, 2)}
F						

3.2 Projections on the TNR

Since the TNR captures all pairwise temporal relations between activities, it generalises existing models of event logs. These models are typically defined as dependency graphs, in which the edges encode a particular temporal relation. Prominent examples used in discovery algorithms such as the Inductive Miner [10], α-Miner [12], or the Heuristic Miner [13] include the direct-follows graph and the concurrency graph. In the direct-follows graph, assuming that it is grounded in completion times of transactional events, a directed edge between activities x and y encodes that there exists a case $\xi \in L$ with two events $e_1, e_2 \in \xi$, such that $e_1.a = x$, $e_2.a = y$, $e_1.c < e_2.c$, and there is no event $e_3 \in \xi$ with $e_1.c < e_3.c < e_2.c$. The concurrency graph, in turn, contains an undirected edge for each pair of activities x and y, for which there exists a case $\xi \in L$ and events $e_1, e_2 \in \xi$ with $e_1.a = x$, $e_2.a = y$, and $e_1.s \leq e_2.s \leq e_1.c \leq e_2.c$.

These graphs may be derived from the TNR by projections. A TNR projection is a function that maps a TNR $G = (V, E, \lambda)$ to another TNR $G' = (V', E', \lambda')$, such that $V = V'$, $E \subseteq E'$, whereas the labelling λ' of G' is not constrained.

We first illustrate the derivation of the directly-follows graph [10], assuming that it is grounded in the completion times of activities. This requires two projections:

(1) For each edge, the temporal evidence for the relations *precedes*, *meets*, *overlaps*, and *contains* is aggregated and considered as part of the *precedes* relation. That is, the frequencies of all these relations are summed up and yield the new frequency of the *precedes* relation. We then remove all edges having a frequency of 0 for *precedes*.

(2) On the TNR that contains only edges with temporal evidence related to *precedes*, we conduct a transitive reduction. We are left with the directly-follows graph.

In the same manner, we can also derive the concurrency graph as used by the life-cycle variant of the Inductive Miner [4]. To this end, the *overlaps*, *is finished by*, *contains*, *starts*, and *equals* relations are aggregated, yielding a new *overlaps* relation. Then, all edges not having temporal evidence related to *overlaps* are removed.

4   Inductive Mining with the TNR

In this section, we show how the TNR can be used to enhance discovery of process models via inductive mining. We first introduce how the TNR is used to make processing delays explicit, before elaborating on the actual construction of a process tree (Sect. 4.1). Then, we propose probabilistic variant mining based on the TNR to handle noisy event logs, while preserving performance details (Sect. 4.2).

4.1   Delay-Aware Inductive Mining

Delay Unfoldings on the TNR. The TNR indicates processing delays by means of the *precedes* interval relation. If the TNR contains an edge between activities x and y with temporal evidence for *precedes*, it means that there is a case in the log in which the start and completion times of two transactional events that represent the occurrence of x and y are ordered, but the occurrence of y does not start immediately after the occurrence of x completes—there is a processing delay between x and y.

To make such processing delays explicit, we define a transformation of the TNR, referred to as delay unfolding. In essence, it inserts a *delay activity* between any two activities for which there is an edge with temporal evidence for the *precedes* interval relation. This delay activity is then linked to the respective activities in terms of temporal evidence for the *meets* relation, with the intuition being that this activity represents the gap between the occurrences of the original activities.

However, an activity x will be in the *precedes* relation with any other activity that starts after x completes. Therefore, we insert a delay activity between two activities x and y solely if there does not exist an activity z, whose occurrence can be seen as the reason for the time gap between the completion of x and the start of y. The situation when a delay-driven gap does not exist between activities x and y would be characterised by one of the following cases:

- There is an activity z that starts after or with the completion of x, while y starts after or with the completion of z, both are manifested as relations $\mathcal{R}_{after} = \{precedes, \; meets\}$; or
- there is an activity z that starts before the completion of x (temporal evidence is given as $\mathcal{R}_{over} = \{overlaps, \; is \; finished \; by, \; contains, \; starts, \; equals\}$), while y starts after or with the start of z (all relations in $\mathcal{R}$).

Using the above sets of temporal relations, we formally define the transformation of delay unfolding as follows:

Definition 3 (Delay Unfolding). *Given a TNR $G = (V, E, \lambda)$, the delay unfolding yields a new TNR $G' = (V', E', \lambda')$, such that:*

- *$V' = V \cup V_\delta$, where V_δ contains a node $\delta_{(x,y)}$ for each edge $d = (x, y) \in E$ with temporal evidence $(precedes, f) \in \lambda(d)$, $f > 0$, if there do not exist edges $d_x = (x, z), d_y = (z, y) \in E$ with temporal evidences $(R_x, f_x) \in \lambda(d_x)$, $(R_y, f_y) \in \lambda(d_y)$, $f_x, f_y > 0$, and either $R_x, R_y \in \mathcal{R}_{after}$, or $R_x \in \mathcal{R}_{over}$ and $R_y \in \mathcal{R}$;*
- *$E' = E \cup E_\delta$, where $E_\delta = \{(x, \delta_{(x,y)}), (\delta_{(x,y)}, y) \mid \delta_{(x,y)} \in V_\delta\}$ connects the new nodes from V_δ with the source and target of the respective edges; and*
- *$\lambda'(d) = \{(R, f) \in \lambda(d) \mid R \neq precedes\}$ for $d \in E$, is the original temporal evidence other than precedes for all edges in the original TNR; and $\lambda'(d) = \{(meets, f)\}$ for $(x, \delta_{(x,y)}), (\delta_{(x,y)}, y) \in E_\delta$, $d = (x, y) \in E$ and $(precedes, f) \in \lambda(d)$, with f being the temporal evidence of the new edges.*

Figure 4 illustrates the delay unfolding for a part of the example in Table 1. Here, a delay activity δ is introduced between A and B, representing the time gap indicated by the temporal evidence for the *precedes* relation. For all other edges with evidence for *precedes*, there are other possible reasons for the respective time gaps. Considering the edge between B and D as an example: while there is temporal evidence for a delayed start of D after B (the *precedes* relation), the temporal evidence of edges between B and C, and C and D, indicates that this delay may stem from the execution of C.

Model Construction. Once the delay unfolding on the TNR has made processing delays explicit, a timed process tree is derived. Here, we exploit the idea of inductive mining (IM), which is a constructive approach to process discovery [4, 10, 14]. In essence, inductive mining proceeds as follows: given a directly-follows graph, it recursively identifies cuts in the graph that separate its components, induce a partition of the log, and yield the control-flow operators of the process tree. For each identified component, this procedure is repeated until trivial components (single activities) are obtained.

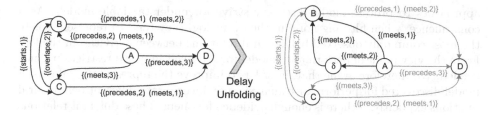

Fig. 4. Delay unfolding for a part of the TNR for the log in Table 1.

We adopt this general approach for the TNR after delay unfolding and rely on the existing algorithms to detect cuts, while integrating the handling of processing delays:

(1) By means of TNR projections, a directly-follows graph and concurrency graph is derived from the TNR as outlined in Sect. 3.2. Then, a process tree is built using the IM for transactional event logs [4], which relies on the concurrency graph to distinguish interleaved from concurrent execution of activities. The resulting process tree contains the delay activities introduced as part of the delay unfolding.

(2) For all activities $a \in \mathcal{A}$ of the obtained process tree, we fit a cumulative distribution function CDF D_a to model the duration. For activities that are observed in the log, we fit the distribution based on all observed durations of events, $\{e.c - e.s \mid e \in \bigcup_{\xi \in L} \xi \land e.a = a\}$. For a delay activity $\delta_{(x,y)}$, the samples to which the distribution is fitted are given as $\{e_2.s - e_1.c \mid e_1, e_2 \in \xi \land \xi \in L \land e_1.a = x \land e_2.a = y\}$, i.e., the durations between the completion of the preceding activity and the start of the succeeding activity. Every exclusive choice operator is enriched with the corresponding occurrence probabilities.

(3) Any delay activity $\delta_{(x,y)}$ is labelled as a distinguished silent activity, $\delta_{(x,y)}.a = \tau_i$ for a unique $i \in \mathbb{N}$, to capture that it does not carry any application semantics.

For the above example, the structure of the timed process trees that results from this procedure is shown in Fig. 5. The model features two silent activities τ_1 and τ_2 that represent common processing delays in handling claims, as outlined already for a single case in Fig. 1b. The CDFs assigned to these silent activities model the durations of these delays.

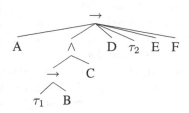

Fig. 5. Process tree with delay activities.

4.2 Probabilistic Variant Mining

In this part, we consider another angle to utilise the TNR in model discovery: it can guide the handling of noise in the event log. We observe that common approaches to process model discovery strive for understandable models. As a consequence, when filtering noise in the event log, they tend to represent solely the most commonly observed behavioural relations between activities in the log. From the viewpoint of performance modelling, this self-imposed restriction is not needed. We therefore argue that, in order to improve the representational bias, a model discovered for performance analysis may incorporate different behavioural relations explicitly, if there is enough evidence for them. These different relations may then be weighted following a probabilistic model.

Below, we introduce *probabilistic variant mining* (PVM), which uses the TNR to handle noise in inductive mining of process trees. PVM comprises two steps: (1) a *preprocessing* step, where a discrete-valued cumulative distribution function is computed for behavioural relations based on the TNR; (2) a *variant construction* step, where this distribution function is used to introduce choices between subtrees in the resulting model.

Preprocessing. Consider a TNR $G = (V, E, \lambda)$ that represents an event log L. The PVF relies on the frequencies defined by the temporal evidence to compute a discrete-valued cumulative distribution function (dv-CDF) over the interval relations, $F(d) : \mathcal{R} \to [0,1], d \in E$. Let $\lambda(d) = \{(R_1, f_1), \ldots, (R_k, f_k)\}$ be the temporal evidence of edge $d \in E$. Without loss of generality, we order the elements (R_i, f_i), $1 \leq i \leq k$, such that $f_1 \leq f_2 \leq \ldots f_k$. The dv-CDF of edge e for the i-th relation is then defined as:

$$F(d)(R_i) = F(d)(R_{i-1}) + \frac{f_i}{\sum_{j=1}^{k} f_j}, \quad \text{with } F(d)(f_0) \text{ defined as } 0.$$

Variant Construction. Following the general approach of inductive mining of process trees (as recalled in Sect. 4.1, yet potentially without handling of processing delays), the control-flow operators of a process tree are identified by iteratively detecting cuts in a dependency graph. If such cuts cannot be detected immediately, edges of the graph may be considered as noise and filtered according to a user-defined noise threshold [14]. In any case, cuts are detected based on a deterministic projection of the dependency graph.

In contrast, PVM defines a probabilistic means to identify cuts based on different TNR projections to obtain a direct-follows graph. In addition to the construction of a direct follows graph outlined in Sect. 3.2, we also consider a construction solely from relations that define an empty intersection of intervals (i.e., the set of relations $\mathcal{R}_{after}$ as defined above).

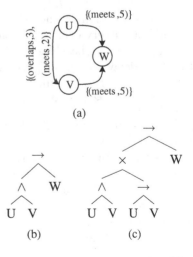

(a)

(b)   (c)

Fig. 6. Example TNR (a); trees obtained with traditional inductive mining (b) and with PVM (b).

If the resulting graphs give rise to different cuts, PVM constructs an exclusive choice operator in the process tree, which embeds all subtrees obtained from the different cuts. Hence, no information is lost in the process. The branching probabilities of this exclusive choice are assigned based on the dv-CDF function F as determined in the preprocessing step: each choice is assigned the aggregated probability of the respective pairwise relations between activities.

For illustration, consider the TNR given in Fig. 6a, which defines temporal evidence for U and V for both, *overlaps* and *precedes*. For this setting, the Inductive Miner for transactional event logs [4] would construct the model given in Fig. 6b, meaning that the relations between B and C are interpreted as concurrent execution. With PVM, it will be noticed that the graph created from relations that define an empty intersection of intervals also yields a sequence cut. Thus, the result would be the model in Fig. 6c; with the branching probabilities of the $\times$ operator set to $\frac{3}{5}$ for the concurrent ($\wedge$) case, and $\frac{2}{5}$ for the sequence case ($\rightarrow$). Clearly, this model is less understandable for users. Yet, focusing on performance modelling, it more accurately captures the behaviour encoded in the TNR and, thus, shall improve the model from the performance perspective.

5   Performance Fitness and Theoretical Guarantees

To evaluate algorithms for performance-driven process discovery, such as the one introduced earlier, this section presents a framework for measuring performance fitness. First, we define performance measures, loss functions, and the notion of *performance replay*, which enable us to quantify the distance between an event log and a process tree (Sect. 5.1). We then prove that TNR-based inductive mining comes with guarantees on the performance fitness with respect to the discovered model (Sect. 5.2).

5.1   Framework for Measuring Performance Fitness

We consider performance fitness with respect to a performance measure (e.g., the sojourn time of a case) and a loss function (e.g., the bias). This pair is defined as follows.

Definition 4 (Performance Measure, Loss Function). *A performance measure* $\psi : 2^{\mathcal{E}} \setminus \emptyset \rightarrow \mathbb{R}_0^+$ *is a function that maps a set of events to a real-value. A loss function* $l : \mathbb{R}_0^+ \times \mathbb{R}_0^+ \rightarrow \mathbb{R}_0^+$ *maps a pair of performance measures to a real-value.*

Having defined ψ and l, we aim at a distance measure between an event log and a model. To this end, we introduce the procedure of *performance replay* between a case $\xi = \{e_1, \ldots, e_n\} \in L$ and a process tree T. It is based on the assumption that T contains all activities referenced in events of cases $\xi \in L$. In other words, we do not consider replay for noise-filtered process trees. Performance replay involves the following steps:

(1) We first match the events of ξ to the process tree T. While there may be multiple leaf nodes to which an event could be matched, we note that matching needs to ensure consistency of the temporal relation between events and the semantics of the process tree operators. That is, this matching can be seen as the reverse operation of constructing a process tree from the TNR, see Sect. 4.1. As a result, we obtain a subtree T_ξ that contains a set of operators (without exclusive choice) and leaf nodes, so that each event in ξ is matched to one of the leaf nodes.

(2) The subtree T_ξ is then played-out into a replayed case $\xi_T = \{e_1^T, \ldots, e_n^T \mid e_i^T.a = e_i.a\}$. It contains the same number of events as ξ and every replayed event e_i^T refers to the same activity as the corresponding event e_i in ξ. Activities that appear in the subtree T_ξ, but are without matching events in ξ, are omitted from ξ_T via a projection of T_ξ on these activities.

(3) For the replayed case, we then construct durations. That is, for event e_i^T, the duration x_i^T is sampled from the CDF D_a of the respective activity $a = e_i^T.a$.

(4) Based on the durations, the start and completion times of the replayed case are set. To this end, we follow the partial order induced by the operators in T_ξ. The first events have a start time of $e_i^T.s = 0$ and a completion time of the sampled duration $e_i^T.c = e_i^T.s + x_i^T$. Subsequent events start at the maximal completion time of preceding events (in terms of the partial order induced by T_ξ), while their completion time is again determined based on their start time and sampled duration.

As an example, we consider the replay of a case $\xi = \{U_0^{10}, V_0^7, W_{10}^{15}\}$ (subscripts and superscripts correspond to start and end times, respectively) on the process tree in Fig. 6b. The replay procedure will match the events to the subtree representing concurrent execution of U and V, which yields a replayed case $\xi_T = \{U, V, W\}$. The durations for U, V, W are then sampled, e.g., as $x_U = 7, x_V = 1, x_W = 4$. Respecting the partial order induced by the subtree, the start and completion times of U, V are set first, before the start time of W is set to be the maximum of completion times of U and V. As a result, the following replayed case is created $\xi_T = \{U_0^7, V_0^1, W_7^{11}\}$.

Performance replay is based on a stochastic model. As such, the *replayed event log*, defined as $L_T = \{\xi_T \mid \forall\, \xi \in L\}$, is different for every enactment of the replay procedure for L over T. To make the event log and the model comparable, we thus use the following statistical approach. We conduct performance replay K times, independently, and consider every replayed log L_T to be a sample from T. This results in a sequence of K replayed logs indexed $L_T^{(1)}, \ldots, L_T^{(K)}$. Then, we define performance fitness between an event log L and a process tree T, as a statistical comparison between L and the sequence of K replayed logs $L_T^{(1)}, \ldots, L_T^{(K)}$.

To simplify notation, we assume that the event log L contains N cases with each case denoted by $\xi_i, i = 1, \ldots, N$; the corresponding replayed cases in the k-th event log are denoted by $\xi_T^{(i,k)}$. We are now ready to define the *performance fitness score*, which quantifies the distance between an event log and a process tree.

Definition 5 (Performance Fitness Score (PFS)). *Given an event log L, a sequence of replayed logs $\{L_T^{(k)}\}, k = 1, \ldots, K$, a performance measure ψ, and a loss function l, the performance fitness score (PFS) between L and T is given by,*

$$S_{K,N}(L, T, \psi, l) = \frac{1}{N} \sum_{i=1}^{N} \frac{1}{K} \sum_{k=1}^{K} l(\psi(\xi_i), \psi(\xi_T^{(i,k)})). \tag{1}$$

5.2 Guarantees on Performance Fitness for TNR-based Inductive Mining

It turns out that inductive mining with TNR-based delay unfolding and probabilistic variant mining comes with guarantees on the performance fitness of the discovered model. To capture these guarantees, we first need the notion of the *expected performance fitness score*, defined as follows:

Definition 6 (Expected Performance Fitness Score (E-PFS)). *Let ξ be a case of an event log L and let ξ_T be the corresponding replayed case. The expected performance fitness score (E-PFS) is*

$$S(L, T, \psi, l) = \mathbb{E}[l(\psi(\xi), \psi(\xi_T))]. \tag{2}$$

The randomness in $l(\psi(\xi), \psi(\xi_T))$ stems from the fact that we observe an arbitrary case from the event log. Assuming that $\psi(\xi_i)$ are independent and identically distributed (i.id.) samples from $\psi(\xi)$ and that $\psi(\xi_T^{(i,k)})$ are i.id. samples from $\psi(\xi_T)$, the PFS, $S_{K,N}(L, T, \psi, l)$, is an unbiased estimator of $S(L, T, \psi, l)$, since it estimates an expected value by using the sampled mean [15].

Below, we instantiate the performance fitness framework with the sojourn time of a case $\xi = \{e_1, \ldots, e_n\} \in L$, i.e., $\psi'(\xi) = \max_{e \in \xi} e.c - \min_{e \in \xi} e.s$, as a performance measure; and the bias, $l'(x, y) = x - y$, as the loss function. Then, we show that a process tree discovered by inductive mining using TNR-based techniques results in an unbiased process tree (in terms of the E-PFS) with respect to the originating event log.

Theorem 1. *Let L be an event log and let T be a process tree discovered from L using TNR-based delay unfolding and PVM. For performance fitness in terms of sojourn times of a case and the bias loss, the corresponding E-PFS is unbiased: $S(L, T, \psi', l') = 0$.*

Theorem 1 implies that $S_{K,N}(L, T, \psi', l') \to 0$ as the two sample sizes, N (number of cases in the event log) and K (number of replays), increase. Below, we provide a proof sketch, while the full proof can be found in an extended version of this paper.[1]

Proof Sketch 1. *We wish to prove that for an arbitrary ξ,*

$$\mathbb{E}[\psi'(\xi) - \psi'(\xi_T)] = \mathbb{E}[\psi'(\xi)] - \mathbb{E}[\psi'(\xi_T)] = 0. \tag{3}$$

---

[1] https://hu.berlin/TNR-extended.

To this end, we write the expression for the difference between the two sojourn times $\psi'(\xi) - \psi'(\xi_T)$ and show that both can be written as the sum of three components: sequential activity durations, sequential delays, and concurrent set durations. Since the replayed case and the log case respect the operators in T and their durations come from the same probability distributions, we show that the expectation of these three components are equal for both ξ and ξ_T, which proves Eq. 3, and the Theorem.

Furthermore, by using similar arguments, we can show that any process tree T that does not explicitly consider delays or is derived by threshold-based noise filtering will have a positive bias in terms of sojourn times, when delays have a positive expected value.

6   Evaluation

The above theoretical guarantees are based on several assumptions that may not hold in practice. For example, Theorem 1 holds if the execution times of activities are independent and identically distributed (i.id.), which must not be the case in real-life business processes. Hence, this section evaluates the usefulness of the TNR from a practical point of view. Our experiments show, based on three real-world datasets, that detecting processing delays and probabilistic variant mining improve performance analysis based on the discovered models. In the experiments, we compare the presented approach against models discovered by state-of-the-art inductive miners [4]. Below, we provide details on the datasets (Sect. 6.1), outline the experimental setting and procedure (Sect. 6.2), before presenting the main results (Sect. 6.3).

6.1   Datasets

We utilise three data sources: two event logs that stem from two different processes of DayHospital, a large cancer outpatient hospital in the United States, and an event log that comes from the Rambam hospital, a general hospital in Haifa, Israel.[2]

The first dataset, named 'Consult', corresponds to a patient consultation process in DayHospital. This process involves several procedures including blood draw, physical examination, vital signs, and consulting with a health provider. The process is typically very sequential. The second dataset, named 'Chemo', comes from a chemotherapy treatment process in DayHospital. It is a hybrid manufacturing-service process. Specifically, in order for a patient to receive chemotherapy, they must go through activities such as blood draw, examination, and chemotherapy infusion. During this process the relevant chemotherapeutic drugs is manufactured. The chemotherapy process exhibits concurrency due to the need to serve the patient and produce their medications. For the DayHospital experiments, we used six months of data (03/2014–09/2014).

---

[2] Data is available at http://seeserver.iem.technion.ac.il/databases/HomeHospital/.

The third dataset, named 'Rambam', originates from a general hospital comprising multiple departments, such as emergency and internal ward. The available data includes process data in terms of department names, start times, and completion times, thereby providing a high-level description of a patient's journey through the hospital as reflected in the information system. For a thorough description of the process and its corresponding data see [16]. For the Rambam hospital, we used a month's worth of data (04/2014).

6.2 Experimental Setting and Procedure

Below, we instantiate the performance fitness framework by setting its building blocks (performance measure ψ and loss function l), and describe the experimental procedure.

Setting. As our performance measure ψ, we chose the sojourn time, which is the total time a case spends in the process. As the loss function, we considered the squared loss, i.e., $l(x, y) = (x - y)^2$. We mine the baseline models via the Inductive Miner–life cycle (IMlc) [4]. Our evaluation scenarios comprise three controlled variables: (1) the dataset, (2) whether to detect delays (TNR approach vs. IMlc), and (3) whether to use probabilistic variant mining (PVM vs. noise filtering).

Procedure. The experimental procedure involved the quantification of the performance fitness score (PFS), namely $S_{K,N}(L, T, \psi, l)$, per scenario. In our case, the PFS corresponds to the average loss in sojourn times. To this end, we discovered and enriched models in correspondence to the 12 aforementioned scenarios. For example, to assess the scenario of the 'Consult' dataset, with delay detection, and probabilistic variant mining, we construct a process tree based on the TNR by applying both delay unfolding and PVM. Next, the model was simulated, case-by-case, with 30 runs per scenario, and the sojourn time of each case was recorded. Lastly, we quantified the empirical root-mean squared error (RMSE) to asses the loss.

6.3 Results

Table 3 summarises our experimental results in terms of the RMSE measure for sojourn time estimation over the 12 scenarios. The 'Delays' column gets values of 'Detected' (if delay unfolding was used), or 'NDetected', otherwise. The 'PVM' value of the 'Noise Handling' column corresponds to probabilistic variant mining (Sect. 4.2), while 'Happy-Path' corresponds to a 20% noise filtering of variants based on the IMlc. We show the average RMSE across 30 runs and the average sojourn times of cases (in minutes for DayHospital, in hours for Rambam). To scale the accuracy of estimation, we also provide the percentage of error out of the average sojourn time.

For DayHospital, we observe that PVM dominates the deterministic approach by IMlc (40%–60% improvement in estimation error). This points toward the ability of the TNR-based approach to improve performance measurement.

Table 3. Experimental results: scenarios and estimated performance fitness score

Dataset	Noise handling	Delays	Avg sojourn time	PFS (RMSE)	Prop. to AVG
Consult	PVM	Detected	73	65	89%
	PVM	NDetected	73	69	95%
	Happy-Path	Detected	66	86	130%
	Happy-Path	NDetected	66	85	129%
Chemo	PVM	Detected	240	172	72%
	PVM	NDetected	240	178	74%
	Happy-Path	Detected	222	233	105%
	Happy-Path	NDetected	222	229	103%
Rambam	PVM	Detected	113	233	206%
	PVM	NDetected	113	234	207%
	Happy-Path	Detected	96	249	259%
	Happy-Path	NDetected	96	250	260%

Further, delay detection improves performance fitness only when combined with probabilistic mining (up to 6% improvement in accuracy); it does not significantly improve estimation when a 'Happy-Path' model is discovered, as the data is filtered first and the TNR receives a biased version of the event log. This result is explained by the fact that deterministic noise filtering removes unlikely paths, which are typically longer and involve more delays. Thus, the TNR is less likely to contain delays. Rambam hospital scenarios suffer from low estimation accuracy across all scenarios due to large variance in patient sojourn times. Yet, the results described for DayHospital are repeated across the Rambam scenarios.

7   Related Work

Our work falls under the field of process mining [3], as it provides a temporal representation for extracting process models from event data. Closest to our work are discovery algorithms that rely on transactional data to extract control-flow relations, such as Tsinghua-α [8], Inductive Miner–life cycle [4], and the Heuristic Miner++ [17].

In this paper, we focus on discovery of timed models for performance analysis of business processes. This topic was the subject of numerous works in process mining [5,6,18]. Commonly, the proposed techniques treat control-flow discovery, and temporal aspects of the event log separately, which results in sub-optimal results when computing performance measures. To overcome this limitation, we propose a new formalism, namely the Temporal Network Representation (TNR), which is based on Allen's interval algebra [9]. Additional formalisms to represent temporal aspects of information systems have been proposed in the literature. These include: generalised interval algebra, temporal networks, fuzzy temporal knowledge, and time ontology methods, see [19] for an extensive survey. In this work, we selected Allen's algebra as it provides a complete representation of all pairwise relations between time intervals that is simple and transparent [20].

Another line of work, namely predictive process monitoring, aims at online forecasts of key performance indicators (KPIs), e.g., the remaining time, and the next event [21]. Our work is mainly concerned with the ability to reconstruct performance measures as they were observed in the event log, and thus focuses on post-mortem analysis.

This paper proposes two techniques that enhance performance-driven discovery based on the TNR, namely delay unfolding and probabilistic variant mining. Related to the former, a method for detecting (unrecorded) queueing delays from event logs with missing temporal information is presented in [22]; it models resource availability via an activity-life cycle representation. The TNR provides a more general approach to delay detection, as it does not require a priory knowledge of resource behaviour.

Probabilistic variant mining relates to recent attempts to handle noise in process discovery, based either on event log filtering [10,13,23], or model abstraction [24,25]. While these works rely on deterministic reasoning, requiring a user to decide on over or under representation of process variants in the model, the TNR enables for a probabilistic framework, which better reflects process heterogeneity in the discovered model.

To assess performance-oriented fitness, we proposed the performance fitness score (PFS) that is derived using a stochastic replay of events logs over process trees. The idea to calculate fitness based on a replay procedure was first introduced in [26], and later extended based on the notion of *alignments* [27]. However, these two approaches do not address the situations when the discovered models are stochastic, i.e., comprising time distributions and branching probabilities. The *second-pass* approach presented in [5] involves simulating a (discovered) stochastic model to obtain a sample of event logs. These simulated logs are then compared to the originating event log that was used for discovery, to check whether the discovered model is a good representation of the event log. However, the comparison is performed without a formal notion of model-log distance (or similarity). In our approach, we propose the PFS as a statistic that measures the distance between event logs and process trees. Further, we show that under probabilistic assumptions, the TNR produces unbiased model with respect to the PFS.

8   Conclusion

Targeting the improvement of business processes based on performance analysis, we introduced a novel model of event logs, namely the temporal network representation (TNR). It captures the temporal relations between activities, which is an essential aspect of performance-oriented process model discovery. To demonstrate the effectiveness of the TNR, we introduced two applications. First, we proposed delay unfolding as a means to detect unrecorded processing delays. Second, we presented probabilistic variant mining (PVM) to preserve performance information, while handling noise in event logs. Further, we developed

a framework for assessing performance fitness of discovered models. We have
shown that under this framework, TNR-based inductive mining is guaranteed
to result in unbiased models with respect to the original event log. We evaluated the approach with three real-world datasets from the healthcare domain.
We have been able to show an up-to 40% percent improvement in sojourn time
estimation, when combining delay detection and PVM.

In future work, we aim at enriching the TNR with workload information to
separate delays stemming from resource queueing and those originating from
synchronisation.

References

1. Dumas, M., Rosa, M.L., Mendling, J., Reijers, H.A.: Fundamentals of Business
   Process Management. Springer, Heidelberg (2013)
2. Senderovich, A., Weidlich, M., Yedidsion, L., Gal, A., Mandelbaum, A., Kadish, S.,
   Bunnell, C.A.: Conformance checking and performance improvement in scheduled
   processes: a queueing-network perspective. Inf. Syst. **62**, 185–206 (2016)
3. van der Aalst, W.M.P.: Process Mining: Discovery, Conformance and Enhancement
   of Business Processes. Springer, Heidelberg (2011)
4. Leemans, S.J.J., Fahland, D., van der Aalst, W.M.P.: Using life cycle information
   in process discovery. In: Reichert, M., Reijers, H.A. (eds.) BPM 2015. LNBIP, vol.
   256, pp. 204–217. Springer, Cham (2016). doi:10.1007/978-3-319-42887-1_17
5. Rozinat, A., Mans, R., Song, M., van der Aalst, W.M.P.: Discovering simulation
   models. Inf. Syst. **34**(3), 305–327 (2009)
6. Rogge-Solti, A., Weske, M.: Prediction of remaining service execution time using
   stochastic petri nets with arbitrary firing delays. In: Basu, S., Pautasso, C., Zhang,
   L., Fu, X. (eds.) ICSOC 2013. LNCS, vol. 8274, pp. 389–403. Springer, Heidelberg
   (2013). doi:10.1007/978-3-642-45005-1_27
7. Senderovich, A., Weidlich, M., Gal, A., Mandelbaum, A.: Queue mining – predicting delays in service processes. In: Jarke, M., Mylopoulos, J., Quix, C., Rolland,
   C., Manolopoulos, Y., Mouratidis, H., Horkoff, J. (eds.) CAiSE 2014. LNCS, vol.
   8484, pp. 42–57. Springer, Cham (2014). doi:10.1007/978-3-319-07881-6_4
8. Wen, L., Wang, J., van der Aalst, W.M., Huang, B., Sun, J.: A novel approach for
   process mining based on event types. J. Intell. Inf. Syst. **32**(2), 163–190 (2009)
9. Allen, J.F.: Maintaining knowledge about temporal intervals. CACM **26**(11), 832–
   843 (1983)
10. Leemans, S.J.J., Fahland, D., van der Aalst, W.M.P.: Discovering block-structured
    process models from event logs - a constructive approach. In: Colom, J.-M., Desel,
    J. (eds.) PETRI NETS 2013. LNCS, vol. 7927, pp. 311–329. Springer, Heidelberg
    (2013). doi:10.1007/978-3-642-38697-8_17
11. Alspaugh, T.A.: Software support for calculations in Allen's interval algebra (2005)
12. Van der Aalst, W., Weijters, T., Maruster, L.: Workflow mining: discovering process
    models from event logs. IEEE Trans. Knowl. Data Eng. **16**(9), 1128–1142 (2004)
13. Weijters, A., van Der Aalst, W.M., De Medeiros, A.A.: Process mining with the
    heuristics miner-algorithm. Technische Universiteit Eindhoven, Technical report
    WP 166, pp. 1–34 (2006)
14. Leemans, S.J.J., Fahland, D., van der Aalst, W.M.P.: Discovering block-structured
    process models from event logs containing infrequent behaviour. In: Lohmann, N.,
    Song, M., Wohed, P. (eds.) BPM 2013. LNBIP, vol. 171, pp. 66–78. Springer, Cham
    (2014). doi:10.1007/978-3-319-06257-0_6

15. Bickel, P.J., Doksum, K.A.: Mathematical Statistics: Basic Ideas and Selected Topics, vol. 2. CRC Press, Boca Raton (2015)
16. Armony, M., et al.: On patient flow in hospitals: a data-based queueing-science perspective. Stoch. Syst. **5**(1), 146–194 (2015)
17. Burattin, A.: Heuristics miner for time interval. In: Process Mining Techniques in Business Environments: Theoretical Aspects, Algorithms, Techniques and Open Challenges in Process Mining. LNBIP, vol. 207, pp. 85–95. Springer, Cham (2015). doi:10.1007/978-3-319-17482-2_11
18. van der Aalst, W.M.P., Schonenberg, M., Song, M.: Time prediction based on process mining. Inf. Syst. **36**(2), 450–475 (2011)
19. Vila, L.: A survey on temporal reasoning in artificial intelligence. AI Commun. **7**(1), 4–28 (1994)
20. Freksa, C.: Temporal reasoning based on semi-intervals. Artif. Intell. **54**(1–2), 199–227 (1992)
21. Maggi, F.M., Di Francescomarino, C., Dumas, M., Ghidini, C.: Predictive monitoring of business processes. In: Jarke, M., Mylopoulos, J., Quix, C., Rolland, C., Manolopoulos, Y., Mouratidis, H., Horkoff, J. (eds.) CAiSE 2014. LNCS, vol. 8484, pp. 457–472. Springer, Cham (2014). doi:10.1007/978-3-319-07881-6_31
22. Senderovich, A., Leemans, S.J.J., Harel, S., Gal, A., Mandelbaum, A., van der Aalst, W.M.P.: Discovering queues from event logs with varying levels of information. In: Reichert, M., Reijers, H.A. (eds.) BPM 2015. LNBIP, vol. 256, pp. 154–166. Springer, Cham (2016). doi:10.1007/978-3-319-42887-1_13
23. Günther, C.W., van der Aalst, W.M.P.: Fuzzy mining – adaptive process simplification based on multi-perspective metrics. In: Alonso, G., Dadam, P., Rosemann, M. (eds.) BPM 2007. LNCS, vol. 4714, pp. 328–343. Springer, Heidelberg (2007). doi:10.1007/978-3-540-75183-0_24
24. Fahland, D., Van Der Aalst, W.M.: Simplifying discovered process models in a controlled manner. Inf. Syst. **38**(4), 585–605 (2013)
25. Senderovich, A., Shleyfman, A., Weidlich, M., Gal, A., Mandelbaum, A.: P^3-Folder: optimal model simplification for improving accuracy in process performance prediction. In: La Rosa, M., Loos, P., Pastor, O. (eds.) BPM 2016. LNCS, vol. 9850, pp. 418–436. Springer, Cham (2016). doi:10.1007/978-3-319-45348-4_24
26. Rozinat, A., van der Aalst, W.M.: Conformance checking of processes based on monitoring real behavior. Inf. Syst. **33**(1), 64–95 (2008)
27. Adriansyah, A., Munoz-Gama, J., Carmona, J., Dongen, B.F., van der Aalst, W.M.P.: Alignment based precision checking. In: Rosa, M., Soffer, P. (eds.) BPM 2012. LNBIP, vol. 132, pp. 137–149. Springer, Heidelberg (2013). doi:10.1007/978-3-642-36285-9_15

Synthesizing Petri Nets from Hasse Diagrams

Robin Bergenthum[(⊠)]

FernUniversität in Hagen, Hagen, Germany
robin.bergenthum@fernuni-hagen.de

Abstract. Synthesis aims at producing a process model from specified sample executions. A user can specify a set of executions of a system in a specification language that is much simpler than a process modeling language. The intended process model is then constructed automatically.

Synthesis algorithms have been extensively explored for cases where the specification language is a reachability graph or a sequential language. Concerning synthesis from partial languages, however, there is a significant gap between theory and practical application. In the literature, we find two different synthesis methods for partial languages, but both have poor runtime even in reasonably sized practical examples. In this paper, we introduce a new and more efficient synthesis algorithm for partial languages based on Hasse diagrams.

1 Introduction

Complex business processes are often modeled by means of Petri nets [1,3,17,29,30]. Petri nets have formal semantics, an intuitive graphical representation, and are able to express concurrency among the occurrence of actions. Petri nets are the formal basis for many workflow modeling languages. However, constructing a Petri net model for a real world process is a costly and error-prone task [1,3,28].

Fortunately, whenever we model a system, there are often some associated descriptions or even specifications of the desired processes. There may be log-files of recorded behavior, example runs, or product specifications describing use cases. Such specifications can be formalized by a set of words, a reachability graph, or a partial language. Yet, only partial languages are able to explicitly express concurrency between events. Thus, partial languages have drawn much attention recently [5,20].

If a specification is incomplete or contains so-called noise, there are algorithms developed in the area of process mining [1,2] to still automatically generate a suitable business process model. If a specification is complete (i.e. is the desired behavior), we can synthesize a model. The synthesis problem is to compute a process model so that: (A) the specification is a subset of the language of the generated model and (B) the generated model has minimal additional behavior.

To showcase a typical use case of synthesis-based model generation, we assume a coffee brewing process together with a domain expert on this process called Robin. Robin has been brewing coffee for years, but just recently received a training in process modeling to document standard processes in his department. Robin observes a sample execution of his process and records the following

© Springer International Publishing AG 2017
J. Carmona et al. (Eds.): BPM 2017, LNCS 10445, pp. 22–39, 2017.
DOI: 10.1007/978-3-319-65000-5_2

sequence of events: *grind beans, unlock machine, empty strainer, clean coffee pot, get water with coffee pot, fill kettle, fill strainer, assemble and turn on*. In a second sample, he uses a glass pot (instead of the coffee pot) to fetch water from the kitchen. With this in mind, Robin builds a naive Petri net-like process model of his process depicted in Fig. 1. All actions of the process are ordered in a sequence and there is an *XOR*-split modeling the choice between the two transitions *get water with coffee pot* and *get water with glass pot*.

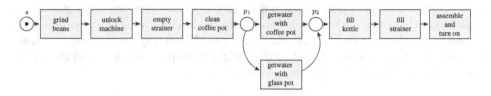

Fig. 1. Petri net-like process model of the coffee brewing process.

We (modeling experts) try to rate the validity of Robin's model. We realize that most of the modeled dependencies may be superfluous. For example, unlocking the coffee machine usually does not depend on grinding beans. Cleaning the coffee pot only depends on unlocking the machine. What is worse, some of these dependencies may change if we consider different executions of the process. In a scenario where we use the coffee pot to get water from the kitchen, the event *fill kettle* depends on the sequence of actions *unlock coffee machine, clean coffee pot*, and *get water with coffee pot*. In another scenario where we use the additional glass pot to fetch water, the event *fill kettle* only depends on unlocking the machine and getting water. Thus, the relation between the occurrences of *fill kettle* and *clean coffee pot* has changed.

There are many possible pitfalls in this small example. Even if Robin understands the concepts that may cause trouble here, he most likely will not be able to adapt his model accordingly. To tackle this problem, we apply a synthesis-based model generation approach. We revisit both initially observed sample executions and depicted them in Fig. 2. In a next step, we ask Robin to delete unnecessary dependencies between observed events. With his expert knowledge on the process at hand, he for example starts to delete the dependency between *grind beans* and *unlock machine* in the first sequence, thus creating a partial order step by step.

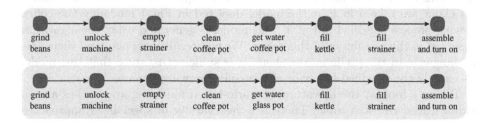

Fig. 2. Observed events of the coffee brewing process.

Robin continues to delete all superfluous dependencies from both observed sequences and finally comes up with the two labeled Hasse diagrams depicted in Fig. 3. Every labeled Hasse diagram specifies a different run of the coffee brewing process. In the first run, we grind beans and unlock the coffee machine. Once the machine is unlocked, we empty the strainer and clean the coffee pot. After it is cleaned, we fetch water from the kitchen using the coffee pot. When the strainer and the kettle are filled, we assemble and turn on the coffee machine. Every arc models a dependency between the occurrence of the related actions and unordered events occur concurrently. In the second run, we use a glass pot (instead of the coffee pot) to fetch water from the kitchen. This activity does not depend on unlocking or cleaning the coffee pot. We can get water right at the beginning of this sample run. Altogether, Fig. 3 depicts a complete and intuitive specification of our coffee brewing process.

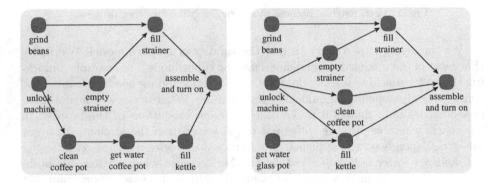

Fig. 3. Two labeled Hasse diagrams, i.e. a specification of the coffee brewing process.

The last step in a synthesis-based model generation approach is to construct a model matching Robin's specification, thus solving the synthesis problem. Such a model offers an integrated view of the process at hand, is more compact, can be analyzed by well-known Petri net algorithms, and can serve as an input for workflow engines.

In this paper, we present a new synthesis technique to automatically transform a specification into a valid process model. The main benefit is that single executions are much easier to model than the complex system itself. To confirm this claim, we take a look at the model depicted in Fig. 4 modeling the coffee brewing process specified in Fig. 3. This model is generated using the algorithm presented in the remainder of this paper. We will recall Petri nets and their partial language in the preliminaries; here, it is sufficient to state that this model has exactly the specified behavior of the coffee brewing process.

Taking a look at the literature, scenario-based modeling approaches are an acknowledged research topic. There is a vast variety of specialized approaches

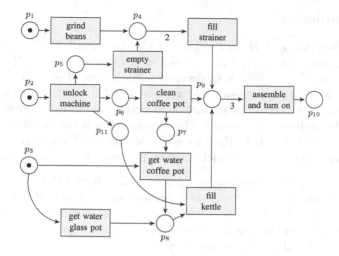

Fig. 4. A marked p/t-net modeling a coffee brewing process.

using different specification languages and process models (see for example [15,16,22,23]). All these approaches have in common that they assume the specification to be valid and rather complete. This is the main requirement for applying precise generation algorithms like synthesis or folding methods. The theory of the related synthesis algorithms is called region theory [4,21]. Region theory has been extensively explored for reachability graphs and sequential languages. There are many non-trivial theoretical results, notions, case studies, as well as tools like ProM [19], Genet [14], or Viptool [10] (see for example [11–13,31,32] for some recent publications). Concerning region theory for partial languages, however, there is a significant gap between theory and practical application. There are two different notions of regions for partial languages: tokenflow regions [9] and transition regions [8,9]. Yet, both related algorithms perform poorly even in reasonably sized practical examples [8]. In Sect. 3, we introduce a synthesis algorithm based on a new concept called compact regions. The name stems from the fact that while tokenflow regions relate to occurrence nets and transition regions relate to step sequences, the new notion relates to compact tokenflows [6,7] and Hasse Diagrams, i.e. a much more compact representation of a partial language. We show that the concept of compact regions introduced here leads to a much faster synthesis algorithm.

The paper is organized as follows: Sect. 2 introduces Petri nets, their partial language, and the synthesis problem. In Sect. 3, we recall the concept of compact tokenflows and introduce compact regions. We prove that compact regions solve the synthesis problem for Petri nets and partial languages. At the end of Sect. 3, we deduce our synthesis algorithm from the new definition of compact regions. In Sect. 4, we discuss the runtime of the new algorithm. We compare the algorithm to its predecessors i.e. algorithms based on transition regions and (ordinary) tokenflow regions. We implement all synthesis techniques in a tool called MoPeBs Eagle Owl and present runtime tests.

2   Preliminaries

Let f be a function and B be a subset of the domain of f. We write $f|_B$ to denote the restriction of f to B. We call a function $m : A \to \mathbb{N}$ a multiset and write $m = \sum_{a \in A} m(a) \cdot a$ to denote multiplicities of elements in m. Let $m' : A \to \mathbb{N}$ be another multiset. We write $m \geq m'$ if $\forall a \in A : m(a) \geq m'(a)$ holds. We denote the transitive closure of an acyclic and finite relation $<$ by $<^*$. We denote the skeleton of $<$ by $<^\circ$. The skeleton of $<$ is the smallest relation $\lhd$ such that $\lhd^* = <^*$ holds. Let $(V, <)$ be some acyclic and finite graph, $(V, <^\circ)$ is called its Hasse diagram. We model business processes by p/t-nets [3,18,29,30].

Definition 1. *A place/transition net (p/t-net) is a tuple (P, T, W) where P is a finite set of places, T is a finite set of transitions such that $P \cap T = \emptyset$ holds, and $W : (P \times T) \cup (T \times P) \to \mathbb{N}$ is a multiset of arcs. A marking of (P, T, W) is a multiset $m : P \to \mathbb{N}$. Let m_0 be a marking, we call the tuple $N = (P, T, W, m_0)$ a marked p/t-net and m_0 the initial marking of N.*

Figure 4 depicts a p/t-net modeling a coffee brewing process. Transitions are rectangles, places are circles, the multiset of arcs is represented by weighted arcs, and the initial marking is represented by black dots called tokens. There is a simple firing rule for transitions of a p/t-net: let t be a transition of a marked p/t-net (P, T, W, m_0). We denote $\circ t = \sum_{p \in P} W(p, t) \cdot p$ the weighted preset of t. We denote $t\circ = \sum_{p \in P} W(t, p) \cdot p$ the weighted postset of t. A transition t is enabled (can fire) at marking m if $m \geq \circ t$ holds. Once transition t fires, the marking changes from m to $m' = m - \circ t + t\circ$. In our example p/t-net, the transitions *grind beans*, *unlock machine*, and *get water with glass pot* can fire at the initial marking. If *unlock coffee machine* fires, this removes the token from the place in its preset and produces two new tokens: one token in the preset of *empty strainer* and another token in the preset of *clean coffee pot*. As soon as *get water glass pot* fires, the token from the lower left place is removed and one token is produced in the preset of *fill kettle*. Note: firing *get water glass pot* disables transition *get water coffee pot* for the rest of this process. Concerning arc weights, the transition *assemble and turn on* is enabled if there are at least three tokens in the rightmost place p_9. Repeatedly processing the firing rule produces firing sequences. These firing sequences are the most basic behavioral model of Petri nets. Let N be a marked p/t-net, the set of all initially enabled firing sequences of N is the sequential language of N.

Petri nets and most business process modeling languages are able to express concurrency of the occurrences of transitions. For example, transitions *grind beans* and *unlock machine* can fire independently from one another. Roughly speaking, they can fire without any order while not sharing resources. However, firing sequences are not able to capture or describe such behavior. The common behavioral model for partially ordered behavior of Petri nets is a so-called process nets [25]. A process net is a Petri net modeling only one single partially ordered run of a marked p/t-net. For a formal definition of process nets we refer to [25,30]. Here, as an example, we depict a process net of our coffee brewing Petri net of Fig. 4.

Every place of a process net relates to a token of the related p/t-net. For example in Fig. 5, there are three places labeled p_9 in the preset of the transition *assemble and turn on*. The set of process nets of a p/t-net is called its unfolding. Events, loops, tokens, and conflicts are unfolded to present single executions of the related net.

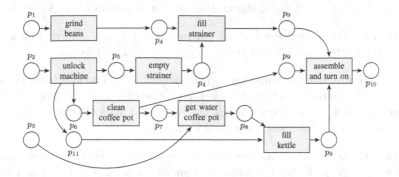

Fig. 5. One process net of the coffee brewing process.

If we abstract from the places of a process net related to a p/t-net N, we have a set of events arranged in a partial order. Just like some valid firing sequence, this partially ordered set of events is enabled in the p/t-net. In other words, we can replay such a partial order by firing transitions of N where unordered parts of the partial order can fire concurrently. The set of labeled partial orders induced by all processes of N is the partial language of N. For example, the left half of Fig. 3 depicts the labeled Hasse diagram representing the partial order underlying Fig. 5. Thus, this labeled partial order is in the partial language of the p/t-net depicted in Fig. 4.

Definition 2. *Let T be a set of labels. A labeled partial order (lpo) is a triple $lpo = (V, <, l)$ where V is a finite set of events, $< \subseteq V \times V$ is a transitive and irreflexive relation, and the labeling function $l : V \to T$ assigns a label to every event.*

Definition 3. *Let $K = (C, E, F, \rho)$ be a process net of a marked p/t-net (P, T, W, m_0) where C is a set of conditions, E is a set of events, F is a set of arcs, and $\rho : (C \cup E) \to (P \cup T)$ is a labeling function. The lpo $(E, F^*|_{E \times E}, \rho|_E)$ is the process lpo of K.*

Let N be a marked p/t-net and $L^\Pi(N)$ be the set of all process lpos of N. $L(N) = \{(E, <, l)|(E, <, l) \text{ an } lpo, (E, <', l) \in L^\Pi(N), <' \subseteq <\}$ is the partial language of N.

As we already pointed out in the introduction, our goal is to synthesize a p/t-net from a specification describing partially ordered behavior. The most suitable manner to represent a partial language is by means of labeled Hasse diagrams (see Fig. 2). A labeled Hasse diagram is a finite set of events ordered by the skeleton of a partial order. Clearly, the transitive closure of a Hasse diagram is an lpo. Thus, the prefix- and sequentialisation-closure of a set of labeled Hasse diagrams is a partial language.

Definition 4. *A triple $run = (V, <, l)$ is a labeled Hasse diagram if $(V, <^*, l)$ is an lpo and $<^\circ = <$ holds. A finite set of labeled Hasse diagrams is a specification. Let $run = (V, <, l)$ be a labeled Hasse diagram, we define $run^* = (V, <^*, l)$.*

Definition 5. *Let N be a marked p/t-net and $S = \{run_1, \ldots, run_n\}$ be a specification. We write $S \subseteq L(N)$ iff $\{run_1^*, \ldots, run_n^*\} \subseteq L(N)$ holds.*

Finally, we are able to define the synthesis problem. The synthesis problem is to construct a p/t-net such that its behavior matches a specification. If there is no such p/t-net, we construct a p/t-net such that its behavior includes the specification and has minimal additional behavior.

Definition 6. *Let S be a specification, the synthesis problem is to compute a marked p/t-net N such that the following conditions hold: $S \subseteq L(N)$ and for all marked p/t-nets $N' : L(N)\backslash L(N') \neq \emptyset \implies S \nsubseteq L(N')$.*

3   Compact Regions and Synthesis Algorithm

The algorithm presented in this paper is based on the theory of regions [21]. For an introduction to region theory, we refer the reader to [4]. As stated in the introduction, the input to our algorithm is a set of labeled Hasse diagrams (see Fig. 3). The first step is to construct a transition for every label to get an initial p/t-net without places. The language of this net includes arbitrary behavior because all the transitions have an empty preset and can fire in any order. Obviously, we need to add places and arcs to restrict the behavior of this initial net. To solve the synthesis problem, we are only allowed to add places and connected arcs that do not inhibit our specification.

Definition 7. *Let S be a specification and $N = (P, T, W, m_0)$ be a marked p/t-net. A place $p \in P$ is called feasible for S iff $S \subseteq L((\{p\}, T, W|_{(\{p\} \times T) \cup (T \times \{p\})}, m_0(p)))$ holds. Let S be a specification and $N = (\{p\}, T, W, m_0)$ be a marked one-place p/t-net. We call N feasible for S iff p is feasible for S.*

If we are able to identify feasible places, we can add these to our initially placeless p/t-net. These places restrict the behavior, yet such a net will still be able to execute all the labeled Hasse diagrams of the specification.

Remark 1. *Let S be a specification and let a set of p/t-nets $\{(\{p_1\}, T, W_1, m_1), \ldots, (\{p_n\}, T, W_n, m_n)\}$ be feasible for S. Let $N = (\bigcup_i \{p_i\}, T, \sum_i W_i, \sum_i m_i)$ be the union of all feasible nets, every place of N is feasible and $S \subseteq L(N)$ holds.*

Theoretically, we could restrict the behavior of the initial p/t-net by adding the set of all feasible places. This would guarantee that the behavior of the net cannot be restricted further without excluding some executions of the specification. This is a fundamental theorem of region theory (see for example [4]).

Theorem 1. *Let S be a specification and T be its set of labels. The p/t-net which is the union of all p/t-nets feasible for S is a solution of the synthesis problem.*

Practically, we need to construct a finite p/t-net with the same behavior as the union-of-all-feasible-places p/t-net. For partial languages, there are two strategies to tackle this problem: We can either calculate a basis of all feasible places and add them to the initial set of transitions. This is always possible and the basis is always finite. According to the firing rules of Petri nets, this net behaves like the infinite p/t-net. In other words, the finite basis p/t-net also solves the synthesis problem. Or we can use the technique of so-called wrong continuations. Roughly speaking, the set of wrong continuations is the border between the specified and all other behaviors. The set of wrong continuations is finite as long as the specification is finite as well. For each wrong continuation, we add one feasible place, thus excluding the wrong continuation from the language of the constructed net. The resulting finite p/t-net solves the synthesis problem as well.

The next step of our algorithm is to characterize the set of all feasible places. To develop an efficient synthesis algorithm, we rely on the behavioral model of compact tokenflows [6,7]. A compact tokenflow is a distribution of tokens along the Hasse diagram of a labeled partial order. A labeled Hasse diagram is in the partial language of a p/t-net if there is a compact tokenflow distributing tokens such that three conditions hold: first, every event receives enough tokens, second, no event has to pass too many tokens, and third, the initial marking is not exceeded. Tokens must be received from the particular presets of events. Thus, we ensure that consumed tokens are available before the actual event occurs. If a transition produces tokens, the related events are allowed to produce tokenflow in the Hasse diagram and pass these tokens to their particular postsets. If an event receives tokens, it consumes the tokenflow needed and passes the redundant tokenflow to later events. Tokens of the initial marking are free for all, i.e. any event can consume tokens from the initial marking.

Definition 8. *Let $N = (P, T, W, m_0)$ be a marked p/t-net and run $= (V, <, l)$ be a labeled Hasse diagram such that $l(V) \subseteq T$ holds. A compact tokenflow is a function $x : (V \cup <) \rightarrow \mathbb{N}$. x is compact valid for $p \in P$ iff the following conditions hold:*

(i) $\forall v \in V: x(v) + \sum_{v' < v} x(v', v) \geq W(p, l(v))$,
(ii) $\forall v \in V: \sum_{v < v'} x(v, v') \leq x(v) + \sum_{v' < v} x(v', v) - W(p, l(v)) + W(l(v), p)$,
(iii) $\sum_{v \in V} x(v) \leq m_0(p)$.

run is compact valid for N iff there is a compact valid tokenflow for every $p \in P$.

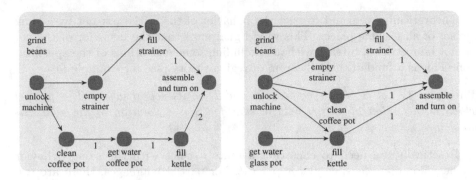

Fig. 6. Two valid compact tokenflows for place p_9 of the marked p/t-net of Fig. 4.

For an example of a compact tokenflow, we consider place p_9 of Fig. 4 and depict two related compact valid tokenflows for the two labeled Hasse diagrams introduced earlier in Fig. 6. We take a look at Fig. 4 and only consider p_9 and its related arcs. We see that all events labeled *assemble and turn on* need to receive three tokens, whereas all events labeled *fill strainer, clean coffee pot*, or *fill kettle* can produce one token. In the Hasse diagrams of Fig. 5, tokenflow is depicted as integers on the related arcs and events (the integer 0 is not shown). According to the depicted tokenflow, in the first diagram *fill strainer* and *clean coffee pot* create one token each. The event *get water coffee pot* cannot create tokens for p_9, but receives a token from *clean coffee pot* and passes this token to *fill kettle*. The event *fill kettle* receives one token and produces another one. Thus, *fill kettle* passes two tokens to *assemble and turn on*. Altogether, *assemble and turn on* receives three tokens and all the conditions for a valid compact tokenflow hold for p_9. All in all, we need to construct eleven compact tokenflows related to the eleven places of the p/t-net of Fig. 4 to deduce that this labeled Hasse diagram is in the language of the p/t-net. In the second Hasse diagram of Fig. 6, three tokens directly reach the event labeled *assemble and turn on*. Again, all conditions for a valid compact tokenflow hold for p_9. Compared to the notion of process nets [25] and to the notion of (ordinary) tokenflows [26, 27], the main advantage of compact tokenflows is that they only consider the Hasse diagrams of the specification. Process nets as well as ordinary tokenflows need to consider the complete (i.e. transitive) relation. Previous work [6, 7] proves that these three notions are equivalent, i.e. they all define the same partial language. For the proof, we refer to [6] but state the following theorem.

Theorem 2. *The language of a marked p/t-net is well-defined by the set of compact valid labeled Hasse diagrams.*

In our algorithm we take advantage of compact tokenflows and define a new notion of regions, i.e. compact regions, for partial languages.

Definition 9. *Let $S = \{(V_1, <_1, l_1), \ldots, (V_n, <_n, l_n)\}$ be a specification, T be its set of labels, and p be a place. We denote V_i' the set of events with an empty prefix* *in* $(V_i, <_i, l_i)$. *A function* $r : (\bigcup_i (V_i' \cup <_i) \cup (T \times \{p\}) \cup (\{p\} \times T) \cup \{p\}) \to \mathbb{N}$ *is a compact region for* S *iff* $\forall i \in \mathbb{N} : r|_{\{V_i' \cup <_i\}}$ *is compact valid for* p *in* $(\{p\}, T, r|_{(T \times \{p\}) \cup (\{p\} \times T)}, r(p))$.

We reconsider our sample brewing process depicted in Fig. 5. We assume, both Hasse diagrams are the input to our synthesis algorithm. The main idea of the algorithm is to construct compact regions. In this example, we implement the domain of a compact region by 41 non-negative integer unknowns. The first 19 unknowns represent a place. A place may have one weighted arc leading to each of the nine transitions related to the labels of our example, one weighted arc coming from each of the nine transitions, and an additional unknown for its initial marking. The next ten unknowns represent a compact tokenflow of the first Hasse diagram. One unknown for each of the eight arcs and two additional unknowns for each of the two minimal events. The last twelve unknowns represent a compact tokenflow of the second Hasse diagram. Again, one unknown for each of the nine arcs and three additional unknowns for each of the three minimal events. Only if all 41 values of these unknowns relate to two compact tokenflows valid for the defined place, this vector is a compact region. For example, assume a fixed ordering of all 41 unknowns, the vector $(0, 1, 3, 0, 0, 1, 0, 0, 0, 0, 0, 1, 0, 0, 0, 0, 0, 0, 0, 0, 0, 0, 0, 0, 1, 1, 1, 2, 0, 0, 0, 0, 0, 0, 1, 1, 1, 0,$ $0, 0, 0, 0)$ may be a compact region. Assuming a correct ordering of all unknowns, the first third of this vector defines the arc weights of p_9 depicted in Fig. 3. The second third defines the compact tokenflow on the left side of Fig. 5. The last third defines the compact tokenflow on the right side of Fig. 5.

To state the correctness of our synthesis algorithm we have to prove that if the compact tokenflows are valid for the defined place, a region defines a feasible place.

Theorem 3. *Let* S *be a specification and* T *be its set of labels. Every compact region* r *for* S *defines a feasible p/t-net* $N_r = (\{p\}, T, W, m_0)$ *and vice versa.*

Proof. Let r be a compact region. For every labeled Hasse diagram in S, there is a valid compact tokenflow $r|_{\{V_i' \cup <_i\}}$ of p in $N_r = (\{p\}, T, r|_{(T \times \{p\}) \cup (\{p\} \times T)}, r(p))$. $S \subseteq L(N_r)$ holds and so N_r is feasible for S.

Let $N = (\{p\}, T, W, m_0)$ be a feasible p/t-net such that $S \subseteq L(N)$ holds. There is a valid compact tokenflow r_i for every labeled Hasse diagram of S. Without loss of generality, every r_i is zero on events with a non-empty prefix. This holds because as long as some valid compact tokenflow is positive for some event e with a non-empty prefix, move this tokenflow to an event e' in its direct prefix and adopt the compact tokenflow on the arc (e', e) accordingly. The union $r = \bigcup_i r_i \cup W \cup m_0$ is a compact region. $\qquad\square$

Every region is a vector of numbers respecting the conditions (i), (ii), and (iii) of Definition 9. With this in mind, we are able to express all feasible p/t-nets by a single inequality system. Again, in this system, there is an unknown for every element in the domain of a compact region, i.e. one unknown for every minimal event, another unknown for every arc, two unknowns for every label, and a single unknown for the initial marking. The inequality system is built from the inequalities defined in Definition 8. According to (i) and (ii), there are two inequalities for every event of the specification. According to (iii), there is another inequality for every labeled Hasse diagram. The set of positive integer solutions of this inequality system is the set of all feasible nets. We call this inequality system the compact region inequality system. Every solution of this system defines one feasible place. Altogether, the compact region inequality system has 41 unknowns as well as 34 inequalities in our coffee brewing example.

Finally, we depict the complete synthesis algorithm using compact regions in Algorithm 1. Input is a set of Hasse diagrams H. We construct a Petri net with an empty set of places and a transition for every label, calculate the compact region inequality system, and the set of wrong continuations of H. For every wrong continuation c we check if it is still executable in the net constructed so far. If it is executable, we need to exclude the wrong continuation from the behavior of the net. This must be done with a feasible place, i.e. a compact region. We encode the non-executability of c in an additional inequality for the compact region inequality system. Every solution of this extended system is a region and excludes c. If this system has a solution, we add the related one-place net to our initially constructed set of transitions. If the extended compact region inequality system has no solution, the wrong continuation c cannot be excluded. We assure that the constructed net is a best approximation to H by adding the set of wrong continuations of c to C. Algorithm 1 will terminate because H is finite.

Algorithm 1.
___

1: **Input**: A set of labeled Hasse diagrams H
2:   $(P, T, W, m_0) \leftarrow (\emptyset, T \leftarrow \bigcup_{(V, <, l) \in H} l(V), \emptyset, \emptyset)$
3:   $M \leftarrow$ compactRegionInequalitySystem(H)
4:   $C \leftarrow$ wrongContinuations(H)
5: **while** $C \neq \emptyset$ **do**
6:      $c \leftarrow C$.remove()
7:      **if** c.isExecutable(P, T, W, m_0) **then**
8:          $M' \leftarrow M$.addInequality(c)
9:          $s \leftarrow M'$.solve()
10:         **if** s.isSolution() **then**
11:            (P, T, W, m_0).add$(s$.getOnePlaceNet())
12:         **else**
13:            C.addAll(wrongContinuations(c))
14: **return** (P, T, W, m_0)

---

4   Comparison and Experimental Results

In this section, we first compare compact regions to the already existing concepts of transition regions [4,8] and ordinary tokenflow regions [9]. Secondly, we present a runtime experiment comparing all three related synthesis algorithms.

A transition region of a partial language L is based on the set of step sequences L^{step} of L (see [24]). As an example, we depict three of the numerous step sequences of the first labeled Hasse diagram of our coffee brewing process using two operators: composition of steps $+$ and sequential composition of steps $\cdot$:

$[(grind\ beans) + (unlock\ machine)] \cdot [(empty\ strainer) + (clean\ coffee\ pot)] \cdot$
$[(fill\ strainer) + (get\ water\ coffeepot)] \cdot (fill\ kettle) \cdot (assemble\ and\ turn\ on)$

$[(grind\ beans) + (unlock\ machine)] \cdot [(empty\ strainer) + (clean\ coffee\ pot)] \cdot$
$(get\ water\ coffee\ pot) \cdot [(fill\ strainer) + (fill\ kettle)] \cdot (assemble\ and\ turn\ on)$

$(unlock\ machine) \cdot [(grind\ beans) + (empty\ strainer) + (clean\ coffee\ pot)] \cdot$
$(get\ water\ coffee\ pot) \cdot [(fill\ strainer) + (fill\ kettle)] \cdot (assemble\ and\ turn\ on)$

The two events *grind beans* and *unlock coffee machine* are the first step of the first sequence. The second step is *empty strainer* and *clean coffee pot*. The number of maximal step sequences may grow exponentially with the number of events of a labeled Hasse diagram. Even in our small coffee brewing specification, the size of this language is huge. A transition region defines a place and requires that this place can fire every maximal step sequence of the specified partial language.

Definition 10. Let $S = \{(V_1, <_1, l_1), \ldots, (V_n, <_n, l_n)\}$ be a specification, T be the set of labels of S, and p be a place. Let $L^{step}(S)$ be the language of step sequences of S. A function $r : ((T \times \{p\}) \cup (\{p\} \times T) \cup \{p\}) \to \mathbb{N}$ is a transition region for S iff for all maximal step sequences $(\tau_1 \ldots \tau_n) \in L(S)^{step}$ and all $j \in \{1, \ldots, n\} : r(p) + \sum_{t \in T}((\tau_1 + \ldots + \tau_{j-1})(t) \cdot r(t, p) - (\tau_1 + \ldots + \tau_j)(t) \cdot r(p, t)) \geq 0$ holds.

We follow Definition 10 to define the transition region inequality system. The number of unknowns is $2 \cdot |T| + 1$, i.e. a place. The inequalities of the inequality system are a subset of the set of conditions of Definition 10. In the transition region inequality system, we discard all constraints that are equal to or less strict than other constraints. Let τ and τ' be two steps such that $\tau' \leq \tau$ holds and let $\tau_1 \ldots \tau_{j-1}$ and $\tau'_1 \ldots \tau'_{k-1}$ be two step sequences such that $\bigcup_{i<j} \tau_i$ and $\bigcup_{i<k} \tau'_i$ share the same multiset of labels. If τ can fire after the occurrence of $\tau_1 \ldots \tau_{j-1}$, the step τ' can fire after the occurrence of $\tau'_1 \ldots \tau'_{k-1}$. Thus, we build the transition region inequality system by merging matching presteps to so-called prefix steps. The number of inequalities of the transition region inequality system is equal to the number of prefix step continuations. A prefix step continuation is a prefix step Γ together with a step τ if there is a matching maximal step sequence $\pi_1 \ldots \pi_n$ of S such that $\Gamma = \bigcup_{i<n} \pi_i$ and $\tau = \pi_n$ holds. Altogether, the number of inequalities is approximately the number of cuts of S. In a worst-case scenario, the number of cuts is exponential in the size of our input. However, if we specify little concurrency, the number of cuts is small. The number of inequalities is, for example, equal to the number of all events if every labeled Hasse diagram is totally ordered. We refer the reader to [8] for a more detailed description of the transition region inequality system.

An ordinary tokenflow region of a partial language is based on the set of labeled partial orders of a specification. Let $S = \{(V_1, <_1, l_1), \ldots, (V_n, <_n, l_n)\}$ be a specification of L, obviously, $S^* = \{(V_1, <_1^*, l_1), \ldots, (V_n, <_n^*, l_n)\}$ specifies L using labeled partial orders. If we input a set of labeled Hasse diagrams we have to calculate the transitive closure in a first step of this algorithm. Figure 3 depicts two Hasse diagrams with a total of 17 arcs. The respective partial orders have 31. In the exceptional case where a Hasse diagram is transitive, both characterizations have the same number of arcs; in most cases, the number of arcs of a partial order may increase quadratic with the number of arcs of a Hasse diagram.

The concept of tokenflows is similar to the concept of compact tokenflows. Yet, ordinary tokenflows directly relate to process nets of Petri nets, whereas compact tokenflows are able to abstract from the history of tokens. In some sense, compact tokenflows rather relate to distributed transition systems than to process nets.

The domain of a tokenflow region r is the partial order of the specification, every event, and a place. Thus, the domain of a tokenflow region includes the domain of a compact region. If we specify little concurrency, the number of arcs of the labeled partial order is quadratic in the number of arcs of the Hasse diagram. The number of conditions of a tokenflow region is equal to the number of conditions of a compact region. Just like for compact regions, we define the tokenflow region inequality system. This system has $\sum_{(V,<,l)\in S}(|V| + |>^*|) + 2 \cdot |T| + 1$ unknowns and $\sum_{(V,<,l)\in S}(2 \cdot |V| + 1)$ inequalities.

Summing up, compared to transition regions and tokenflow regions, the new compact regions define by far the smallest region inequality system for partial languages. Since algorithms have to solve these systems multiple times during the synthesis procedure, compact tokenflows lead to the fastest synthesis algorithms for partial languages.

To support the scenario-based modeling approaches with Hasse diagrams we developed our tool called *MoPeBs eagle owl*. In *MoPeBs* we implement three synthesis algorithms, each using a different concept of regions. *MoPeBs* is a lightweight editor embedding Viptool [10] plug-ins. *MoPeBs* uses the Simplex algorithm of *LpSolve* to handle the occurring region inequality systems (http://lpsolve.sourceforge.net).

Figure 7 depicts a screenshot of *MoPeBs eagle owl*. The main application window can handle, save, and load synthesis projects. We see a list of specified tasks, two files specifying two different Hasse diagrams, *CoffeePot.lpo* and *GlassPot.lpo*, and three files relating to different p/t-net models. The list of all .*lpo*-files is the specification, i.e. the input of the synthesis algorithms. Every file in the list of p/t-net models was synthesized using a different concept of regions. In the bottom left-hand corner of the main application window, there are three buttons: *transition regions*, *tokenflow regions*, and *compact regions*. Every button starts the related synthesis algorithm. The second window of Fig. 7 depicts the MoPeBs editor showing the Hasse diagram of *CoffeePot.lpo*. Of course, *MoPeBs* can edit, save, and load Hasse diagrams.

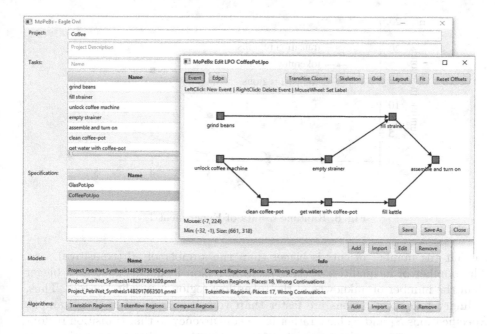

Fig. 7. A screenshot of MoPeBs eagle owl.

We perform the following experiments to measure how well the different synthesis approaches scale with respect to the size of the input and to compare the overall runtime. We use MoPeBs and an Intel Core i5 3.30 GHz (4 CPUs) machine with 8 GB RAM running a Windows 10 operating system. *MoPeBs eagle owl* is available on the *MoPeBs* homepage at www.fernuni-hagen.de/sttp/forschung/mopebs.shtml.

Experiment 1. *We consider five specifications S_1, S_2, S_3, S_4, and S_5. Specification S_1 is the sample specification depicted in* Fig. 3. *Every other specification is a sequential composition of copies of these two labeled Hasse diagrams. S_2 is the sequential composition of twice the first labeled Hasse diagram and twice the second labeled Hasse diagram. S_3, S_4, and S_5 are three, four, and five copies. We solve the synthesis problem using compact regions, tokenflow regions, and transition regions. We depict the mean of the runtimes of 20 runs of each algorithm in seconds if the algorithm terminates within 15 min in* Fig. 8.

In Experiment 1, the Hasse diagrams grow in length. Specification S_1 has 24 events and 17 arcs, Specification S_5 has 80 events and 105 arcs. Algorithm 1, which uses compact tokenflows, outperforms Algorithm 2 and Algorithm 3 in every test. This is not surprising because only compact regions are tailored to relate to small region inequality systems. As pointed out in the first part of this section, the compact region inequality system is much easier to solve than the tokenflow and the transition region inequality systems. If we compare Algorithm 2 and Algorithm 3, the specifications are rather short at first,

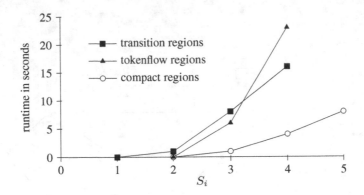

Fig. 8. Runtime results of Experiment 1.

so that the number of cuts is bigger than the number of all (transitive) arcs. The number of inequalities of the transition region inequality system is bigger than the number of unknowns of the tokenflow region inequality system. Thus, ordinary tokenflow regions are faster than transition regions in this example. Specifications S_4 and S_5 have rather little concurrency so that transition regions outperform ordinary tokenflow regions as soon as the number of (transitive) arcs exceeds the number of cuts. Compact regions are fast, independent of the level of concurrency. Considering S_5, only Algorithm 1 is able to solve the synthesis problem within 15 min.

Experiment 2. *We consider four specifications* $X_1, X_2, X_3,$ *and* X_4. *Specification* X_1 *is three Hasse diagrams of the partial language of the so-called repair example from www.processmining.org. Every other specification is a parallel composition of copies of these diagrams, i.e. the specifications grow in width. The maximal size of a cut in* X_1 *is two, four in* X_2, *six in* X_3, *and eight in* X_4. *We solve the synthesis problem using compact regions, tokenflow regions, and tran-*

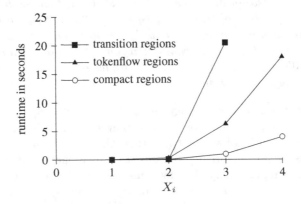

Fig. 9. Runtime results of Experiment 2.

sition regions. We depict the mean of the runtimes of 20 runs of each algorithm in seconds if the algorithm terminates within 15 min in Fig. 9.

In Experiment 2, the Hasse diagrams of the specifications grow in width. Specification X_1 has 30 events, X_2 has 45 events, X_3 has 72 events, and X_4 has 90 events. The length of the longest path in every specifications is 8 (at most one loop of the repair example). Just like in Experiment 1, Algorithm 1 outperforms Algorithm 2 and Algorithm 3 in every test. Again, the compact region inequality system is much easier to solve than the tokenflow and the transition region inequality systems. If we compare Algorithm 2 and Algorithm 3, the number of cuts is big and the number of transitive arcs is small. Thus, ordinary tokenflow regions are faster than transition regions in this example. In both experiments compact regions outperform both older algorithms independent from the structure of the specification.

5 Conclusion and Future Work

We presented an approach to generate a process model from a set of Hasse diagrams specifying a set of sample executions. Using our approach, a user can specify a set of executions in a very intuitive and simple specification language and get the complex Petri net model for free.

We presented a new concept of regions for partial languages. The definition is based on the semantics of compact tokenflows. The domain of a compact tokenflow is the number of arcs and the set of initial events of a labeled Hasse diagram. Both numbers grow neither like the number of events nor like the number of arcs of a labeled partial order. We compared compact regions to tokenflow and transition regions referring to the size of the related region inequality systems.

Furthermore, we presented a synthesis algorithm and experimental results of its implementation in the tool called *MoPeBs eagle owl*. *MoPeBs* supports the sample based modelling approach for Hasse Diagramms. We compared the runtime of the new algorithm to the runtime of both existing synthesis algorithms for partial languages.

An important topic for future research will be to develop a concept of wrong continuations using a concept of tokenflows. Right now, the definition of wrong continuations is based on the step language of a partial language. Even though the size of the compact region inequality system is reasonable, the huge number of wrong continuations corrupts the synthesis algorithm if a specification has many concurrent events.

References

1. van der Aalst, W.M.P., Dongen, B.F.: Discovering petri nets from event logs. In: Jensen, K., Aalst, W.M.P., Balbo, G., Koutny, M., Wolf, K. (eds.) Transactions on Petri Nets and Other Models of Concurrency VII. LNCS, vol. 7480, pp. 372–422. Springer, Heidelberg (2013). doi:10.1007/978-3-642-38143-0_10

2. van der Aalst, W.M.P., Weijters, T., Maruster, L.: Workflow mining: discovering process models from event logs. IEEE Trans. Knowl. Data Eng. **16**(9), 1128–1142 (2004)

3. van der Aalst, W.M.P.: The application of petri nets to workflow management. J. Circ. Syst. Comput. **8**(1), 21–66 (1998)

4. Badouel, E., Bernardinello, L., Darondeau, P.: Petri Net Synthesis. Texts in Theoretical Computer Science. Springer, Heidelberg (2015)

5. van Beest, N., Dumas, M., Garca-Bauelos, L., La Rosa, M.: Log delta analysis: interpretable differencing of business process event logs. Eprint no. 83018. Queensland University of Technology (2015)

6. Bergenthum, R., Lorenz, R.: Verification of scenarios in petri nets using compact tokenflows. Fundam. Informaticae **137**, 117–142 (2015). IOS Press

7. Bergenthum, R.: Faster verification of partially ordered runs in petri nets using compact tokenflows. In: Colom, J.-M., Desel, J. (eds.) PETRI NETS 2013. LNCS, vol. 7927, pp. 330–348. Springer, Heidelberg (2013). doi:10.1007/978-3-642-38697-8_18

8. Bergenthum, R., Desel, J., Mauser, S.: Comparison of different algorithms to synthesize a petri net from a partial language. In: Jensen, K., Billington, J., Koutny, M. (eds.) Transactions on Petri Nets and Other Models of Concurrency III. LNCS, vol. 5800, pp. 216–243. Springer, Heidelberg (2009). doi:10.1007/978-3-642-04856-2_9

9. Bergenthum, R., Desel, J., Lorenz, R., Mauser, S.: Synthesis of petri nets from finite partial languages. Fundam. Informaticae **88**, 437–468 (2008). IOS Press

10. Bergenthum, R., Desel, J., Lorenz, R., Mauser, S.: Synthesis of petri nets from scenarios with viptool. In: Hee, K.M., Valk, R. (eds.) PETRI NETS 2008. LNCS, vol. 5062, pp. 388–398. Springer, Heidelberg (2008). doi:10.1007/978-3-540-68746-7_25

11. Bergenthum, R., Desel, J., Lorenz, R., Mauser, S.: Process mining based on regions of languages. In: Alonso, G., Dadam, P., Rosemann, M. (eds.) BPM 2007. LNCS, vol. 4714, pp. 375–383. Springer, Heidelberg (2007). doi:10.1007/978-3-540-75183-0_27

12. Carmona, J.: Projection approaches to process mining using region-based techniques. Data Min. Knowl. Discov. **24**(1), 218–246 (2012)

13. Carmona, J., Cortadella, J., Kishinevsky, M.: New region-based algorithms for deriving bounded petri nets. IEEE Trans. Comput. **59**(3), 371–384 (2010)

14. Carmona, J., Cortadella, J., Kishinevsky, M.: Genet: a tool for the synthesis and mining of petri nets. Appl. Concurrency Syst. Des. **2009**, 181–185 (2009)

15. Desel, J., Juhás, G., Lorenz, R., Neumair, C.: Modelling and validation with viptool. In: Aalst, W.M.P., Weske, M. (eds.) BPM 2003. LNCS, vol. 2678, pp. 380–389. Springer, Heidelberg (2003). doi:10.1007/3-540-44895-0_26

16. Desel, J., Erwin, T.: Quantitative engineering of business processes with *VIPbusiness*. In: Ehrig, H., Reisig, W., Rozenberg, G., Weber, H. (eds.) Petri Net Technology for Communication-Based Systems. LNCS, vol. 2472, pp. 219–242. Springer, Heidelberg (2003). doi:10.1007/978-3-540-40022-6_11

17. Desel, J., Juhás, G.: "What is a petri net?" Informal answers for the informed reader. In: Ehrig, H., Padberg, J., Juhás, G., Rozenberg, G. (eds.) Unifying Petri Nets. LNCS, vol. 2128, pp. 1–25. Springer, Heidelberg (2001). doi:10.1007/3-540-45541-8_1

18. Desel, J., Reisig, W.: Place/transition Petri nets. In: Reisig, W., Rozenberg, G. (eds.) ACPN 1996. LNCS, vol. 1491, pp. 122–173. Springer, Heidelberg (1998). doi:10.1007/3-540-65306-6_15

19. van Dongen, B.F., Medeiros, A.K.A., Verbeek, H.M.W., Weijters, A.J.M.M., Aalst, W.M.P.: The ProM framework: a new era in process mining tool support. In: Ciardo, G., Darondeau, P. (eds.) ICATPN 2005. LNCS, vol. 3536, pp. 444–454. Springer, Heidelberg (2005). doi:10.1007/11494744_25

20. Dumas, M., García-Bañuelos, L.: Process mining reloaded: event structures as a unified representation of process models and event logs. In: Devillers, R., Valmari, A. (eds.) PETRI NETS 2015. LNCS, vol. 9115, pp. 33–48. Springer, Cham (2015). doi:10.1007/978-3-319-19488-2_2

21. Ehrenfeucht, A., Rozenberg, G.: Partial (set) 2-structures. part i: basic notions and the representation problem, part ii: state spaces of concurrent systems. Acta Inf. **27**(4), 315–368 (1990)

22. Fahland, D.: Scenario-based process modeling with Greta. In: BPM Demonstration Track 2010, vol. 615. CEUR (2010)

23. Fahland, D.: Oclets – scenario-based modeling with Petri nets. In: Franceschinis, G., Wolf, K. (eds.) PETRI NETS 2009. LNCS, vol. 5606, pp. 223–242. Springer, Heidelberg (2009). doi:10.1007/978-3-642-02424-5_14

24. Grabowski, J.: On partial languages. Fundam. Informaticae **4**, 427–498 (1981). IOS Press

25. Goltz, U., Reisig, W.: Processes of place/transition-nets. In: Diaz, J. (ed.) ICALP 1983. LNCS, vol. 154, pp. 264–277. Springer, Heidelberg (1983). doi:10.1007/BFb0036914

26. Juhás, G., Lorenz, R., Desel, J.: Can I execute my scenario in your net? In: Ciardo, G., Darondeau, P. (eds.) ICATPN 2005. LNCS, vol. 3536, pp. 289–308. Springer, Heidelberg (2005). doi:10.1007/11494744_17

27. Lorenz, R., Juhás, G., Bergenthum, R., Desel, J., Mauser, S.: Executability of scenarios in Petri nets. Theoret. Comput. Sci. **410**(12–13), 1190–1216 (2009). Elsevier

28. Mayr, H.C., Kop, C., Esberger, D.: Business process modeling and requirements modeling. In: ICDS 2007, pp. 8–14. IEEE Computer Society (2007)

29. Peterson, J.L.: Petri Net Theory and the Modeling of Systems. Prentice-Hall, Englewood Cliffs (1981)

30. Reisig, W.: Understanding Petri Nets - Modeling Techniques, Analysis Methods, Case Studies. Springer, Heidelberg (2013)

31. Solé, M., Carmona, J.: Region-based foldings in process discovery. IEEE Trans. Knowl. Data Eng. **25**(1), 192–205 (2013)

32. van Zelst, S.J., van Dongen, B.F., van der Aalst, W.M.P.: ILP-based process discovery using hybrid regions. In: ATAED 2015, vol. 1371, pp. 47–61. CEUR (2015)

PE-BPMN: Privacy-Enhanced Business Process Model and Notation

Pille Pullonen[1(✉)], Raimundas Matulevičius[2], and Dan Bogdanov[1]

[1] Cybernetica AS, Tartu, Estonia
{pille.pullonen,dan}@cyber.ee
[2] University of Tartu, Tartu, Estonia
rma@ut.ee

Abstract. Privacy Enhancing Technologies (PETs) play an important role in preventing privacy leakage of data along information flows. Although business process modelling is well-suited for expressing stakeholder collaboration and process support by technical solutions, little is done to visualise and analyse privacy leakages in the processes. We propose PE-BPMN – privacy-enhanced extensions to the BPMN language for capturing data leakages. We demonstrate its feasibility in the mobile app scenario where private data leakages are determined. Our approach helps system builders make decisions on the privacy solutions at the early stages of development and lets auditors analyse existing systems.

Keywords: Privacy · Business process model and notation (BPMN) · Privacy enhancing technology · Data leakage

1 Introduction

The importance of privacy is continuously growing. A new General Data Protection Regulation is entering into force in EU [13] and the new Privacy Shield agreement will be affecting businesses in US with restrictions compared to the Safe Harbour agreement [28]. Furthermore, companies are starting to use privacy as a sales argument, e.g., adding differential privacy to their services [16].

Organisations wishing to cope with new restrictions or to deploy new privacy enhancing technologies (PETs), need to understand the privacy properties and assumptions of their current and newly developed systems. There exist regulatory standards (e.g., [1,18]) and approaches (e.g., [22]) for risk-oriented privacy management. However, little is done [20] to assess privacy properties within business processes, or to address unintentional privacy leakages when some input data objects or derived data objects are sent to other parties.

In this paper, we primarily focus on privacy analysis in the *honest-but-curious adversary* cases. For example, if a bank officer generates a report about the payment transactions, data about the payer, recipient, etc. should not leak to the officer or any other unintended party. The PETs are applied to enforce privacy requirements. We consider *how business process model and notation* (BPMN)

© Springer International Publishing AG 2017
J. Carmona et al. (Eds.): BPM 2017, LNCS 10445, pp. 40–56, 2017.
DOI: 10.1007/978-3-319-65000-5_3

[12,23] *could support analysis of private data leakage*. Based on the PET classification [9], we propose PE-BPMN – a BPMN extension with privacy enhancing technologies and characterize the goals of using PETs. Feasibility of PE-BPMN is illustrated using the mobile application scenario where emergency data are gathered using a mobile phone app.

The rest of the paper is structured as follows: in Sect. 2 we discuss some related studies; Sect. 3 overviews the PET classification, which is used to describe PE-BPMN in Sect. 4. Section 5 illustrates PE-BPMN application and development. Finally, Sect. 6 concludes the paper and provides directions for future work.

2 Related Work

Literature suggests several studies where BPMN is extended towards security and privacy modelling. Extensions to modelling secure business processes through understanding the security requirements are proposed in [24]. Menzel *et al.* have presented enhancements towards trust modelling [21]. In [8], BPMN is enriched with information assurance and security modelling capabilities. In [3], BPMN is aligned to the domain model of security risk management. Schleicher *et al.* have defined the concept of compliance scope used to restrict certain areas of a business process [25]. The above studies introduce security extensions to BPMN, which do not necessarily model all privacy risks. Our proposal focuses on the private data leakages and protecting against them within the organisation.

In [2] authors are using the principle of the Petri-nets reachability to detect places where information leaks occur in the business processes. In the current study we focus on the BPMN modelling language and extend it with the abilities to introduce PETs in order to mitigate information leakage.

In [4], the BPMN collaboration and choreography models are used to detail message exchange and identity contract negotiation. In [7], BPMN is extended with *access control, separation of duty, binding of duty* and *need to know* principles. *Privacy-aware* BPMN is presented in [20]. Similarly to [7], privacy concerns are captured by annotating the BPMN model with *access control, separation of tasks, binding of tasks, user consent* and *necessity to know* icons. Although these studies focus on the privacy requirements and their potential implementation, they (*i*) basically consider only business process entailment constraints, and (*ii*) do not analyse *privacy leakage cases* nor take PETs into account.

In [11] a quantification of private data leakage is discussed using differential privacy. In this paper we expand this principle and illustrate how privacy leakage in the business processes could be determined using other PETs.

3 Classification of Privacy Enhancing Technologies

We adapt the results of the recent survey [9] on existing PETs to introduce a PET classification (see Table 1). In this classification, PETs are grouped according to their application *goals* to aid choosing PETs and expanded with *targets*, which should be met within different privacy areas. It should be noted that the same

PET could appear at different categories; for example, encryption is used for data protection and secure communication. This classification could be extended with more sub-categories for other PETs. For example, the computation on protected inputs can be divided to distributed and single party techniques. Added details can help choose the right PET in a decision tree manner.

Table 1. Classification of privacy enhancing technologies

Goal	Target	Examples of technology
Communication protection	Secure	Client-Server encryption, TLS, IPSec, End-to-End encryption, PGP, OTR
	Anonymous	Proxies and VPN, onion routing, mix-networks, broadcast
Data protection	Integrity	Message authentication codes, signatures
	Confidentiality	Encryption, secret sharing
Entity authentication	Identity based	Username and password, single-sign-on
	Attribute based	Credential used only once, zero-knowledge proofs
Privacy-Aware computation	Confidential inputs	Homomorphic encryption, secure multiparty computation, private information retrieval
	Privacy adding	Differential privacy, k-anonymity, cell suppression, noise addition, aggregation, anonymisation
Human-Data interaction	Transparency of data usage	Information flow detection, logging, declarations about information usage
	Intervenability	Information granularity adjustment, access control

Communication protection protects the content and the parties. Security means that the protected contents (e.g., using client-server encryption, TLS, etc.) can travel without external parties seeing or modifying them. Anonymity ensures that the interacting parties can not be deduced by an observer.

Data protection ensures integrity and confidentiality of the data. For example, signatures or message authentication codes can not be modified by external parties who do not have access to respective keys. Encrypted data remains confidential unless a party has the decryption key. Data protected by secret sharing raises an additional constraint that it must be stored in a distributed manner.

Entity Authentication is a procedure for proving that user corresponds to the claimed attributes. Identity authentication requires some identity provider to verify all accesses (e.g., based on a fixed account). Attribute based methods deal with proving one's membership to some group, without identifying herself.

Privacy-Aware Computations focus on the utility of private data. Computations on confidential inputs allow one to securely process various operations without removing the protection mechanisms. For example, these computations use homomorphic properties of encryption or secret sharing. Privacy adding computations can add a layer of privacy to their outputs instead of fully protecting the inputs. For example, differential privacy adds some noise to the query reply so that it is hard to infer something about single entries in the database.

Human-data interaction is a field that combines technical means and policies with user experience. In essence, the users allowing some processing of their data should be knowledgeable about how and why their data is used. In addition, they may be able to regulate the data processing.

The taxonomy in [9] thoroughly describes commonly used PETs. Another recent systematic comparison of properties of PETs is given in [17] that could be used to enhance the decision tree for PETs. Our taxonomy combines the *aim*, *data* and *aspect* ideas of [17]. Our focus is on privacy goals, but it is nicely complemented with a legal viewpoint of activities that are harmful for privacy [27]. We cover *data collection* and *processing* parts of [27] and discuss how leakage analysis can help to quantify problems in *information dissemination*. There are also attempts at creating guidelines for choosing PETs, for example [9,14,19].

4   Extending BPMN with Privacy Enhancing Technologies

4.1   Abstract Syntax and Semantics

Figure 1 presents extensions of BPMN abstract syntax [23] with the PET concepts.

The BPMN Data Flow is extended with Communication Protection. In common secure channels, the message is hidden and can not be modified during transit. Thus, secure channels are straightforward to model in the sense that the communication and privacy risks occur between different pools. We introduce SecureChannel privacy class as a specialisation of Communication Protection.

Most privacy related technologies result in specific tasks, thus BPMN Task is extended with abstract *PET-Task*. Figure 1 illustrates four specialisations of PET-Task based on Table 1: Data Protection, Entity Authentication, Privacy-Aware Computation and Human-Data Interaction. We focus on Data Protection and Privacy-aware computation technologies that are used in the scenario discussed in Sect. 5.

Secret sharing is a specialisation of Confidentiality. It splits private values among participants so that some predefined groups of parties can collaboratively restore the secret [5,26]. Secret sharing consists of two major tasks: producing the shares (i.e., SSsharing) from a secret and restoring the secret from the shares (i.e., SSreconstruction). Secret sharing is most useful if it is homomorphic and allows Secure Multiparty Computation (i.e., SScomputation).

Encryption is another specialisation of Confidentiality. For example, public key cryptosystems [10] protect data using encryption (i.e. PKencryption) and allow to open it using decryption (i.e. PKdecryption). More specifically, encryption

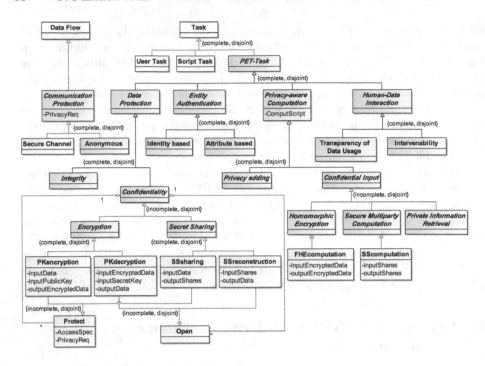

Fig. 1. Extension of the BPMN abstract syntax

requires the input data and a public key to produce a ciphertext and decryption needs a ciphertext and the respective secret key to restore the encrypted data. Some cryptosystems also have homomorphic properties, for example fully homomorphic encryption (FHE) [15] allows to compute any functionality on encrypted data to provide an encrypted output. Hence, FHEcomputation is an extension of Privacy-Aware Computation on confidential input, where the confidentiality is ensured by encryption. We assume that the encrypted values reveal nothing about the inputs that they protect except to parties that also hold the secret key.

Figure 1 is not complete as other privacy technologies can be added from Table 1. However, it gives an example to both single task and multi-task technologies as well as secure communication, making extensions straightforward.

4.2   Concrete Syntax

Extension of the BPMN concrete syntax to add PETs is done using *stereotypes* with the general stereotypes given in Table 2. The *stereotype* characterises the changed type of the BPMN construct. *Parameter* (it has similarity to the *tag* in UML) is the variable, that gives additional details about the execution of the action. Some goals result in a series of tasks, for example data protection allows adding protection with Protect and removing it with Open.

Table 2. Stereotypes for PET goals

Group	Stereotype	Parameter
Communication	SecureChannel	Privacy requirements
Data protection	Protect	Access specification, privacy requirements
	Open	
Entity authentication	AuthenticationProof	
	AuthenticationVerification	
Privacy aware computation	PETcomputation	Computation script
Human-Data interaction	CheckingPermissions	

For concrete stereotypes (see Table 3), the *input* describes the types of data needed to perform the task, and the *output* is the result. The input and output types can be used to typecheck the models and help the user. Note that, based on the PET classification, we also obtain a hierarchy of the stereotypes. We discuss the concrete syntax of PE-BPMN needed in the example scenario in Sect. 5.

Table 3. Example stereotypes

Stereotype	General stereotype	Input	Output
SecureChannel	SecureChannel	Data	Data
SSsharing	Protect	Data	Shares
AddSSsharing	SSsharing	Data	Additive shares
FunSSsharing	SSsharing	Data	Function shares
PKencryption	Protect	Data, public key	Encrypted data
SSreconstruction	Open	Shares	Data
AddSSreconstruction	SSreconstruction	Additive shares	Data
FunSSreconstruction	SSreconstruction	Function shares	Data
PKdecryption	Open	Encrypted data, secret key	Data
SScomputation	PETComputation	Shares	Shares
AddSScomputation	SScomputation	Additive shares	Additive shares
FunSScomputation	SScomputation	Function shares	Additive shares
FHEcomputation	PETComputation	Encrypted data	Encrypted data

Secret Sharing needs specializations of Protect and Open, Fig. 2. SSsharing splits the input *data* to the number of *shares*, determined by the *access specification*. SSreconstruction inverts SSsharing: it restores input *shares* to public *data*. SScomputation tasks define the computations on shares specified by the *script*.

We need two *secret sharing* specialisations in Sect. 5 – *additive secret sharing* (AddSS) and *function secret sharing* (FunSS). In *AddSS* each participant P_i gets a share x_i of secret x so that it can be reconstructed as $x = \sum x_i$. *FunSS* [6]

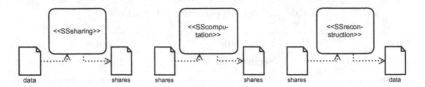

Fig. 2. Secret sharing concrete syntax

can be thought of as an additive secret sharing scheme for functions. In FunSS, the secret is a function f and the shares are also functions f_i. Moreover, if all participants evaluate their functions f_i on a common input x then $f_i(x)$ are the additive shares of $f(x) = \sum f_i(x)$. In both schemes, the secret is revealed only when all parties join their shares.

Encryption specifies data protection tasks in Fig. 3. Specifically, PKencryption encrypts data with a public key (PK) that is paired with a secret key used to open the secret with PKdecryption. Fully homomorphic encryption (**FHE**) defines FHEcomputation to process encrypted values according to *script*.

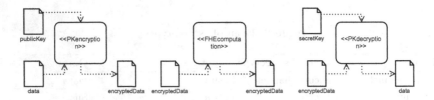

Fig. 3. Encryption and FHE concrete syntax

Secure Channel is an extension of the BPMN DataFlow construct with stereotype SecureChannel, Fig. 4. It means that *data* are sent by activity A and received by activity B without interference. Additional *privacy requirements* (e.g., anonymity requirement) can be used to change the properties of the channel.

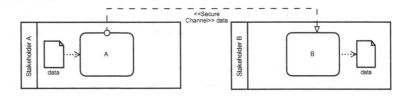

Fig. 4. Secure channel concrete syntax

4.3   Use-Cases for PE-BPMN

The purpose of general stereotypes is twofold: they can be used to specify the basis for privacy analysis or used to iteratively move to concrete technologies. The general stereotypes specify the overall goals and the concrete PET stereotypes also contribute the limitations of the technology. We discuss how PE-BPMN supports choosing PETs, validating the models and characterising leakages.

Choosing PETs. Concrete technologies introduce different trade-offs or limitations even if they belong to the same groups in Table 1. Therefore, considering privacy in business process starts with fixing the general goal and stereotype and then iteratively making specific decisions to fix the PETs. Choosing PETs from the general stereotypes can be done with the help of decision trees or other specifications of the PET properties. For example, the decision should take into account the necessary efficiency or computation capabilities. It is useful to model the PETs in BPMN and not leave the choice to later stages of development as they may introduce new stakeholders. For example, secure multiparty computation techniques are often applicable in theory but, in many processes, it is hard to find stakeholders that are willing to participate in the computation.

Leakage Analysis. We foresee the use of PE-BPMN for the analysis in the *honest-but-curious* security model where we are only interested in what leaks in the process when followed as designed. For each stakeholder (i.e., *swimlane*) and each *data object*, we can determine, to which extent does the stakeholder learn the data. The stakeholders learn all data sent to them and protected contents of objects they can open. In addition, everything sent over an insecure network is considered to leak to the external telecommunication service provider.

Data privacy analysis focuses more on the PETs that provide confidentiality and less on integrity and authenticity. Some cryptographic means can be abstracted as providing perfect privacy (data confidentiality protection, computation on confidential inputs, secure communication) and others as limiting data leakage but not removing it fully (privacy adding computation, human-data interaction). The latter group requires careful treatment to find ways to quantify the leakage. The first group enables a black-or-white leakage analysis where we consider the protected contents of a data object not to leak to parties only holding the protected object and not meeting the access specification.

In the most direct analysis, we identify the data protection mechanism applied for each data object and stakeholder. For example, [2] provides a tool to find leaking data flows in business processes without PETs that could be used as a basis for further development. The next step is to quantify how much the derived public data objects leak about the private inputs. The basis for any quantitative analysis is the data dependency and the computation scripts or metadata. For example, [11] provides a version of quantitative analysis for differentially private tasks. Analysing other technologies that require quantification involves detailed study of the properties of these technologies and specific PET stereotypes. Our PE-BPMN notation enables to denote both privacy-adding and full protection mechanisms in BPMN and enables to combine the analysis.

Protection mechanisms can only be considered secure, if their underlying assumptions are satisfied. A version of the analysis lists all underlying assumptions based on the PETs information. For example, that cryptographic keys are generated by correct parties and some participants do not collude. Such analysis can also benefit from the integrity and authenticity providing PETs that can be used to lift some assumptions (e.g. authenticity of the key). Furthermore, models using PE-BPMN could be a good input documents for auditing the systems.

5   Applying PE-BPMN

To illustrate feasibility of the PE-BPMN to various processes, we describe an extract of the mobile app RapidGather and summarise other experiments. This app is developed by the Privacy-Enhanced Android Research and Legacy Systems (PEARLS) team in DARPA Brandeis program[1]. It enables a rapid response to an imminent threat. In case of an event, emergency officers would use the RapidGather infrastructure to collect data from RapidGather app and to analyse them at the command center. RapidGather has many scenarios deploying different privacy enhancing technologies. These include location analysis, private machine learning using photos from the mobile device and computing a reputation for each device using secure hardware. We also model procedures such as uploading the application to the app store or installing it to the phone. In addition to RapidGather, we are exploring other scenarios in the Brandeis program, for example, Internet of things setting with data streams, that need a different level of detail on the inner workings of the computation. Use-cases for PE-BPMN are all characterized as processes with multiple stakeholders and private data.

5.1   Lessons Learned

The idea of using various levels of generalisation of stereotypes arose from the different requirements of the teams in Brandeis, whereas our initial approach was focused on concrete technologies. The abstract syntax and classification are expressive enough to allow a wide selection of technologies. Elsewhere the approach was applied in a commercial project to assess the suitability of a secure querying system in a healthcare service. Adjusting the querying system to the stakeholders' business processes highlighted unforeseen (and unacceptable) data leakages due to the setup invalidating the underlying assumptions of the querying system. This helped to shape the idea of the stereotypes and to explore the balance between the approaches of the systems analyst and security engineer. We have observed that using PE-BPMN requires a different focus from the analyst: (i) analysis requires more details of the data objects that need to be explicit, and

---

[1] DARPA Brandeis—http://www.darpa.mil/program/brandeis.

(*ii*) fine-grained separation of stakeholders yields better analysis. More specifically the suggested approach helps to identify and fix data leakages during system design, supports communication and documentation of PETs usage, and stresses on limitations of PETs usage (e.g. separation of duty).

We have also noticed some limitations to the current validation and ideas of PE-BPMN. For example, there may be better means of recording PETs information on the model. Additionally our proposal is based on the assumption that cryptographic techniques have standardised usage (i.e., fixed tasks).

5.2   RapidGather Location Analysis

We show details of a location analysis process in RapidGather scenario. The goal is twofold: (*i*) to demonstrate the PE-BPMN modelling applicability, and (*ii*) to illustrate privacy analysis means in business processes with PETs.

Scenario Description: Modelling Without PETs. As illustrated in Fig. 5, the RapidGather app initiates the collection of location data (see A1). The Android OS is responsible for preprocessing (see A2 and A3) and submitting data to the Compute server. The Compute server processes (see A5) and stores (see A6) the data as a heatmap which characterizes the emergency. Next, data are analysed by the Command center employees (see Fig. 6). Once the request for the movement heatmap is submitted to the Compute server (see A8), the heatmap is extracted and provided to the Command center for inspection (see A11).

The main privacy concern in this case is the danger of leaking mobile device location through communication or computation. We need to ensure that the

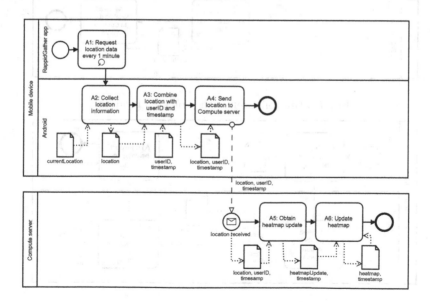

Fig. 5. Data collection

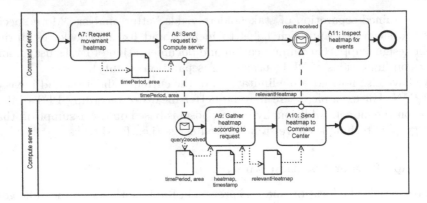

Fig. 6. Data analysis

compute server does not learn the location of any specific device and command center is only able to see the aggregated heatmap. We apply PE-BPMN to address the privacy concerns and illustrate the model changes as we add PETs.

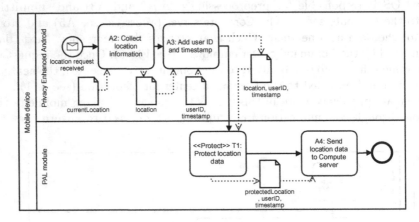

Fig. 7. Data collection with protected location

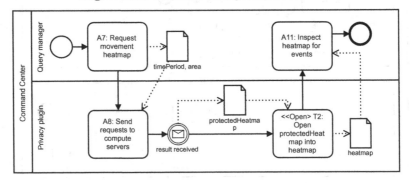

Fig. 8. Data analysis results from protected location

Choosing the PETs. Tracking which data object is seen by which party identifies that the location data might be revealed to the telecommunication party and computing server as it is sent in clear. The former can be solved with SecureChannel. For the latter, there are two initial choices, either to ensure full confidentiality or loose the precision to reduce the leakage. Note that many of the privacy adding computations can not be easily used as they work on databases but we have a single location. The remaining choice is to add a task T1 with Protect stereotype and confidentiality parameter before sending the location to the compute server as on Fig. 7. This introduces the requirement that the following computations are PETcomputation and the output must be *opened* as task T2 on Fig. 8 to comply with the input and output types.

It is possible to leave the choice of PETs at this level to show the desired properties. It is also possible to narrow the general stereotypes to concrete technologies with the help of the PET classification. This choice depends on the stage of the system development and the capabilities of the analyst. There are many considerations besides the goal of the PET to choose the exact technology. In this scenario, the process on the mobile phone should be efficient to save the battery and limit data usage. Also, the overall heatmap updates in the compute server should be fast to allow timely updates. Applicable confidentiality mechanisms include *encryption* and *secret sharing* with respective computation technologies.

Encryption allows using one compute server to perform computations privately (see data collection in Fig. 9 and analysis in Fig. 10). However, the required computations are broad, meaning that it would require FHE as no special purpose encryption supports these operations. The main trouble is that FHE com-

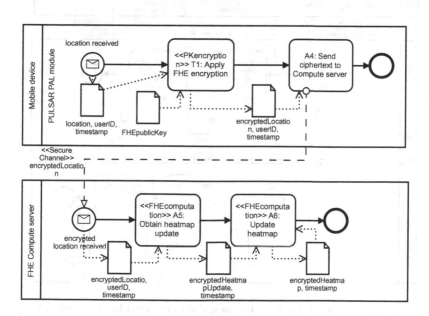

Fig. 9. Data collection with FHE protected location

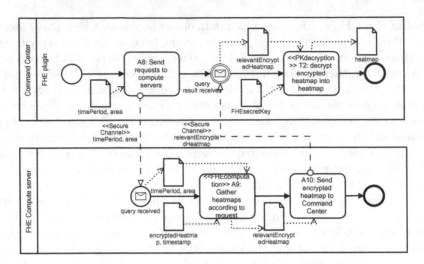

Fig. 10. Data analysis results from FHE protected location

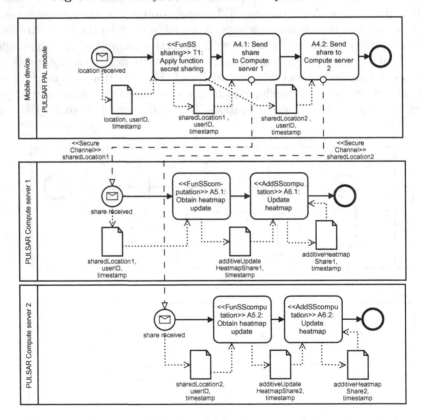

Fig. 11. Data collection with location data protected by secret sharing

putation is not efficient yet and especially, it requires significant computations in the phone to encrypt the data. In addition, FHE requires key distribution.

Secret sharing with SScomputation is an alternative approach finding the heatmap. In this case, the protection mechanism produces shares that are distributed to the computing parties. A new stakeholder is required to deploy the second computing server. The compute servers collaboratively use computation protocols to obtain the final heatmap that is reconstructed at the command center. To stress details that can be documented in PE-BPMN, Figs. 11 and 12 combine AddSS and FunSS sharing schemes as used in RapidGather. In comparison to the FHE, the secret sharing solution requires more communication.

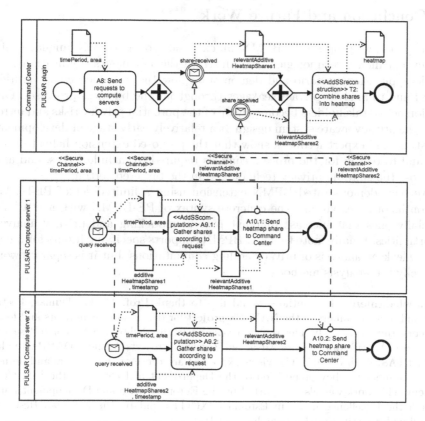

Fig. 12. Data analysis results from location data protected by secret sharing

Privacy Analysis. The first layer of leakage analysis is simplified by data object names (e.g. shared or encrypted vs unprotected) in the examples, e.g. encrypted location does not leak the location. In both scenarios with PETs, the location data is protected before it is given to the compute servers and the compute servers only process it without revealing the data. Therefore, nothing about the location leaks to the compute servers. All communication is secured and, therefore, the telecommunications companies can not learn the contents. However, the heatmap of the private location is revealed to the command center. It is a further question to quantify how much it leaks about the locations. Such quantification can be approached by finding the sensitivity of the computation and applying differential privacy to reduce the leakage, as in [11].

The assumptions specify that, for secret sharing, the two computing parties must be non-colluding. For FHE, the encryption key must be generated correctly and location privacy is only preserved if encryption uses the command center's public key and the secret key is kept securely by the command center.

6   Conclusion and Future Work

In this paper, we propose PE-BPMN, an extension to the BPMN language with privacy enhancing technologies. We illustrate the language's application in a mobile application scenario and demonstrate how the approach could be used to find and to analyse privacy leakages throughout the business processes. Our solution helps stakeholders become aware of the potential privacy risks and introduces the privacy aware system design at a relatively early stage of development. The study and expert feedback show that the proposed extension helps to visualise and to reason for the process changes required to include PETs, and aids to choose the suitable privacy technologies in the targeted setting.

We have demonstrated BPMN extension using a limited set of PETs. We are continuously expanding the concrete syntax of PE-BPMN with new PETs, especially ones that increase different process modelling and privacy analysis opportunities. Finally, note that PE-BPMN supports specifying inputs for quantitative leakage analysis of privacy adding computations, but it is separate work to specify the analysis methods.

Acknowledgment. The authors would like to thank Prof. Marlon Dumas, Peeter Laud and other members of the NAPLES project for discussions, comments and feedback concerning this study. This research was, in part, funded by the Air Force Research laboratory (AFRL) and Defense Advanced Research Projects Agency (DARPA) under contract FA8750-16-C-0011. The views expressed are those of the authors and do not reflect the official policy or position of the Department of Defense or the U.S. Government. This work was also supported by the European Regional Development Fund through the Excellence in IT in Estonia (EXCITE) and by the Estonian Research Council under Institutional Research Grant IUT27-1.

References

1. Privacy management reference model and methodology (PMRM) version 1.0. OASIS Committee Specification 02 (2016). http://docs.oasis-open.org/pmrm/PMRM/v1.0/cs02/PMRM-v1.0-cs02.html
2. Accorsi, R., Lehmann, A., Lohmann, N.: Information leak detection in business process models. Inf. Syst. **47**(C), 244–257 (2015)

3. Altuhhova, O., Matulevičius, R., Ahmed, N.: An extension of business process model and notification for security risk management. IJISMD **4**(4), 93–113 (2013)
4. Ayed, G.B., Ghernaouti-Helie, S.: Processes view modeling of identity-related privacy business interoperability: considering user-supremacy federated identity technical model and identity contract negotiation. In: 2012 Proceedings of the ASONAM (2012)
5. Blakley, G.R.: Safeguarding cryptographic keys. In: Proceedings of the 1979 AFIPS National Computer Conference, pp. 313–317. AFIPS Press (1979)
6. Boyle, E., Gilboa, N., Ishai, Y.: Function secret sharing. In: Oswald, E., Fischlin, M. (eds.) EUROCRYPT 2015. LNCS, vol. 9057, pp. 337–367. Springer, Heidelberg (2015). doi:10.1007/978-3-662-46803-6_12
7. Brucker, A.D., Hang, I., Lückemeyer, G., Ruparel, R.: SecureBPMN: modeling and enforcing access control requirements in business processes. In: Proceedings of the SACMAT 2012, pp. 123–126. ACM (2012)
8. Cherdantseva, Y., Hilton, J., Rana, O.: Towards SecureBPMN - aligning BPMN with the information assurance and security domain. In: Mendling, J., Weidlich, M. (eds.) BPMN 2012. LNBIP, vol. 125, pp. 107–115. Springer, Heidelberg (2012). doi:10.1007/978-3-642-33155-8_9
9. Danezis, G., Domingo-Ferrer, J., Hansen, M., Hoepman, J.-H., Metayer, D.L., Tirtea, R., Schiffner, S.: Privacy and data protection by design-from policy to engineering. Technical report, European Union Agency for Network and Information Security (2015)
10. Diffie, W., Hellman, M.: New directions in cryptography. IEEE Trans. Inf. Theor. **22**(6), 644–654 (2006)
11. Dumas, M., García-Bañuelos, L., Laud, P.: Differential privacy analysis of data processing workflows. In: Kordy, B., Ekstedt, M., Kim, D.S. (eds.) GraMSec 2016. LNCS, vol. 9987, pp. 62–79. Springer, Cham (2016). doi:10.1007/978-3-319-46263-9_4
12. Dumas, M., La Rosa, M., Mendling, J., Reijers, H.: Fundamentals of Business Process Management. Springer, Heidelberg (2013)
13. Regulation on the protection of natural persons with regard to the processing of personal data and on the free movement of such data, and repealing Directive 95/46/EC (General Data Protection Regulation) (2016). http://data.europa.eu/eli/reg/2016/679/oj
14. Joint Task Force and Transformation Initiative: Security and privacy controls for federal information systems and organizations. NIST Special Publication, 800:53 (2013)
15. Gentry, C.: Fully homomorphic encryption using ideal lattices. In: Proceedings of the Forty-first Annual ACM Symposium on Theory of Computing, STOC 2009, pp. 169–178. ACM, New York (2009)
16. Greenberg, A.: Apple's 'Differential Privacy' is about collecting your data-but not your data. Wired (2016)
17. Heurix, J., Zimmermann, P., Neubauer, T., Fenz, S.: A taxonomy for privacy enhancing technologies. Comput. Secur. **53**, 1–17 (2015)
18. ISO/IEC DIS 29134: Information technology - security techniques - privacy impact assessment - guidelines. Technical report, International Organization for Standardization (2016)
19. Koorn, R., van Gils, H., ter Hart, J., Overbeek, P., Tellegen, R., Borking, J.: Privacy Enhancing Technologies, White Paper for Decision Makers. Ministry of the Interior and Kingdom Relations, The Netherlands (2004)

20. Ladha, W., Mehandjiev, N., Sampaio, P.: Modelling of privacy-aware business processes in BPMN to protect personal data. In: Proceedings of the 29th Annual ACM Symposium on Applied Computing, pp. 1399–1405 (2014)
21. Menzel, M., Thomas, I., Meinel, C.: Security requirements specification in service-oriented business process management. In: ARES 2009, pp. 41–49 (2009)
22. Mouratidis, H., Kalloniatis, C., Islam, S., Hudic, A., Zechner, L.: Model based process to support security and privacy requirements engineering. Int. J. Secur. Softw. Eng. **3**(3), 1–22 (2012)
23. OMG: Business Process Model and Notation (BPMN). http://www.omg.org/spec/BPMN/2.0/
24. Rodriguez, A., Fernandez-Medina, E., Piattini, M.: A BPMN extension for the modeling of security requirements in business processes. IEICE Trans. Inf. Syst. **90**(4), 745–752 (2007)
25. Schleicher, D., Leymann, F., Schumm, D., Weidmann, M.: Compliance scopes: extending the BPMN 2.0 meta model to specify compliance requirements. In: SOCA 2010, pp. 1–8 (2010)
26. Shamir, A.: How to share a secret. Commun. ACM **22**(11), 612–613 (1979)
27. Solove, D.J.: A taxonomy of privacy. Univ. Pa. Law Rev. **154**, 477–564 (2006)
28. Weiss, M.A., Archick, K.: US-EU Data Privacy: From Safe Harbor to Privacy Shield. Congressional Research Service (2016)

Process Mining 1

Learning Hybrid Process Models from Events
Process Discovery Without Faking Confidence

Wil M.P. van der Aalst[1,2]([⊠]), Riccardo De Masellis[2],
Chiara Di Francescomarino[2], and Chiara Ghidini[2]

[1] Eindhoven University of Technology, PO Box 513, Eindhoven, The Netherlands
w.m.p.v.d.aalst@tue.nl
[2] FBK-IRST, Via Sommarive 18, 38050 Trento, Italy
{r.demasellis,dfmchiara,ghidini}@fbk.eu

Abstract. Process discovery techniques return process models that are
either formal (precisely describing the possible behaviors) or informal
(merely a "picture" not allowing for any form of formal reasoning). Formal models are able to classify traces (i.e., sequences of events) as fitting
or non-fitting. Most process mining approaches described in the literature produce such models. This is in stark contrast with the over 25
available commercial process mining tools that only discover informal
process models that remain *deliberately vague* on the precise set of possible traces. There are two main reasons why vendors resort to such models:
scalability and *simplicity*. In this paper, we propose to combine the best
of both worlds: discovering *hybrid process models* that have formal and
informal elements. As a proof of concept we present a discovery technique
based on *hybrid Petri nets*. These models allow for formal reasoning, but
also reveal information that cannot be captured in mainstream formal
models. A novel discovery algorithm returning hybrid Petri nets has been
implemented in ProM and has been applied to several real-life event logs.
The results clearly demonstrate the advantages of remaining "vague"
when there is not enough "evidence" in the data or standard modeling
constructs do not "fit". Moreover, the approach is scalable enough to be
incorporated in industrial-strength process mining tools.

Keywords: Process mining · Process discovery · Petri nets · BPM

1 Introduction

The increased interest in process mining illustrates that Business Process Management (BPM) is rapidly becoming more data-driven [1]. Evidence-based BPM
based on process mining helps to create a common ground for business process
improvement and information systems development. The uptake of process mining is reflected by the growing number of commercial process mining tools available today. There are over 25 commercial products supporting process mining
(Celonis, Disco, Minit, myInvenio, ProcessGold, QPR, etc.). All support process
discovery and can be used to improve compliance and performance problems.

© Springer International Publishing AG 2017
J. Carmona et al. (Eds.): BPM 2017, LNCS 10445, pp. 59–76, 2017.
DOI: 10.1007/978-3-319-65000-5_4

For example, without any modeling, it is possible to learn process models clearly showing the main bottlenecks and deviating behaviors.

These commercial tools are based on variants of techniques like the heuristic miner [17] and the fuzzy miner [8] developed over a decade ago [1]. All return process models that *lack formal semantics* and thus cannot be used as a *classifier* for traces. Classifying traces into *fitting* (behavior allowed by the model) and *non-fitting* (not possible according to the model) is however important for more advanced types of process mining. Informal models ("boxes and arcs") provide valuable insights, but cannot be used to draw reliable conclusions. Therefore, most discovery algorithms described in the literature (e.g., the α-algorithm [3], the region-based approaches [6,15,18], and the inductive mining approaches [11–13]) produce formal models (Petri nets, transition systems, automata, process trees, etc.) having clear semantics.

So why did vendors of commercial process mining tools opt for informal models? Some of the main drivers for this choice include:

- *Simplicity*: Formal models may be hard to understand. End-users need to be able to interpret process mining results: Petri nets with smartly constructed places and BPMN with many gateways are quickly perceived as too complex.
- *Vagueness*: Formal models act as binary classifiers: traces are fitting or non-fitting. For real-life processes this is often not so clear cut. The model capturing 80 percent of all traces may be simple and more valuable than the model that allows for all outliers and deviations seen in the event log. Hence, "vagueness" may be desirable to show relationships that cannot be interpreted in a precise manner.
- *Scalability*: Commercial process mining tools need to be able to handle logs with millions of events and still be used in an interactive manner. Many of the more sophisticated discovery algorithms producing formal models (e.g., region-based approaches [6,15,18]) do not scale well.

The state-of-the-art commercial products show that simplicity, vagueness and scalability can be combined effectively. Obviously, vagueness and simplicity may also pose problems. People may not trust process mining results when a precise interpretation of the generated model is impossible. When an activity has multiple outgoing arcs, i.e., multiple preceding activities, one would like to know whether these are concurrent or in a choice relation. Which combinations of output arcs can be combined? Showing frequencies on nodes (activities) and arcs may further add to the confusion when "numbers do not add up".

We propose *hybrid process models* as a way to combine the best of both worlds. Such models show informal dependencies (like in commercial tools) that are deliberately vague and at the same time provide formal semantics for the parts that are clear-cut. Whenever there is enough structure and evidence in the data, explicit routing constructs are used. If dependencies are weak or too complex, then they are not left out, but depicted in an informal manner.

We use *hybrid Petri nets*, a new class for Petri nets with informal annotations, as a concrete representation of hybrid process models. However, the ideas, concepts, and algorithms are generic and could also be used in the context of

BPMN, UML activity diagrams, etc. Our proposed *discovery technique* has two phases. First we discover a *causal graph* based on the event log. Based on different (threshold) parameters we scan the event log for possible causalities. In the second phase we try to learn places based on explicit quality criteria. Places added can be interpreted in a precise manner and have a guaranteed quality. Causal relations that cannot or should not be expressed in terms of places are added as sure or unsure arcs. The resulting hybrid Petri net can be used as a starting point for other types of process mining.

The approach has been implemented in *ProM* and has been tested on various event logs and processes. These applications of our approach show that hybrid process models are useful and combine the best of both worlds: *simplicity, vagueness, and scalability can be combined with partly formal models that allow for reasoning and provide formal guarantees.*

The remainder is organized as follows. We first present a running example (Sect. 2) and some preliminaries (Sect. 3). Section 4 defines hybrid Petri nets. The actual two-phase discovery approach is presented in Sect. 5. Section 6 describes the *ProM* plug-ins developed to support the discovery of hybrid process models. Section 7 evaluates the approach. Section 8 discusses related work and Sect. 9 concludes the paper.

2   Motivating Example

Figure 1 illustrates the trade-offs using example data from an order handling process. All five models have been produced for the same event log containing 12,666 cases, 80,609 events, and eight unique activities. Each case has a corresponding trace, i.e., a sequence of events. Models (a), (b), and (c) are expressed in terms of a Petri net and have formal semantics. Model (a) was created using the ILP miner with default settings; it is precise and each of the 12,666 cases perfectly fits the model. However, model (a) is difficult to read. For larger event logs, having more activities and infrequent paths, the ILP miner is not able to

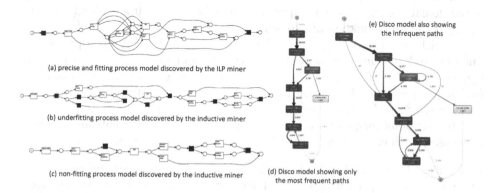

(a) precise and fitting process model discovered by the ILP miner

(b) underfitting process model discovered by the inductive miner

(c) non-fitting process model discovered by the inductive miner

(d) Disco model showing only the most frequent paths

(e) Disco model also showing the infrequent paths

Fig. 1. Five process models discovered for an event log recorded for 12,666 orders (labels are not intended to be readable).

produce meaningful models (the approach becomes intractable and/or produces incomprehensible models). Models (b) and (c) were created using the inductive miner (IMf [12]) with different settings for the noise threshold (0.0 respectively 0.2). Model (b) is underfitting, but able to replay all cases. Model (c) focuses on the mainstream behavior only, but only 9,440 of the 12,666 cases fit perfectly. In 3,189 cases there are multiple reminders and in 37 cases the payment is done before sending the invoice. All other cases conform to model (c). Models (d) and (e) were created using the commercial process mining tool *Disco* (Fluxicon) using different settings. These models are informal. Model (d) shows only the most frequent paths and model (e) shows all possible paths. For such informal models it is impossible to determine the exact nature of splits and joins. Commercial tools have problems dealing with loops and concurrency. For example, for each of the 12,666 cases, activities *make delivery* and *confirm payment* happened at most once, but not in a fixed order. However, these concurrent activities are put into a loop in models (d) and (e). This problem is not specific for *Disco* or this event log: all commercial tools suffer from this problem.

We would like to combine the left-hand side and the right-hand side of Fig. 1 by using formal semantics when the behavior is clear and easy to express and resorting to informal annotations when things are blurry or inexact.

3   Preliminaries

In this section we introduce basic concepts, including multisets, operations on sequences, event logs and Petri nets.

$\mathcal{B}(A)$ is the set of all multisets over some set A. For some multiset $X \in \mathcal{B}(A)$, $X(a)$ denotes the number of times element $a \in A$ appears in X. Some examples: $X = []$, $Y = [x, x, y]$, and $Z = [x^3, y^2, z]$ are multisets over $A = \{x, y, z\}$. X is the empty multiset, Y has three elements ($Y(x) = 2$, $Y(y) = 1$, and $Y(z) = 0$), and Z has six elements. Note that the ordering of elements is irrelevant.

$\sigma = \langle a_1, a_2, \ldots, a_n \rangle \in A^*$ denotes a sequence over A. $\sigma(i) = a_i$ denotes the i-th element of the sequence. $|\sigma| = n$ is the length of σ and $dom(\sigma) = \{1, \ldots, |\sigma|\}$ is the domain of σ. $\langle \rangle$ is the empty sequence, i.e., $|\langle \rangle| = 0$ and $dom(\langle \rangle) = \emptyset$. $\sigma_1 \cdot \sigma_2$ is the concatenation of two sequences.

Let A be a set and $X \subseteq A$ one of its subsets. $\lceil_X \in A^* \to X^*$ is a projection function and is defined recursively: $\langle \rangle \lceil_X = \langle \rangle$ and for $\sigma \in A^*$ and $a \in A$: $(\langle a \rangle \cdot \sigma) \lceil_X = \sigma \lceil_X$ if $a \notin X$ and $(\langle a \rangle \cdot \sigma) \lceil_X = \langle a \rangle \cdot \sigma \lceil_X$ if $a \in X$. For example, $\langle a, b, a \rangle \lceil_{\{a,c\}} = \langle a, a \rangle$. Projection can also be applied to multisets of sequences, e.g., $[\langle a, b, a \rangle^5, \langle a, d, a \rangle^5, \langle a, c, e \rangle^3] \lceil_{\{a,c\}} = [\langle a, a \rangle^{10}, \langle a, c \rangle^3]$.

Starting point for process discovery is an event log where events are grouped into cases. Each case is represented by a trace, e.g., $\langle \triangleright, a, b, c, d, \square \rangle$.

Definition 1 (Event Log). *An event log $L \in \mathcal{B}(A^*)$ is a non-empty multiset of traces over some activity set A. A trace $\sigma \in L$ is a sequence of activities. There is a special start activity $\triangleright$ and a special end activity $\square$. We require that $\{\triangleright, \square\} \subseteq A$ and each trace $\sigma \in L$ has the structure $\sigma = \langle \triangleright, a_1, a_2, \ldots, a_n, \square \rangle$*

and $\{\triangleright, \square\} \cap \{a_1, a_2, \ldots, a_n\} = \emptyset$. $\mathcal{U}_L$ *is the set of all event logs satisfying these requirements.*

An event log captures the observed behavior that is used to learn a process model. An example log is $L_1 = [\langle \triangleright, a, b, c, d, \square \rangle^{45}, \langle \triangleright, a, c, b, d, \square \rangle^{35}, \langle \triangleright, a, e, d, \square \rangle^{20}]$ containing 100 traces ($|L_1| = 100$) and 580 events ($\sum_{\sigma \in L_1} |\sigma| = 580$). In reality, each event has a timestamp and may have any number of additional attributes. For example, an event may refer to a customer, a product, the person executing the event, associated costs, etc. Here we abstract from these notions and simply represent an event by its activity name.

A *Petri net* is a bipartite graph composed of places (represented by circles) and transitions (represented by squares).

Definition 2 (Petri Net). *A Petri net is a tuple $N = (P, T, F)$ with P the set of places, T the set of transitions, $P \cap T = \emptyset$, and $F \subseteq (P \times T) \cup (T \times P)$ the flow relation.*

Transitions represent activities and places are added to model causal relations. $\bullet x = \{y \mid (y, x) \in F\}$ and $x \bullet = \{y \mid (x, y) \in F\}$ define input and output sets of places and transitions. Places can be used to causally connect transitions as is reflected by relation $\widehat{F}$: $(t_1, t_2) \in \widehat{F}$ if t_1 and t_2 are connected through a place p, i.e., $p \in t_1 \bullet$ and $p \in \bullet t_2$.

Definition 3 ($\widehat{F}$). *Let $N = (P, T, F)$ be a Petri net. $\widehat{F} = \{(t_1, t_2) \in T \times T \mid \exists_{p \in P} \{(t_1, p), (p, t_2)\} \subseteq F\}$ are all pairs of transitions connected through places.*

The state of a Petri net, called *marking*, is a multiset of places indicating how many *tokens* each place contains. Tokens are shown as block dots inside places.

Definition 4 (Marking). *Let $N = (P, T, F)$ be a Petri net. A marking M is a multiset of places, i.e., $M \in \mathcal{B}(P)$.*

A transition $t \in T$ is *enabled* in marking M of net N, denoted as $(N, M)[t\rangle$, if each of its input places ($p \in \bullet t$) contains at least one token. An enabled transition t may *fire*, i.e., one token is removed from each of the input places ($p \in \bullet t$) and one token is produced for each of the output places ($p \in t \bullet$).

$(N, M)[t\rangle(N, M')$ denotes that t is enabled in M and firing t results in marking M'. Let $\sigma = \langle t_1, t_2, \ldots, t_n \rangle \in T^*$ be a sequence of transitions, sometimes referred to as a *trace*. $(N, M)[\sigma\rangle(N, M')$ denotes that there is a set of markings $M_0, M_1, \ldots, M_n$ such that $M_0 = M$, $M_n = M'$, and $(N, M_i)[t_{i+1}\rangle(N, M_{i+1})$ for $0 \leq i < n$.

A *system net* has an initial and a final marking. The *behavior* of a system net corresponds to the set of traces starting in the initial marking M_{init} and ending in the final marking M_{final}.

Definition 5 (System Net Behavior). *A system net is a triplet $SN = (N, M_{init}, M_{final})$ where $N = (P, T, F)$ is a Petri net, $M_{init} \in \mathcal{B}(P)$ is the initial marking, and $M_{final} \in \mathcal{B}(P)$ is the final marking. $behav(SN) = \{\sigma \mid (N, M_{init})[\sigma\rangle(N, M_{final})\}$ is the set of traces possible according to the model.*

Note that a system net classifies traces σ into *fitting* ($\sigma \in behav(SN)$) and *non-fitting* ($\sigma \notin behav(SN)$).

4   Hybrid Petri Nets

A *formal process model* is able to make firm statements about the inclusion or exclusion of traces, e.g., trace $\langle \triangleright, a, b, c, d, \Box \rangle$ fits the model or not. *Informal process models* are unable to make such precise statements about traces. Events logs only show example behavior: (1) logs are typically incomplete (e.g., the data only shows a fraction of all possible interleavings, combinations of choices, or unfoldings) and (2) logs may contain infrequent exceptional behavior where the model should abstract from. Therefore, it is impossible to make conclusive decisions based on event logs. More observations may lead to a higher certainty and the desire to make a formal statement (e.g., "after a there is a choice between b and c"). However, fewer observations and complex dependencies create the desire to remain "vague". Models (a), (b) and (c) in Fig. 1 have formal semantics as described in Definition 5. (The initial and final markings are defined but not indicated explicitly: the source places are initially marked and the sink places are the only places marked in the final markings.) Models (d) and (e) in Fig. 1 are informal and therefore unable to classify traces into fitting and non-fitting.

In essence process models describe *causalities* between activities. Depending on the evidence in the data these causalities can be seen as stronger ("sure") or weaker ("unsure"). The strength of a causal relation expresses the level of confidence. A strong causality between two activities a and b suggests that one is quite sure that activity a causes activity b to happen later in time. This does not mean that a is always followed by b. The occurrence of b may depend on other factors, e.g., b requires c to happen concurrently or a only increases the likelihood of b.

The strength of a causality and the formality of a modeling construct are orthogonal as shown in Fig. 2. Even when one is not sure, one can still use a formally specified modeling construct. Moreover, both notions may be local, e.g., parts of the process model are more certain or modeled precisely whereas other parts are less clear and therefore kept vague.

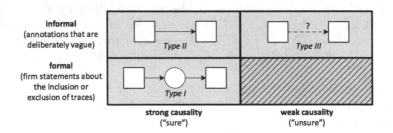

Fig. 2. The strength of a causality and the formality of a modeling construct are orthogonal. However, it makes less sense to express a weak causality in a formal manner.

As Fig. 2 suggests it seems undesirable to express a weak causality using a formal construct. Moreover, depending on the *representational bias* of the modeling notation, strong causalities may not be expressed easily. The modeling notation may not support concurrency, duplicate activities, unstructured models, long-term dependencies, OR-joins, etc. Attempts to express behavior incompatible with representational bias of the modeling notation in a formal model are doomed to fail. Hence, *things that cannot be expressed easily in an exact manner can only be captured using annotations that are deliberately vague and non-executable.* Instead, we aim to combine the best of both worlds, i.e., marrying the left-hand side and the right-hand side of Fig. 1 by combining both formal and informal notations.

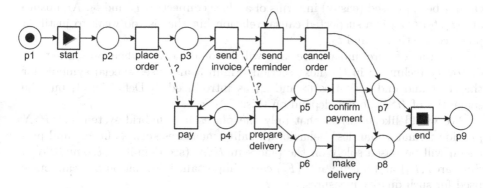

Fig. 3. A hybrid system net with $M_{init} = [p1]$ and $M_{final} = [p9]$. This hybrid model was discovered using the approach presented in Sect. 5.

Although the ideas are *generic* and also apply to other notations (BPMN, UML activity diagrams, etc.), we operationalize the notion of hybrid process models by defining and using so-called *hybrid Petri nets.* Unlike conventional Petri nets, we use different types of arcs to indicate the level of *certainty.*

Figure 3 shows an example of a hybrid Petri net discovered based on the event log also used to create the models in Fig. 1. Strong causalities are expressed through conventional places and arcs and *sure arcs* (arcs directly connecting transitions). Weak causalities are expressed using *unsure arcs* (dashed arcs with a question mark). Figure 2 shows the three types of arcs.

Definition 6 (Hybrid Petri Net). *A hybrid Petri net is a tuple* $HPN = (P, T, F_1, F_2, F_3)$ *where* (P, T, F_1) *is a Petri net,* $F_2 \subseteq T \times T$, *and* $F_3 \subseteq T \times T$ *such that* $\widehat{F_1}$, F_2, *and* F_3 *are pairwise disjoint. Arcs of Type I (*$(p,t) \in F_1$ *or* $(t,p) \in F_1$*) are the normal arcs connecting a place to a transition or vice versa. Arcs of Type II (*$(t_1, t_2) \in F_2$*) are arcs indicating a strong causality between two transitions (sure arcs). Arcs of Type III (*$(t_1, t_2) \in F_3$*) are arcs indicating a weak causality between two transitions (unsure arcs).*

Transitions, places, and normal (*Type I*) arcs have formal semantics as defined in Sect. 3. Again we define an initial and final marking to reason about the set of traces possible. Therefore, we define the notion of a *hybrid system net.*

Definition 7 (Hybrid System Net). *A hybrid system net is a triplet $HSN = (HPN, M_{init}, M_{final})$ where $HPN = (P, T, F_1, F_2, F_3)$ is a hybrid Petri net, $M_{init} \in \mathcal{B}(P)$ is the initial marking, and $M_{final} \in \mathcal{B}(P)$ is the final marking. $\mathcal{U}_{HSN}$ is the set of all possible hybrid system nets. behav(HSN) is defined as in* Definition 5 *while ignoring the sure and unsure arcs (i.e., remove F_2 and F_3).*

Only normal (*Type I*) arcs have formal semantics; the other two types of arcs are informal and do not include or exclude traces. Recall that Petri net without any places allows for any behavior and adding a place can only restrict behavior. A sure arc $(t_1, t_2) \in F_2$ should be interpreted as a strong causal relationship that cannot be expressed (easily) in terms of a place connecting t_1 and t_2. An unsure arc $(t_1, t_2) \in F_3$ is a suspected causal relationship that is too weak to justify a place connecting t_1 and t_2.

The role of sure and unsure arcs will become clearer when presenting the discovery technique in the next section. Figure 3 also uses special symbols for the start and end activities ($\triangleright$ and $\square$) as introduced in Definition 1, but the semantics of *HSN* do not depend on this.

We would like to stress that only the places in a hybrid system net *HSN* provide formal semantics. Behavioral quality measures such as fitness and precision will be based solely on the places in *HSN* (see definition *behav(HSN)*). Sure arcs (F_2) and unsure arcs (F_3) carry important information but cannot be used for such quality measures.

5   Discovering Hybrid Process Models

We aim to discover hybrid process models. As a target format we have chosen hybrid system nets that have three types of arcs. We use a *two-step approach*. First, we discover a *causal graph* (Sect. 5.1). Based on the causalities identified, we generate candidate places. These places are subsequently evaluated using replay techniques (Sect. 5.2). Strong causalities that cannot be expressed in terms of places are added to the *hybrid system net* as sure arcs. Moreover, the resulting hybrid model may also express weak causal relations as unsure arcs.

5.1   Discovering Causal Graphs

A causal graph is a directed graph with activities as nodes. There is always a unique start activity ($\triangleright$) and end activity ($\square$). There are two kinds of causal relations: *strong* and *weak*. These correspond to the two columns in Fig. 2.

Definition 8 (Causal Graph). *A causal graph is a triplet $G = (A, R_S, R_W)$ where A is the set of activities including start and end activities (i.e., $\{\triangleright, \square\} \subseteq A$), $R_S \subseteq A \times A$ is the set of strong causal relations, $R_W \subseteq A \times A$ is the set of weak causal relations, and $R_S \cap R_W = \emptyset$ (relations are disjoint). $\mathcal{U}_G$ is the set of all causal graphs.*

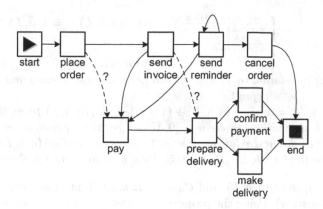

Fig. 4. A causal graph: nodes correspond to activities and arcs correspond to causal relations.

Figure 4 shows a causal graph derived from the event log also used to discover the models in Fig. 1. The dashed arcs with question marks correspond to weak causal relations. The other arcs correspond to strong causal relations.

Definition 9 (Causal Graph Discovery). *A causal graph discovery function $disc_{cg} \in \mathcal{U}_L \rightarrow \mathcal{U}_G$ is a function that constructs a causal graph $disc_{cg}(L) = (A, R_S, R_W)$ for any event log $L \in \mathcal{U}_L$ over A.*

There are many algorithms possible to construct a causal graph from a log. As an example, we use a variant of the approach used by the heuristic miner [1,17]. We tailored the approach to hybrid discovery (i.e., different types of arcs) while aiming for parameters that are intuitive and can be used interactively (e.g., thresholds can be changed seamlessly while instantly showing the resulting graph). Note that we clearly separate the identification of causalities from the discovery of process logic (see Sect. 5.2).

Definition 10 (Log-Based Properties). *Let $L \in \mathcal{U}_L$ be an event log over A and $\{a, b\} \subseteq A$.*

- *$\#(a, L) = \sum_{\sigma \in L} |\{i \in dom(\sigma) \mid \sigma(i) = a\}|$ counts the number of a's in log L.*
- *$\#(X, L) = \sum_{x \in X} \#(x, L)$ counts the number of $X \subseteq A$ activities in L.*
- *$\#(a, b, L) = \sum_{\sigma \in L} |\{i \in dom(\sigma) \setminus \{|\sigma|\} \mid \sigma(i) = a \wedge \sigma(i+1) = b\}|$ counts the number of times a is directly followed by b in event log L.*
- *$\#(*, b, L) = \sum_{\sigma \in L} |\{i \in dom(\sigma) \setminus \{|\sigma|\} \mid \sigma(i+1) = b\}|$ counts the number of times b is preceded by some activity.*
- *$\#(a, *, L) = \sum_{\sigma \in L} |\{i \in dom(\sigma) \setminus \{|\sigma|\} \mid \sigma(i) = a\}|$ counts the number of times a is succeeded by some activity.*
- *$Rel1(a, b, L) = \dfrac{\#(a, b, L) + \#(a, b, L)}{\#(a, *, L) + \#(*, b, L)}$ counts the strength of relation (a, b) relative to the split and join behavior of activities a and b.*

$$- \ Rel2_c(a, b, L) \ = \ \begin{cases} \frac{\#(a,b,L)-\#(b,a,L)}{\#(a,b,L)+\#(b,a,L)+c} & \text{if } \#(a, b, L) - \#(b, a, L) > 0 \\ \frac{\#(a,b,L)}{\#(a,b,L)+c} & \text{if } a = b \\ 0 & \text{otherwise} \end{cases} \quad counts$$

the strength of relation (a, b) taking into account concurrency and loops using parameter $c \in \mathbb{R}^+$ (default $c = 1$).

− $Caus_{c,w}(a, b, L) = w \cdot Rel1(a, b, L) + (1 - w) \cdot Rel2_c(a, b, L)$ takes the weighted average of both relations where $w \in [0, 1]$ is a parameter indicating the relative importance of the first relation. If $w = 1$, we only use $Rel1(a, b, L)$. If $w = 0$, we only use $Rel2_c(a, b, L)$. If $w = 0.5$, then both have an equal weight.

$Rel1(a, b, L)$, $Rel2_c(a, b, L)$, and $Caus_{c,w}(a, b, L)$ all produce values between 0 (weak) and 1 (strong). Using the properties in Definition 10, we define a concrete function $disc_{cg}$ to create causal graphs. All activities that occur at least t_{freq} times in the event log are included as nodes. The strength of relations between remaining activities (based on $Caus_{c,w}$) are used to infer causal relations. t_{R_S} and t_{R_W} are thresholds for strong respectively weak causal relations. Parameter w determines the relative importance of $Rel1$ and $Rel2_c$. Parameter c is typically set to 1.

Definition 11 (Concrete Causal Graph Discovery Technique). Let $L \in \mathcal{U}_L$ be an event log over A and let $t_{freq} \in \mathbb{N}^+$, $c \in \mathbb{R}^+$, $w \in [0, 1]$, $t_{R_S} \in [0, 1]$, $t_{R_W} \in [0, 1]$ be parameters such that $t_{R_S} \geq t_{R_W}$. The corresponding causal graph is $G = disc_{cg}(L) = (A', R_S, R_W)$ where

− $A' = \{a \in A \mid \#(a, L) \geq t_{freq}\} \cup \{\triangleright, \square\}$ is the set of activities that meet the threshold (the start and end activities are always included).
− $R_S = \{(a, b) \in A' \times A' \mid Caus_{c,w}(a, b, L{\restriction}_{A'}) \geq t_{R_S}\}$ is the set of strong causal relations.
− $R_W = \{(a, b) \in A' \times A' \mid t_{R_S} > Caus_{c,w}(a, b, L{\restriction}_{A'}) \geq t_{R_W}\}$ is the set of weak causal relations.

Figure 4 shows a causal graph constructed using parameters $t_{freq} = 1000$, $c = 1$, $w = 0.2$, $t_{R_S} = 0.8$, and $t_{R_W} = 0.75$.

5.2   Discovering Hybrid System Nets

In the second step of the approach we use the causal graph to create a hybrid system net (that turns strong causalities into formal constraints if possible).

Definition 12 (Hybrid System Net Discovery). A hybrid system net discovery function $disc_{hsn} \in (\mathcal{U}_L \times \mathcal{U}_G) \rightarrow \mathcal{U}_{HSN}$ is a function that for any event log L and causal graph G discovers a hybrid system net $disc_{hsn}(L, G) \in \mathcal{U}_{HSN}$.

Just like there are many algorithms possible to create a causal graph, there are also multiple ways to construct a hybrid system net from an event log and causal graph. The minimal consistency requirements can be defined as follows.

Definition 13 (Consistent). *Let $L \in \mathcal{U}_L$ be an event log, let $G = (A, R_S, R_W) \in \mathcal{U}_G$ be a causal graph, and let $HSN = (HPN, M_{init}, M_{final}) \in \mathcal{U}_{HSN}$ with $HPN = (P, T, F_1, F_2, F_3)$ be a hybrid system net. $L, G,$ and SN are consistent if and only if: $T = A \subseteq \bigcup_{\sigma \in L}\{a \in \sigma\}, \{p_\triangleright, p_\square\} \subseteq P, F_1 \cap ((\{p_\triangleright, p_\square\} \times T) \cup (T \times \{p_\triangleright, p_\square\})) = \{(p_\triangleright, \triangleright), (\square, p_\square)\}, M_{init} = [p_\triangleright] \text{ and } M_{final} = [p_\square], \text{ for all } p \in P \setminus \{p_\triangleright, p_\square\}: \bullet p \neq \emptyset$ and $p\bullet \neq \emptyset$, $R_S = \widehat{F_1} \cup F_2$, $\widehat{F_1} \cap F_2 = \emptyset$, and $R_W = F_3$.*

An event log L, causal graph G, and hybrid system net HSN are consistent if (1) L and G refer to the same set of activities all appearing in the event log, (2) there is a source place $p_\triangleright$ marked in the initial place and enabling start activity $\triangleright$, (3) there is a sink place $p_\square$ marked in the final marking and connected to end activity $\square$, (4) all other places connect activities, (5) there is a one-to-one correspondence between strong causal relations (R_S) and connections through places ($\widehat{F_1}$) or sure arcs (F_2), and (6) there is a one-to-one correspondence between weak causal relations (R_W) and unsure arcs (F_3).

Consider two activities $a_1, a_2 \in A$ that are frequent enough to be included in the model. These can be related in three different ways: $(a_1, a_2) \in \widehat{F_1}$ if there is a place connecting a_1 and a_2, $(a_1, a_2) \in F_2$ if there is no place connecting a_1 and a_2 but there is a strong causal relation between a_1 and a_2 (represented by a sure arc), $(a_1, a_2) \in F_3$ if there is a weak causal relation between a_1 and a_2 (represented by an unsure arc).

Any discovery function $disc_{hsn} \in (\mathcal{U}_L \times \mathcal{U}_G) \to \mathcal{U}_{HSN}$ should ensure consistency. In fact, Definition 13 provides hints on how to discover a hybrid system net.

Assume a place $p = (I, O)$ with input transitions $\bullet p = I$ and output transitions $p \bullet = O$ is added. $R_S = \widehat{F_1} \cup F_2$ implies that $\widehat{F_1} \subseteq R_S$. Hence, $I \times O \subseteq R_S$, i.e., place $p = (I, O)$ can only connect transitions having strong causal relations. Moreover, I and O should not be empty. These observations based on Definition 13 lead to the following definition of candidate places.

Definition 14 (Candidate Places). *Let $G = (A, R_S, R_W) \in \mathcal{U}_G$ be a causal graph. The candidate places based on G are: $candidates(G) = \{(I, O) \mid I \neq \emptyset \land O \neq \emptyset \land I \times O \subseteq R_S\}$.*

Given a candidate place $p = (I, O)$ we can check whether it allows for a particular trace.

Definition 15 (Replayable trace). *Let $p = (I, O)$ be a place with input set $\bullet p = I$ and output set $p \bullet = O$. A trace $\sigma = \langle a_1, a_2, \dots, a_n \rangle \in A^*$ is perfectly replayable with respect to place p if and only if*

- *for all $k \in \{1, 2, \dots, n\}$: $|\{1 \leq i < k \mid a_i \in I\}| \geq |\{1 \leq i < k \mid a_i \in O\}|$ (place p cannot "go negative" while replaying the trace) and*
- *$|\{1 \leq i \leq n \mid a_i \in I\}| = |\{1 \leq i \leq n \mid a_i \in O\}|$ (place p is empty at end).*

We write $\checkmark(p, \sigma)$ if σ is perfectly replayable with respect to place $p = (I, O)$. $act(p, \sigma) = \exists_{a \in \sigma} a \in (I \cup O)$ denotes whether place $p = (I, O)$ has been activated, i.e., a token was consumed or produced for it in σ.

Note that $\checkmark(p,\sigma)$ if σ is a trace of the system net having only one place p. To evaluate candidate places one can define different scores.

Definition 16 (Candidate Place Scores). *Let $L \in \mathcal{U}_L$ be an event log. For any candidate place $p = (I, O)$ with input set $\bullet p = I$ and output set $p \bullet = O$, we define the following scores:*

- *$score_{freq}(p, L) = \frac{|[\sigma \in L | \checkmark(p,\sigma)]|}{|L|}$ is the fraction of fitting traces,*
- *$score_{rel}(p, L) = \frac{|[\sigma \in L | \checkmark(p,\sigma) \wedge act(p,\sigma)]|}{|[\sigma \in L | act(p,\sigma)]|}$ is the fraction of fitting traces that have been activated, and*
- *$score_{glob}(p, L) = 1 - \frac{|\#(I,L) - \#(O,L)|}{max(\#(I,L), \#(O,L))}$ is a global score only looking at the aggregate frequencies of activities.*

To explain the three scoring functions consider again $L_1 = [\langle \triangleright, a, b, c, d, \square \rangle^{45}, \langle \triangleright, a, c, b, d, \square \rangle^{35}, \langle \triangleright, a, e, d, \square \rangle^{20}]$. Let us consider place $p_1 = (I_1, O_1)$ with $I_1 = \{a\}$ and $O_2 = \{b\}$. $score_{freq}(p_1, L_1) = score_{rel}(p_1, L_1) = {}^{80}/100 = 0.8$ and $score_{glob}(p_1, L_1) = 1 - {}^{|100-80|}/max(100,80) = 0.8$. For place $p_2 = (I_2, O_2)$ with $I_2 = \{a\}$ and $O_2 = \{b, e\}$: $score_{freq}(p_2, L_1) = score_{rel}(p_2, L_1) = score_{glob}(p_2, L_1) = 1$. Hence, all three scoring functions agree and show that the second place is a better candidate. Note that if the candidate place p does not inhibit any of the traces in the log, then all scores are 1 by definition.

Let us now consider event log $L_2 = [\langle c, d \rangle^{1000}, \langle a, b \rangle^{100}, \langle b, a \rangle^{10}, \langle a, a, a, a, \ldots, a \rangle]$ (with the last trace containing 1000 a's) and candidate place $p_1 = (I_1, O_1)$ with $I_1 = \{a\}$ and $O_2 = \{b\}$. $score_{freq}(p_1, L_2) = {}^{1100}/1111 = 0.99$, $score_{rel}(p_1, L_2) = {}^{100}/111 = 0.90$, $score_{glob}(p_1, L_2) = 1 - {}^{|1110-110|}/max(1110,110) = 0.099$. Now the values are very different. Interpreting the scores reveals that $score_{freq}$ is too optimistic. Basically one can add any place connected to low frequent activities, without substantially lowering the $score_{freq}$ score. Hence, $score_{rel}$ is preferable over $score_{freq}$. $score_{glob}$ can be computed very efficiently because traces do not need to be replayed. It can be used to quickly prune the set of candidate places, but the last example shows that one needs to be careful when traces are unbalanced (i.e., I or O activities occur many times in a few traces).

Based on the above discussion we use scoring function $score_{rel}$ in conjunction with a threshold t_{replay}. The causal graph, a set of candidate places, and this threshold can be used to discover a hybrid system net.

Definition 17 (Concrete Discovery Technique). *Let $L \in \mathcal{U}_L$ be an event log and let $G = (A, R_S, R_W) \in \mathcal{U}_G$ be a causal graph. t_{replay} is the threshold for the fraction of fitting traces that have been activated. The discovered hybrid system net $disc_{hsn}(L, G) = (HPN, M_{init}, M_{final})$ with $HPN = (P, T, F_1, F_2, F_3)$ is constructed as follows*

- *$Q = \{p \in candidates(G) \mid score_{rel}(p, L\lceil_A) \geq t_{replay}\}$ is the set of internal places (all candidate places meeting the threshold),*
- *$P = \{p_\triangleright, p_\square\} \cup Q$ is the set of places ($\{p_\triangleright, p_\square\} \cap Q = \emptyset$),*
- *$T = A$ is the set of transitions,*

- $F_1 = \{(p_\rhd, \rhd), (\Box, p_\Box)\} \cup \{(t, (I, O)) \in T \times Q \mid t \in I\} \cup \{((I, O), t) \in Q \times T \mid t \in O\}$ *is the set of normal arcs,*
- $F_2 = R_S \setminus \widehat{F_1}$ *is the set of sure arcs, and*
- $F_3 = R_W$ *is the set of unsure arcs.*

It is easy to check that this concrete $disc_{hsn}$ function indeed ensures consistency. The construction of the discovered hybrid system net is guided by the causal graph. We can construct hybrid system net $disc_{hsn}(L, disc_{cg}(L))$ for any event log L using parameters t_{freq}, c, w, t_{R_S}, t_{R_W}, and t_{replay}. For example, the hybrid model shown in Fig. 3 was discovered using $t_{freq} = 1000$, $c = 1$, $w = 0.2$, $t_{R_S} = 0.8$, $t_{R_W} = 0.75$, and $t_{replay} = 0.9$. Our discovery approach is highly configurable and also provides formal guarantees (e.g., $t_{replay} = 1$ ensures perfect fitness). When there is not enough structure or evidence in the data, the approach is not coerced to return a model that suggests a level of confidence that is not justified.

6 Implementation

Two novel *ProM* plug-ins have been created to support the approach described in this paper.[1] The *Causal Graph Miner* plug-in is used to create a causal graph using the approach described in Definition 11. The user can control the parameters w, t_{freq}, t_{R_S}, and t_{R_W} through sliders and directly see the effects in the resulting graph. The *Hybrid Petri Net Miner* plug-in implements Definition 17 and takes as input an event log and a causal graph. The plug-in returns a discovered hybrid system net. Only places that meet the t_{replay} threshold are added. The replay approach has been optimized to stop replaying a trace when it does not fit.

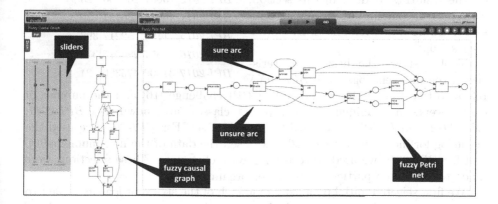

Fig. 5. Screenshots of the *Causal Graph Miner* (left) and the *Hybrid Petri Net Miner* (right) analyzing the running example with parameter settings $t_{freq} = 1000$, $c = 1$, $w = 0.2$, $t_{R_S} = 0.8$, $t_{R_W} = 0.75$, and $t_{replay} = 0.9$.

---

[1] Install *ProM* and the package *HybridMiner* from http://www.promtools.org.

Figure 5 shows the two plug-ins in action for the event log containing 12,666 cases and 80,609 events. The results returned correspond to the causal graph depicted in Fig. 4 and the hybrid system net depicted in Fig. 3. *Both were computed in less than a second on a standard laptop.* Activity *send reminder* may occur repeatedly (or not) after sending the invoice but before payment or cancellation. However, payments may also occur before sending the invoice. The hybrid system net in Fig. 5 (also see Fig. 3 which is better readable) clearly differentiates between (1) the behavior which is dominant and clear and (2) the more vague behavior that cannot be captured formally or is not supported by enough "evidence". The example illustrates the scalability of the approach while supporting simplicity and deliberate vagueness.

7 Evaluation

Process discovery techniques can be evaluated using a range of indicators referring to *fitness* (ability to replay the observed behavior), *precision* (avoiding underfitting), *generalization* (avoiding overfitting), and *simplicity* (is the model easy to understand) [1]. Existing indicators are less suitable for the evaluation of hybrid models explicitly capturing vagueness. Criteria involving fitness, precision, and generalization can also not be measured for the informal models produced by existing commercial process mining tools. When computing traditional quality measures for hybrid system nets we basically ignore the sure and unsure arcs.

We applied our approach to a large number of real-life events logs and analyzed the effects of the different parameters (t_{freq}, c, w, t_{R_S}, t_{R_W}, and t_{replay}) on the resulting models. In this section, we report on our findings using six data sets taken from the well-known *BPI Challenges* [16].[2]

Table 1 shows the basic characteristics of the six event logs used: *BPI-20XX*

Table 1. Six data sets used.

Log	Cases	Events	Activities
BPI-2011	1143	150291	624
BPI-2012	13087	164506	23
BPI-2014	46616	466737	39
BPI-2015	1199	52217	398
BPI-2016	557	286075	312
BPI-2017	31509	475306	24

refers to the year of the corresponding BPI challenge [16] and the number of cases, events, and unique activities (event classes) are shown. For *BPI-2011*, *BPI-2012*, and *BPI-2017* we used the full data set. For *BPI-2014* we used the event log for incidents, for *BPI-2015* we used the data of the first municipality, and for *BPI-2016* we used the event log with click data. These selections were made to focus on a particular process or organization.

We first selected initial parameters for each of the six event logs in Table 1 to create six "reasonable" base models. To create the base models we interactively set the thresholds in such a way that the underlying graph is connected. t_{replay}

---

[2] The reader is invited to redo the experiments using the latest version of *ProM*, the *HybridMiner* package (promtools.org), and the publicly available data sets used here [16].

Table 2. Parameters used to create the base models and their characteristics.

Log	t_{freq}	t_{R_S}	t_{R_W}	w	t_{replay}	$|T|$	$|P|$	$|\widehat{F_1}|$	$|F_2|$	$|F_3|$	Fitness	Precision	Time (ms)
BPI-2011	343	0.81	0.80	0.10	0.80	38	6	4	200	6	0.84	0.04	11772
BPI-2012	3926	0.90	0.89	0.10	0.80	14	8	7	20	1	0.90	0.26	12414
BPI-2014	13985	0.90	0.90	0.10	0.80	10	5	3	13	0	0.93	0.54	21233
BPI-2015	360	0.45	0.40	0.50	0.80	59	26	24	145	75	0.74	0.05	7055
BPI-2016	445	0.50	0.50	0.10	0.80	12	2	0	31	0	0.83	0.10	31428
BPI-2017	9453	0.51	0.50	0.50	0.80	22	8	7	36	12	0.95	0.12	24772

was set in such way that a reasonable number of places remained. Table 2 shows the settings used and some of the characteristics of the resulting hybrid process models.

Obviously different parameter settings lead to different models. For example, if t_{replay} is set to 1, then (by definition) the fitness will be 1. Similarly, the number of unsure arcs is directly affected by t_{R_W}. If $t_{R_W} = t_{R_S}$, then (by definition) there will be no unsure arcs. Column $|T|$ shows the number of retained activities. Columns $|P|$ and $|\widehat{F_1}|$ provide insights in the dominant and clear behavior captured in terms of normal arcs. Columns $|F_2|$ and $|F_3|$ indicate the number of sure and unsure arcs. These numbers give insights in the complexity of the models (simplicity dimension). Fitness and precision are computed using the techniques from [4,5] while ignoring the sure and unsure arcs (i.e., only considering the normal places and arcs).

The fitness values in Table 2 are as expected. It is possible to improve fitness at the cost of having fewer places. The precision values in Table 2 vary widely. Precision is very low for BPI-2011 and BPI-2015. However, for these models there are many sure arcs showing the added value of our hybrid approach. Things that cannot be expressed in terms of reasonable places ($t_{replay} = 0.8$) can still be expressed. Traditional approaches would be forced to accept places that have a lower quality or ignore the causalities observed. For example, the inductive miner would generate underfitting models or models focusing on the mainstream behavior only.

The computation times (last column in Table 2) are in milliseconds. Clearly, the size of the event log and computation time positively correlate. Moreover, the fewer candidate places the faster the second step is performed. These numbers show that the approach is already quite fast compared to other approaches returning a formal model (all models are computed in less than 25 seconds). Implementation-wise there is ample room for improvement, showing that the approach itself is highly scalable.

For each of the six event logs, we used the baseline values for w, t_{R_S}, t_{R_W}, t_{replay}, and t_{freq} (Table 2) as a starting point. (We fixed the value of c to its default value: $c = 1$.) Next, we varied some of the key parameters one-by-one while keeping the baseline values for the other parameters fixed: $t_{replay} \in \{0.7, 0.8, 0.9, 1.0\}$, $w \in \{0.0, 0.25, 0.5, 0.75, 1.0\}$, $t_{R_S} \in \{0.5, 0.6, 0.7, 0.8, 0.9\}$, and $t_{R_W} \in \{0.5, 0.6, 0.7, 0.8, 0.9\}$ (such that $t_{R_S} \geq t_{R_W}$). These results are discussed in detail in a technical report [2]. Given the limited space, we only summarize the main findings here.

- Increasing the value of t_{replay} improves fitness of the model because places that are not perfectly fitting are removed. The precision of the model typically decreases when t_{replay} goes up. Moreover, the removal of places leads to an increase in sure arcs.
- Increasing the value of w has a marginal effect on fitness and precision. For some of the event logs, precision is better for lower values of w (i.e., more weight is given to $Rel2_c(a, b, L)$).
- Increasing the value of t_{R_S} leads to fewer connections through places and sure arcs. This can only improve fitness. However, the effect is moderate and heavily depends on t_{replay}. Precision tends to go down when t_{R_S} goes up.
- Increasing the value of t_{R_W}, by definition, has no effect on precision and fitness and only affects the number of unsure arcs.

In summary, the discovery approach works in a predictable manner. Using the parameters the analyst can influence the characteristics of the discovered model in a fast and reliable manner. It is possible to express "vagueness" in terms of sure and unsure arcs. If there is not enough evidence in the data to justify the addition of many "good" places, then the resulting model will have a low precision. Fitness can be controlled directly by t_{replay}. We refer to the technical report for detailed experimental results [2], but acknowledge that additional evaluations are needed (involving new metrics and groups of users).

8   Related Work

The work reported in this paper was inspired by the work of Herrmann et al. [9,10] who argue that modeling "requires the representation of those parts of knowledge which cannot be stated definitely and have to be modeled vaguely". They propose annotations to make vagueness explicit. In [9,10] the goal is to *model* vagueness, but we aim to automatically *discover* hybrid models supporting both vagueness and formal semantics.

Hybrid process models are related to the partial models considered in software engineering [7,14]. These partial models can be completed into formal models and do not consider data-driven uncertainty. In fact, these partial models are closer to configurable process models representing sets of concrete models

In literature one can find a range of process discovery approaches that produce formal models [1]. The α-algorithm [3] and its variants produce a Petri net. Approaches based on state-based regions [15] and language-based regions [6,18] also discover Petri nets. The more recently developed inductive mining approaches produce process trees that can be easily converted to Petri nets or similar [11–13].

Commercial process mining tools typically produce informal models. These are often based on the first phases of the heuristic miner [17] (dependency graph) or the fuzzy miner [8] (not allowing for any form of formal reasoning).

It is impossible to give a complete overview of all discovery approaches here. However, as far as we know there exist on other discovery approaches that return hybrid models having both formal and informal elements.

9 Conclusion

In this paper we advocated the use of *hybrid models* to combine the best of two worlds: commercial tools producing informal models and discovery approaches providing formal guarantees. We provided a concrete realization of our hybrid discovery approach using *hybrid Petri nets*. The ideas are not limited to Petri nets and could be applied to other types of process models (e.g., BPMN models with explicit gateways for the clear and dominant behavior and additional arcs to capture complex or less dominant behavior). Unlike existing approaches there is no need to straightjacket behavior into a formal model that suggests a level of confidence that is not justified. The explicit representation of vagueness and uncertainty in hybrid process models is analogous to the use of confidence intervals and box-and-whisker diagrams in descriptive statistics.

The approach has been fully implemented and tested on numerous real-life event logs. The results are very promising, but there are still many open questions. In fact, the paper should be seen as the starting point for a new branch of research in BPM and process mining. To evaluate differences between informal, formal, and hybrid models from a user perspective, we need new evaluation criteria taking understandability and perceived complexity into account. Future work will also include "hybrid BPMN and UML activity diagrams" focusing on different model constructs (gateways, swimlanes, artifacts, etc.). Existing techniques (also supported by *ProM*) can already be used to map compliance and performance indicators onto causalities expressed in terms of explicit places. We would like to also provide approximative compliance and performance indicators for sure and unsure arcs. Note that commercial tools show delays and frequencies on arcs, but these indicators may be very misleading as demonstrated in Sect. 11.4.2 of [1]. Finally, we would like to improve performance. The approach has already a good performance. Moreover, there are several ways to further speed-up analysis (e.g., pruning using $score_{glob}$ or user-defined preferences). Moreover, computation can be distributed in a straightforward manner (e.g., using MapReduce).

References

1. van der Aalst, W.M.P.: Process Mining: Data Science in Action. Springer, Berlin (2016)
2. van der Aalst, W.M.P., De Masellis, R., Di Francescomarino, C., Ghidini, C.: Learning Hybrid Process Models From Events: Process Discovery Without Faking Confidence (Experimental Results). ArXiv e-prints 1703.06125 (2017). arXiv.org/abs/1703.06125

3. van der Aalst, W.M.P., Weijters, A.J.M.M., Maruster, L.: Workflow mining: discovering process models from event logs. IEEE Trans. Knowl. Data Eng. **16**(9), 1128–1142 (2004)
4. Adriansyah, A.: Aligning observed and modeled behavior. Ph.D thesis, Eindhoven University of Technology, April 2014
5. Adriansyah, A., Munoz-Gama, J., Carmona, J., van Dongen, B.F., van der Aalst, W.M.P.: Measuring precision of modeled behavior. ISeB **13**(1), 37–67 (2015)
6. Bergenthum, R., Desel, J., Lorenz, R., Mauser, S.: Process mining based on regions of languages. In: Alonso, G., Dadam, P., Rosemann, M. (eds.) BPM 2007. LNCS, vol. 4714, pp. 375–383. Springer, Heidelberg (2007). doi:10.1007/978-3-540-75183-0_27
7. Famelis, M., Salay, R., Chechik, M., Models, P.: Towards modeling and reasoning with uncertainty. In: International Conference on Software Engineering (ICSE 2012), pp. 573–583. IEEE Computer Society (2012)
8. Günther, C.W., van der Aalst, W.M.P.: Fuzzy mining – adaptive process simplification based on multi-perspective metrics. In: Alonso, G., Dadam, P., Rosemann, M. (eds.) BPM 2007. LNCS, vol. 4714, pp. 328–343. Springer, Heidelberg (2007). doi:10.1007/978-3-540-75183-0_24
9. Herrmann, T., Hoffmann, M., Loser, K.U., Moysich, K.: Semistructured models are surprisingly useful for user-centered design. In: De Michelis, G., Giboin, A., Karsenty, L., Dieng, R. (eds.) Designing Cooperative Systems (Coop 2000), pp. 159–174. IOS Press, Amsterdam (2000)
10. Herrmann, T., Loser, K.U.: Vagueness in models of socio-technical systems. Behav. Inf. Technol. **18**(5), 313–323 (1999)
11. Leemans, S.J.J., Fahland, D., van der Aalst, W.M.P.: Discovering block-structured process models from event logs - a constructive approach. In: Colom, J.-M., Desel, J. (eds.) PETRI NETS 2013. LNCS, vol. 7927, pp. 311–329. Springer, Heidelberg (2013). doi:10.1007/978-3-642-38697-8_17
12. Leemans, S.J.J., Fahland, D., van der Aalst, W.M.P.: Discovering block-structured process models from event logs containing infrequent behaviour. In: Lohmann, N., Song, M., Wohed, P. (eds.) BPM 2013. LNBIP, vol. 171, pp. 66–78. Springer, Cham (2014). doi:10.1007/978-3-319-06257-0_6
13. Leemans, S.J.J., Fahland, D., van der Aalst, W.M.P.: Scalable process discovery and conformance checking. Softw. Syst. Model., pp. 1–33 (2016). doi:10.1007/s10270-016-0545-x
14. Salay, R., Chechik, M., Horkoff, J., Di Sandro, A.: Managing requirements uncertainty with partial models. Requirements Eng. **18**(2), 107–128 (2013)
15. Solé, M., Carmona, J.: Process mining from a basis of state regions. In: Lilius, J., Penczek, W. (eds.) PETRI NETS 2010. LNCS, vol. 6128, pp. 226–245. Springer, Heidelberg (2010). doi:10.1007/978-3-642-13675-7_14
16. van Dongen, B.F.: BPI Challenges (2011–2017), Real life Event Logs Collection (2017). data.4tu.nl/repository/collection:event_logs
17. Weijters, A.J.M.M., van der Aalst, W.M.P.: Rediscovering workflow models from event-based data using little thumb. Integr. Comput.-Aided Eng. **10**(2), 151–162 (2003)
18. van der Werf, J.M.E.M., van Dongen, B.F., Hurkens, C.A.J., Serebrenik, A.: Process discovery using integer linear programming. Fundam. Inf. **94**, 387–412 (2010)

Multi Instance Anomaly Detection in Business Process Executions

Kristof Böhmer[✉] and Stefanie Rinderle-Ma

Faculty of Computer Science, University of Vienna, Vienna, Austria
{kristof.boehmer,stefanie.rinderle-ma}@univie.ac.at

Abstract. Processes control critical IT systems and business cases in dynamic environments. Hence, ensuring secure model executions is crucial to prevent misuse and attacks. In general, anomaly detection approaches can be employed to tackle this challenge. Existing ones analyze each process instance individually. Doing so does not consider attacks that combine multiple instances, e.g., by splitting fraudulent fund transactions into multiple instances with smaller "unsuspicious" amounts. The proposed approach aims at detecting such attacks. For this, anomalies between the temporal behavior of a set of historic instances (ex post) and the temporal behavior of running instances are identified. Here, temporal behavior refers to the temporal order between the instances and their events. The proposed approach is implemented and evaluated based on real life process logs from different domains and artificial anomalies.

Keywords: Runtime anomaly detection · Secure business processes · Multiple instances · Temporal anomalies

1 Introduction

Business process anomaly detection identifies *anomalous behavior* in recorded (ex post) or ongoing (real time) process executions in order to expose and prevent fraud, misuse, unknown attacks, and errors. Hence it constitutes a critical IT security defense line in today's interconnected business driven organizations [3,5]. Existing process anomaly detection work analyzes single process instances in order to distinguish if their behavior is anomalous (i.e., *unlikely*) or not. Doing so does not provide protection against all possible attack vectors. For example, assume an attack scenario where the attacker (Trudy) strives to quickly transfer funds from an organization's bank account. Therefore, Trudy could instantiate a single transaction process and transfer all the money at once. Alternatively, Trudy could start multiple transaction processes in parallel and split up the transactions into smaller chunks. The first approach would likely be detected by existing anomaly detection approaches while the second would not.

This is, because in the first case the transferred funds are exceptionally high, i.e., they exceed previously transferred funds and are, therefore, unlikely. In this

© Springer International Publishing AG 2017
J. Carmona et al. (Eds.): BPM 2017, LNCS 10445, pp. 77–93, 2017.
DOI: 10.1007/978-3-319-65000-5_5

case, analyzing each process instance execution individually is sufficient to identify Trudy's attack. In the second case each individual process instance only transfers a small amount of money. Through this, the transferred funds are, likely, comparable to transactions represented in the known historic behavior. Accordingly the executions would *not* be identified as anomalous.

Hence, an anomaly detection approach is required that is able to consider *multiple* instances – from the same or different process models. In the example scenario these are instances which take place before, during, or after one of the fraudulent transaction process executions. Through this the parallel executions are noticed as unusual and the second attack scenario is identified. The assumption behind this is that the massive parallel execution of multiple transaction processes – which was never observed before – is unlikely (i.e., anomalous).

This paper proposes a configurable and unsupervised anomaly detection heuristic for business processes that exploits the temporal dependencies between multiple instances. In detail, for each instance of interest, all temporal relations of preceding, succeeding, and simultaneous process executions are taken into account. The business process instances and executions which are taken into consideration can stem from *various models*, i.e., not all instances need to be spawned from the same process model but, e.g., from multiple models.

This work applies *design science research*, cf. [14]. Doing so multi instance process executions were identified as a problem (i.e., an unprotected attack vector). To tackle this problem artifacts are created and evaluated, here this is a prototypical implementation of the proposed multi instance anomaly detection approach. Stakeholders for the approach are organizations and security experts.

More precisely, we assume a set of process models R, and a set of execution log files L. Hereby, R could be a process repository and L holds all executions of the processes in R. The key idea is to generate an anomaly detection signature G for a process $P \in R$ so that it represents for P the behavior of P's instances and temporally related instances from multiple other models. This is achieved by mining and combining temporal relations from multiple instances, stored in L, that take place during or close to executions of P into a signature G.

Finally, behavior that should be analyzed for anomalies, from P's instances and other temporally related instances, is assumed as given. For example, such behavior can be extracted from logs, for ex post analysis, or be collected directly during model executions – from process execution engines – for real time analysis. To analyze if behavior is anomalous or not it is mapped to G, which enables to calculate its behavior likelihood. If the behavior to analyze for anomalies is found to be unlikely, when comparing it to the logged historic behavior in L, then the behavior is identified as anomalous. The artifacts generated in this work comprise multi instance process behavior mining and signature generation and matching algorithms. The presented approach is evaluated using real life process execution logs from multiple domains along with artificially generated anomalies.

This paper is organized as follows: Related work is discussed in Sect. 2. Prerequisites and the proposed approach are introduced in Sect. 3. The proposed anomaly detection approach (i.e., signature generation and matching) is, in detail, described in Sect. 4. Section 5 holds the evaluation. Finally, conclusions, discussions, and future work is given in Sect. 6.

2   Related Work

Related anomaly detection work was searched for in the *process domain* and in the *security domain*. The results found for the process domain were limited as related work focuses only on single individual process models and instances. Hence, the anomalies which this work is capable of identifying are not supported by existing process anomaly detection work. Our systematic literature review in [6] provides a more detailed analysis. The most comparable work [12] analyzes temporal behavior of individual activities to identify unlikely anomalous execution behavior. However, this work also concentrates only on single instances.

In a broader context, i.e., the security domain in general, several temporal anomaly detection approaches are suggested, cf. [10]. However, according to [10] and our own findings, those approaches are typically domain or data specific (i.e., focus on specific protocols, such as, SIP or network packages) and can, because of this, hardly be generalized, e.g., to analyze process behavior data for anomalies. It can be concluded that an anomaly detection approach specifically tailored for process behavior is a necessity to identify related attacks and anomalies.

Moreover, it was found that existing approaches show, likely, an underwhelming anomaly detection performance when dealing with unexpected behavior. Existing anomaly detection work frequently classifies unexpected behavior, even if it only slightly deviates from, e.g., a signature, as anomalous. This could potentially result in a large amount of false positives [5] in flexible and dynamic execution scenarios. Hence, this work proposes a novel approach to deal with unexpected behavior by assigning it with an artificially calculated likelihood.

Existing approaches which are comparable to the presented artificial likelihood calculation are so called *soft matching* techniques. Soft matching generalizes expected behavior patterns by constructing multiple slightly deviating, but still presumably "correct" patterns (e.g., based on expert knowledge) [2]. Hereby the area of data or behavior which is identified as non-anomalous is widened. Unfortunately, it frequently requires expert knowledge to soften the patterns and through this soft matching lacks in flexibility, compared to the presented automatic approach. Moreover, the presented approach enables to "aggregate" multiple occurrences of slightly unlikely behavior to identify collective anomalies [5] which could be missed by less sensitive detection approaches. This is because this work does not flag each observed process execution behavior solely as anomalous or non-anomalous. Instead a more flexible likelihood is calculated and aggregated over multiple successive process execution events and instances.

3   Prerequisites and General Approach

This paper proposes a multi instance anomaly detection approach that enables to distinguish process execution behavior as anomalous (i.e., unlikely) or not.

For this the behavior is compared with a signature generated from given process execution logs L which represent historic process executions. Generating signatures from logs is beneficiary as logs are frequently generated automatically by today's process execution engines, contain real behavior and executions, and include manual adaptations. Moreover, exploiting execution logs enables to become independent from abstracted and potentially outdated documentation [11].

Let each execution log $l \in L$ hold the associated process model **name** and a bag of execution **Events**, i.e., $l := (n, E)$. Each execution event $e \in l.E$, i.e., $e := (s, c)$ represents an activity execution by its **s**tart and **c**ompletion timestamp s and c respectively, with $s, c \in \mathbb{N}_{>0}$. An exemplary log for model A with two activity executions could be defined as $l_A := (A, \{(s_1, c_1), (s_2, c_2)\})$. We assume that each individual model execution (i.e., each instance) is held by an individual execution log $l \in L$ and that timestamps are defined in a range of $\mathbb{N}_{>0}$.

Such a brief definition is sufficient because the presented approach mainly analyzes temporal relations between models and their instances. For this the following auxiliary functions, inspired by a subset of Allen's interval algebra [1], are defined. The start timestamp of an instance execution is found by $\min(E) := \{e.s | e \in E; \forall e' \in E, e.s \le e'.s\}^0$. Here, $\{\cdots\}^0$ returns the only element held by a set or bag if it is a *singleton* or a random set/bag element if it is not. A similar definition is applied for $\max(E)$ to determined the end timestamp of an instance execution. Through this the duration of an instance is $\text{dur}(E) := \max(E) - \min(E)$. Moreover, $\text{execP}(t, L)$ extracts a bag of process model names that are executed at point t (i.e., a timestamp) based on the logs in L, i.e., $\text{execP}(t, L) := \{l.n | \exists l \in L, \exists e \in l.E; e.s \le t \wedge e.c \ge t\}$. Similarly, $\text{act}(t_1, t_2, L)$ counts the activities that are executed in a specific interval given by t_1, t_2, i.e., $\text{act}(t_1, t_2, L) := |\{e | l \in L, e \in l.E; e.s \ge t_1 \wedge e.c \le t_2\}|$ where $t_1 \le t_2$. Function $\text{next}(t, L)$ determines the process instance start or end timestamp that occurs as close after t as possible, i.e., $\text{next}(t, L) := \{t_1 | t_1, t_2 \in T; \nexists t_2 < t_1; t_1 > t \wedge t_2 > t\}^0$ where $T := \{e.s | l \in L, e \in l.E\} \cup \{e.c | l \in L, e \in l.E\}$. Further, $\text{mid}(t_1, t_2) := t_1 + ((t_2 - t_1)/2)$ calculates the average of two timestamps where $t_1 < t_2$.

Fig. 1. Proposed multi instance anomaly detection approach – overview

Figure 1 provides an overview on the proposed anomaly detection heuristic. The related algorithms are presented in Sect. 4. Firstly, a signature is generated for a process $P \in R$ based on a set L of historic instance executions – both are assumed as given input. The first idea is to extract process execution events in

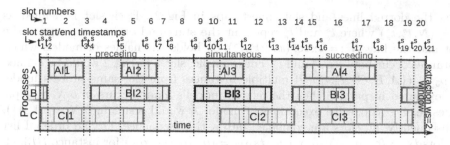

Fig. 2. Running example for window & behavior extraction and noise reduction

L that precede, succeed, or occur simultaneously to executions of P's instances, cf. Fig. 2. The figure depicts three processes – A, B, and C – along with a number of instances (i.e., the rectangles, e.g., $AI1$ to $AI4$) and activity execution events (i.e., the vertical bars in the instance rectangles, e.g., $AI3$ holds 4 activity executions). For the sake of brevity Fig. 2 depicts only a snapshot of all instance executions, i.e., additional instances are stored in L but not depicted.

Assume a signature is generated for process B in Fig. 2. Then the signature generation starts by identifying *relevant* execution events in L (i.e., historic behavior) ①. Relevant events are events which most likely affect B's instances (i.e., events which precede B's instances) or which are affected by B's instances (i.e., events which are succeeding and simultaneous to B's instances). Extraction windows are applied to identify such events in the following.

An individual *extraction window* $w := [wt_1; wt_2]$ is created for each of B's instances ①. Extraction windows enable to determine which of the behavior held by L is relevant (i.e., preceding, succeeding, and simultaneous events) for a specific instance and model and should, therefore, be contained in the generated signature. The beginning and end of the window (i.e., wt_1 and wt_2) is calculated by multiplying the duration of the respective instance (this example uses $BI3$ and $l_{BI3} \in L$) with a user chosen **window** size modifier $ws \in \mathbb{R}_{>0}$. So $wt_1 := \min(l.E) - (ws \cdot \mathbf{dur}(l.E))$ and $wt_2 := \max(l.E) + (ws \cdot \mathbf{dur}(l.E))$. Hence, when assuming $ws = 2$ and $\min(l_{BI3}.E) = t_9^s$, $\max(l_{BI3}.E) = t_{13}^s$ then $w_{BI3} = [t_1^s; t_{21}^s]$.

The size of an extraction window is defined in a direct relation to the duration of the corresponding instance. Moreover, the parameter ws enables to adapt the extraction window size to the density of the analyzed event logs. For example, if a log is very dense (i.e., it holds a large amount of events in a short timespan) then applying extraction windows with a fixed size could result in an overly detailed signature (i.e., overfitting occurs) which could, subsequently, lead to flawed anomaly detection results, cf. [7]. In comparison a sparse log combined with a fixed size window could result in a signature that contains insufficient historic behavior to identify anomalies (i.e., underfitting occurs).

Subsequently step ② is applied to mine all the behavior that occurs in a chosen window in a *time sequence*. Therefore, the window is split into multiple *slots* based on the start and end of the process instances covered by the window (i.e., the dotted lines and slot timestamps in Fig. 2). Each slot is defined as $o := (N, t_s^s, t_e^s)$ where t_s^s and t_e^s represent the start and end timestamps of the slot and the bag $N := \texttt{execP}(\texttt{mid}(t_s^s, t_e^s), L)$ holds the names of models whose instances occur between t_s^s and t_e^s. For example, slot 5 in Fig. 2 would be defined as $o_5 = (\{A, B, C\}, t_5^s, t_6^s)$. If multiple instances from the same process model are executed in parallel then the related model's name occurs in $o.N$ multiple times (e.g., two parallel executions of model A would result in $o.N = \{A, A\}$). Subsequently, all slots are combined into a time sequence, i.e., an ordered list of slots $ts := \langle o_1, o_2, \cdots, o_n \rangle$, e.g., $ts_{BI3} = \langle o_1, o_2, \cdots, o_{20} \rangle$ for instance $BI3$, cf. Fig. 2. Finally, noise in the mined time sequences is addressed, for example, by removing slots which do not cover any instance execution (e.g., slot 8 in Fig. 2).

The signature generation ends by merging the resulting time sequences from all windows (one window for each instance is generated) into one signature ③. The signatures are represented as *likelihood graphs*, which were also already successfully applied in [5] for this purpose. Here, likelihood graphs enable to calculate the likelihood of instance behavior to determine unlikely (i.e., anomalous) ones. For this a likelihood graph encodes which and how instances and models are typically temporally related to each other during their execution (e.g., how instances succeed or precede each other). Moreover, it encodes the likelihood and order of such relations based on the mined time sequences, cf. Fig. 4.

Secondly, the signature is utilized to assess if a given process instance execution behavior, for P's instances, is anomalous or not. Hence, given behavior is filtered ④, and mapped ⑤ to the signatures (i.e., for each process model an individual signature is generated) to determine the likelihood of given instance execution behavior. Of course, some of the instance behavior could be unexpected because it never occurred before and is, accordingly, also not represented by the signatures (i.e., it cannot be mapped to a signature), cf. [5]. In such cases a configurable artificial behavior likelihood is calculated to flexibly deal with noise and slight – likely harmless – deviations from the historic behavior.

Thirdly, the likelihood of the given instance execution behavior is compared to a reference likelihood generated from P's historic instance executions stored in the historic execution log files L, ⑥. If a deviation between both likelihoods (reference likelihood and likelihood of the given instance execution to analyze) is observed then the analyzed given instance execution is identified as anomalous.

4   Multi Instance Anomaly Detection

This section presents the algorithms for the approach set out in Fig. 1.

4.1   Temporal Behavior Mining from Execution Logs

The proposed anomaly detection approach starts with a process model $P \in R$ (i.e., a signature is generated for P) and logs containing historic process execution behavior L. Subsequently, extraction windows are constructed for P's instances, as described before. This enables to mine historic execution behavior that takes place before, during, and after P's executions as *time sequences*.

Mining Time Sequences. Each time sequence is a sequence of slots $o :=$ (N, t_s^s, t_e^s) which are ordered based on their end timestamp, i.e., t_e^s. In the following a signature is generated by merging multiple time sequences. The presented mining approach generates a time sequence (i.e., a sequence of slots) for each extraction window – and through this for each instance. Therefore, the start and end timestamps of each process instance covered by the window are exploited, i.e., the dotted vertical lines in Fig. 2. Algorithm 1 formalizes the mining of a single time sequence for a given window w and the historic execution logs in L. In the following the symbol $\oplus$ denotes the appending of a slot to the end of a sequence.

Algorithm mineTS(*extraction window* $w := [wt_1; wt_2]$, *execution logs* L)
    Result: mined time sequence ts
    $ts := \langle\rangle$; $first := w.wt_1$ // initially ts is empty
    // extract the interval between instance start and end timestamps as slots
    while $second := next(first, L) \wedge second < w.wt_2$ **do**
        $ts := ts \oplus (\text{execP}(mid(first, second), L), first, second)$
        $first := second$ // preserve for next iteration
    // interval from the last instance start or end till the end of the window
    $ts := ts \oplus (\text{execP}(mid(first, w.wt_2), L), first, w.wt_2))$
    return ts // the mined time sequence for the window w and a given log L

Algorithm 1. Mines a time sequence for a given window w and logs L

The time sequence generated for the execution scenario depicted in Fig. 2 is shown at the left side of Fig. 3. For the sake of brevity Fig. 3 only depicts the number of the respective slot and the covered process model names while start and end timestamps are omitted. Only the first eight slots are depicted.

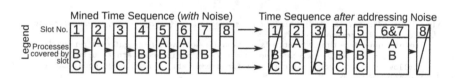

Fig. 3. Time sequence before and after addressing noise, window size is $ws = 2$

Addressing Noise in Time Sequences. The mined time sequences likely contain slots which are not relevant for or even interfering with the following signature generation. Such noise in the time sequences could result in an overfitting of the generated signatures, i.e., the signatures would be "too" specific and detailed, cf. [7]. This could result in false positives, i.e., non-anomalous executions which are incorrectly identified as anomalous. So, the proposed noise reduction heuristic will deal, for a given time sequence ts, with all slots which are *empty* or *volatile*. A slot $o := (N, t_s^s, t_e^s)$ is empty if $o.N = \varnothing$, i.e., if no instance is executed at the timespan (t_s^s to t_e^s) covered by the slot, e.g., slot 8 in Fig. 2 is empty.

Moreover, a slot is volatile if it covers only a low amount of activity executions. Typically volatile slots are placed at the beginning or end of instances and cover only a short time span. Hence even a minor shift in an instance's start or end timestamp can have a large impact on the slot. For example, if instance $BI2$'s duration, cf. Fig. 2, would only be a bit shorter (e.g., when it would end at t_7^s instead of t_8^s) then slot 7 would no longer be present or be part of the mined time sequence. Formally, a slot o is identified as volatile if $\mathsf{act}(o.t_s^s, o.t_e^s, L) < c$, i.e., c controls the minimum number of activities covered by the slot, cf. Algorithm 2.

Empty slots are removed from a time sequence ts, using list comprehension notation, i.e., $ts := \langle o \in ts | o.N \neq \varnothing \rangle$. For volatile slots, in comparison, it is checked if they could be aggregated with *one or more* directly successive slots, which are also volatile, to become non-volatile. If this is not possible then they are also removed, cf. Fig. 3 (right side). For this Algorithm 2 must identify directly connected volatile slots. Hence, the algorithm stores the start of the first volatile slot $volS$ and the end of the most recent successive volatile slot in $volE$. Based on this information $\mathsf{act}(volS, volE, L) > c$ enables to determine if an aggregation of the found successive volatile slots results in a non-volatile slot. Imagine that slot 18 ($\{C\}, t_{18}^s, t_{19}^s$) and 19 ($\{B, C\}, t_{19}^s, t_{20}^s$) in the running example Fig. 2 were found as volatile. This is because c was assumed as 3 and each of both slots covers less than three complete activity executions, i.e., $\mathsf{act}(t_{18}^s, t_{19}^s, L) = 2$ for slot 18 and $\mathsf{act}(t_{19}^s, t_{20}^s, L) = 2$ for slot 19. However, by aggregating both slots a new slot is created that is not volatile. Thus the aggregated slot becomes ($\{B, C\}, t_{18}^s, t_{20}^s$) (i.e., $\mathsf{act}(t_{18}^s, t_{20}^s, L) = 4$) and replaces the old slots 18 and 19. Alternative slot merging approaches, e.g., to identify and merge slot combinations which cover the longest timespan are conceivable and were evaluated. However, as these are computational intense the described greedy heuristic was applied.

Algorithm addressVolatileSlotsInTS(*time sequence* $ts = \langle o_1, o_2, \cdots, o_n \rangle$, *execution logs* L, *slot volatile threshold* $c \in \mathbb{N}_{>0}$)
    Result: noise free time sequence ts_{nv} (volatile slots were aggregated or removed)
    $ts_{nv} := \langle\rangle; volN := \varnothing; volS := 0; volE := 0$ // store intermediate results for the
      following steps, $volS$ and $volE$ are timestamps while $volN$ holds model names
    foreach $o \in ts$ **do**
        if $\mathsf{act}(o.t_s^s, o.t_e^s, L) < c$ // check if o is a volatile slot **then**
            $volE := o.t_e^s; volN := volN \cup o.N$// aggregate slots
            if $volS = 0$ // i.e., first volatile slot found **then**
                $volS := o.t_s^s$ // preserve start time of the first volatile slot found
            else if $\mathsf{act}(volS, volE, L) > c$ // aggregated slot is not volatile **then**
                $ts_{nv} := ts_{nv} \oplus (volN, volS, volE)$ // append aggregated slot on ts_{nv}
                $volN := \varnothing; volS := 0; volE := 0$ // purge preserved data
        else
            // a non-volatile slot was found, purge preserved data because only
            directly successive volatile slots are aggregated
            $ts_{nv} := ts_{nv} \oplus o; volN := \varnothing; volS := 0; volE := 0$
    return ts_{nv}// similar to input time sequence ts but without volatile slots

Algorithm 2. Addressing volatile slots in mined time sequences

---

Consider the right side of Fig. 3. The original time sequence (left side, cf. the running example in Fig. 2) was adapted to remove or address noise (right side). Note, slots which are crossed out were removed because they are empty (slot 8) or volatile slots which could not be aggregated with other volatile slots (slot 1 and 3). Two volatile slots (slot 6 and 7) were replaced by an aggregated non-volatile slot, i.e., "6&7". The slot volatile threshold c was assumed as three.

4.2 Signature Generation from Time Sequences

Subsequently, signatures are generated for each individual process model P based on time sequences TS which were generated for P's historic instances in L. For this the noise free time sequences are merged and transition likelihoods are calculated. Transition likelihoods represent the likelihood that slots which cover specific instances of processes follow each other (based on all time sequences $ts \in TS$). For example, the likelihood that a slot which holds an execution of process model A, B, and C is followed by a slot with execution A and C, cf. slots 16 and 17 in Fig. 2. In the following this likelihood information is utilized to differentiate between likely and unlikely (i.e., anomalous) model instance executions.

This work proposes to represent the mined temporal behavior in three independent signatures (i.e., one for behavior that happens before, during, or after a process model's execution). For this the mined time sequences are split accordingly into three parts – based on P's associated instance starts and ends, cf. Fig. 2. This decreases the size of each signature. Also this was found to increase the anomaly detection performance of the presented approach. The latter is because during each point in time a signature can be applied that specializes on the specific kind of behavior that is currently observed (e.g., behavior that was historically observed after or during an instance execution). Hereby, the applied signature can be more specific than one large generic signature that needs to cover all the historic instance behavior (i.e., before, during, and after) at once.

Each signature is represented as a likelihood graph $G = (V, D)$, cf. [5]. A likelihood graph is a directed cyclic graph that consists of a set of vertexes $v \in V$ and a set of edges $d \in D$ with $D \subseteq V \times V \times [0; 1]$. Each vertex v represents processes covered by a specific $o.N$ for a given slot o. In comparison each edge $d = (v_s, v_e, tl)$ represents the *transition likelihood* $tl \in [0; 1]$ from one "slot" v_s (i.e., a vertex holding the process model names covered by a slot) to another vertex v_e based on the mined time sequences $ts \in TS$.

Algorithm 3 creates a likelihood graph (i.e., a signature) by merging multiple time sequences. For this, the algorithm extracts from each slot, covered by the merged time sequences, the processes covered by that slot and stores them in the set V. Moreover, the set VC is populated with triplets $vc = (v_1, v_2, tc)$ that indicate, based on the analyzed time sequences, that a slot v_2 is preceded by a slot v_1, $tc \in \mathbb{N}_{>0}$ times (i.e., tc denotes the *transition count*). Subsequently these absolute numbers (i.e., tc) are converted into relative transition likelihoods tl and stored into D as edges. V is initialized with a dummy entry v_d that is used as a general entry point for the signature and the following mapping of behavior to it.

Algorithm mergeTimeSequencesIntoSignature(*time sequences* TS, *dummy vertex, i.e., the entry point for the signature* v_d)

> **Result:** likelihood graph (i.e., a signature) $G = (V, D)$ from the behavior in TS
> $V := \{v_d\}$; $D := \varnothing$; $VC := \varnothing$
> **foreach** $ts \in TS$ **where** $|ts| > 0$ **do**
> > $VC := VC \cup \{(v_d, ts_0, 1)\}$ // add dummy vertex, ts_0 identifies the first slot in ts, i.e., ts_i with $0 \geq i < |ts|$ identifies the slot with the index i
> > **for** $i := 0; i < (|ts| - 1); i := i + 1$ **do**
> > > $v_1 := ts_i.N$; $v_2 := ts_{i+1}.N$// ts_i and its successor ts_{i+1} in ts
> > > $V := V \cup \{v_1, v_2\}$;// add slots to signature graph vertex set V
> > > $tcount := 1$// holds how frequently v_1 is followed by v_2 in all sequences
> > > **if** $(v_1, v_2, \cdot) \in VC$ // previous ts contained the same transition **then**
> > > > $tcount := tcount + \{vc.tc | vc \in VC, vc.v_1 = v_1 \wedge vc.v_2 = v_2\}^0$
> > > // purge old information for v_1/v_2, then add updated or new information
> > > $VC := \{vc \in VC | vc.v_1 \neq v_1 \wedge vc.v_2 \neq v_2\} \cup \{(v_1, v_2, tcount)\}$
> **foreach** $vc \in VC$ // convert absolute numbers into likelihoods **do**
> > $s := vc.tc$; $TC := \{vc'.tc | vc' \in VC \wedge vc'.v_1 = vc.v_1\}$
> > // create and add edges to D that connect the signature vertexes in V
> > $D := D \cup \{(vc.v_1, vc.v_2, \frac{s}{\sum_{tc_s \in TC} tc_s})\}$// fraction $\mapsto$ transaction likelihood
> **return** $G = (V, D)$// return signature, it was created for sequences in TS

Algorithm 3. Merge time sequences TS for P into a signature G

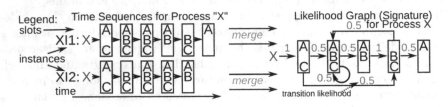

Fig. 4. Merging two mined time sequences into a signature

Figure 4 depicts an example for the proposed signature generation. Two time sequences ($XI1$ and $XI2$, mined for the process "X" – left side) are merged into a likelihood graph signature representation (right side). The depicted time sequences and represented behavior occurred after X's instances (i.e., succeeding behavior). So process X is placed at the start of the time sequences and signature.

4.3 Signature Matching for Execution Event Streams and Logs

The signatures are applied to calculate the likelihood of process execution behavior based on *given execution events*. Today's execution engines store (ex post analysis) or stream (real time analysis) various events. This work is mainly interested in process instance and activity start and end events. Hence, all perceived events are filtered accordingly and mined into time sequences by applying the presented approach. Finally the resulting time sequences are mapped to signatures which were generated for the executed processes based on historic behavior.

To determine if a given process instance execution is unlikely (i.e., anomalous) or not its likelihood is calculated (i.e., *execution likelihood* $l_e \in \mathbb{R}_{>0}$) by mapping it to the signatures. Moreover, comparable executions (i.e., that show a similar temporal execution behavior to the given execution) are identified in the historic logs L and mapped to the same signature as the given execution to generate a *reference likelihood* $l_r \in \mathbb{R}_{>0}$. Finally, both likelihoods are compared. If the execution likelihood is *below* the smallest found reference likelihood then the analyzed execution is identified as unlikely and, because of this, as anomalous.

The *execution likelihood* is calculated by Algorithm 4. Therefore, a signature (i.e., a likelihood graph $G = (V, D)$) is utilized along with a time sequence ts that is mapped to the signature (i.e., the sequence ts, here, represents given instance execution behavior that should be analyzed for anomalies). To calculate the likelihood, the slots held by the time sequence are mapped to the signature one after another while aggregating the transition likelihoods encoded in the edges $d \in D$ that connect all signature vertexes. Finally, when all recorded behavior (i.e., the time sequence ts) was mapped the likelihood is returned.

Algorithm matchSig(*signature* $G = (V, D)$, *dummy vertex* v_d *representing the process which* G *was generated for, time sequence* ts *holding behavior to map, punishment factors* $pNDC, pDP, pOS \in (0; 1]$)

  Result: calculated likelihood $lh \in \mathbb{R}_{>0}$ for ts

  $lh := 1; v_l := v_d$ // behavior likelihood lh and most recent signature vertex v_l

  foreach $o \in ts$ // individually for each slot **do**

    $lh' := \{d.tl \in D | d.v_1 = v_l \wedge d.v_2 = o.N\}$

    $P := \{(v', sim) | v' \in V; sim := \frac{|v' \triangle o.N|}{|v'| + |o.N|}, \forall a \in o.N, \exists b \in v'; a = b\}$

    $MS := \{(v', sim) | v' \in V; sim := \frac{|v' \triangle o.N|}{|v'| + |o.N|}\}$// $\triangle$ notates a symmetric difference

    if $lh' = \varnothing$ // unexpected behavior was found **then**

      if $o.N \in V$ // stage one: exact behavior is present in the signature **then**

        $v_l := o.N; lh := lh \cdot pNDC$

      else if $P \neq \varnothing$ // stage two: present but different parallelism **then**

        $min := \{p_1 | p_1 \in P; \forall p_2 \in P, p_1.sim \leq p_2.sim\}^0$

        $v_l := min.v; lh := lh \cdot (1 - min.sim) \cdot pDP$

      else

        // stage three: fallback if one and two are not applicable

        $min := \{ms_1 | ms_1 \in MS; \forall ms_2 \in MS, ms_1.sim \leq ms_2.sim\}^0$

        $v_l := min.v; lh := lh \cdot (1 - min.sim) \cdot pOS$

    else

      $v_l := o.N; lh := lh \cdot lh'^0$ // if the behavior is expected

  return lh// return likelihood lh of the behavior in ts

Algorithm 4. Likelihood for a time sequence ts based on a signature G

Of course, it is possible that some behavior cannot be mapped successfully. For example, this is the case if behavior occurs in unexpected orders, e.g., instance A is succeed by B but it was expected (i.e., specified in the signature) the other way around. Another reason could be that the parallelism of observed and the expected behavior deviates, e.g., it was expected that a single instance of A is executed, but two concurrent executions of A were observed.

Existing process anomaly detection work typically classifies any unexpected behavior, such as the preceding examples, as anomalous. However, as argued in [5] this is not always beneficial. Process models and model executions are known to occur in flexible dynamic environments, struggling with ad-hoc changes, and the need to cope with multiple frequently changing requirements [9]. Hence, we assume that existing anomaly detection approaches are too strict to be successfully applied in today's flexible and dynamic process execution environments. So, the proposed anomaly detection approach provides the flexibility to deal with unexpected behavior by calculating an artificial likelihood for it.

The flexibility that should be granted by the proposed anomaly detection approach varies between different organizations, processes, and use cases. Hence, the flexibility can be configured in Algorithm 4 based on three punishment factor variables, i.e., $pNDC$, pDP, and pOS. Those enable to punish unexpected behavior by reducing its calculated artificial likelihood. Hence, while the scenario in the initial motivating example is identified as anomalous – because significantly more parallelism is observed than expected based on historic executions – minor, probably harmless, deviations from the historic behavior are "granted" until, e.g., a combination of multiple minor deviations becomes too unlikely.

So when calculating the *execution likelihood* it is checked if the current slot $o \in ts$ (i.e., the behavior to map next) is a direct successor of the last mapped slot v_l. If it is, then the likelihood is extracted from the related transition likelihood hold by the signature in D. Hence, when mapping the short example timesequence $ts := \{(\{X\}), (\{A, C\}), (\{A, B, C\})\}$ (timestamps are omitted) on the signature depicted in Fig. 4 then a likelihood of $1 \cdot 0.5 = 0.5$ is calculated. However, if o is not a successor of the last mapped slot then unexpected behavior was found. For this, it becomes necessary to calculate an artificial likelihood by applying a three staged approach which is discussed in the following.

Stage *one*: It is checked if the unexpected behavior (i.e., slots) is represented in the signature but occurred in an unexpected order. If this is the case then the punishment factor $pNDC$ it utilized as the artificial likelihood. Stage two and three calculate the artificial likelihood based on the similarity of the given instance behavior and the behavior represented in the signature. Stage *two* is applied if the expected processes are executed but with an unexpected parallelism. For example, A, A, B was observed but expected was A, B, i.e., two parallel executions of process A were found but only one was expected. This stage utilizes the punishment factor pDP. Stage *three*, which utilizes the punishment factor pOS, can always be applied and is, because of this, used as a fallback. It is similar to stage two but more relaxed, i.e., it does not enforce that the behavior to map and the related signature behavior must only consist of the same processes.

Imagine that the slot o with $o.N := \{A, A\}$ should be mapped to a signature which only consist of A as the expected behavior. Because, the observed $\{A, A\}$ and the expected $\{A\}$ behavior is different the slot cannot be found in the signature. Hence the proposed approach falls back to stage two of the artificial likelihood calculation. So the likelihood is calculated as $\frac{|\{A\} \triangle \{A, A\}|}{|\{A\}| + |\{A, A\}|} \mapsto \frac{1}{3} = 0.\bar{3}$ so that the final artificial likelihood becomes $(1 - 0.\bar{3}) \cdot 0.8 = 0.5\bar{3}$ when a punishment factor pDP of 0.8 is used.

The *reference likelihood* l_r is calculated based on logged historic executions in L that show comparable behavior to the given behavior (i.e., given instance execution behavior to analyze for anomalies). In this case comparable means that the time sequence describing the given behavior and the time sequences describing the historic behavior hold similar slots. For this the presented approach to measure the similarity between two slots (i.e., for artificial likelihood calculation, cf. Algorithm 4) is generalized and applied on the historic time sequences which were mined from L during the signature generation. The $k \in (0, 1]$ percent most similar historic time sequences are subsequently compared with the signature G using Algorithm 4. Finally the lowest likelihood found during that comparisons is utilized as the reference likelihood l_r. If $l_e < l_r$ then the given behavior (i.e., the behavior that is analyzed for anomalies) is identified as anomalous. This bears two advantages: Executions in L are never identified as anomalous and the flexibility which was historically observed for the process under analysis, and which is because of this stored in L, is taken into account during anomaly detection.

5   Evaluation

The evaluation utilizes real life process execution logs from multiple domains and artificially generated anomalies in order to assess the anomaly detection performance and feasibility of the proposed approach. It was necessary to generate artificial anomalies as information about real anomalies are not provided by the log sources. The utilized logs were taken from the BPI Challenge 2015[1] and 2017[2] (BPIC5 and BPIC7), and Higher Eduction Processes (HEP), cf. [13].

The BPIC5 logs hold 262,628 execution events which origin from 5,649 instances and 398 activities. The logs cover the processing of building permit applications at five (BIPC5_1 to BPIC5_5) Dutch building authorities between 2010 and 2015. In comparison the BPIC7 logs hold 1,202,267 events from 31,509 instances, recorded in 2016 and 2017, which focused on loan application management. The HEP logs contain 28,129 events, 354 execution traces (i.e., instances), and 147 activities – recorded from 2008 to 2011. Each trace holds the interactions of a student with an e-learning platform (e.g., exercise uploads). The interactions are recorded individually for each academic year $\mapsto$ HEP_1 to HEP_3. All logs (i.e., BPIC and HEP) contain sufficient details to apply the proposed approach (e.g., instance execution events and relevant timestamps).

The logs were evenly and randomly separated into training (for signature generation) and test data (for the anomaly detection performance evaluation). A tenth of the test data was mutated by one (out of four) randomly chosen mutators (we regard this amount as being sufficient, cf. [4]). This enables to generate labeled non-anomalous (i.e., non-mutated) and anomalous (i.e., mutated) test data entries, i.e., to determine if both behavior "types" are correctly differentiated by the proposed approach. The applied four mutators generate multi instance anomalies that cannot be detected by existing single business process instance focused anomaly detection work, hence, a comparison with such existing

---

[1] http://www.win.tue.nl/bpi/2015/challenge—DOI:10.4121/uuid: 31a308ef-c844-48da-948c-305d167a0ec1.

[2] http://www.win.tue.nl/bpi/doku.php?id=2017—DOI:10.4121/uuid: 5f3067df-f10b-45da-b98b-86ae4c7a310b.

work is not possible: (*a*) *Parallel Executions* – a process execution is duplicated so that it occurs in parallel; and (*b*) *Sequential Executions* – a process execution is duplicated so that occurs in a sequential order; and (*c*) *Execution Order* – the process execution order is randomly changed; and (*d*) *New or Missing Process* – new process executions are artificially added or recorded executions are removed.

The mutators were adapted and extended from our work in [4] – which was assessed by security experts as being realistic. It was chosen to combine multiple mutators to represent that real life anomalies are diverse and affect different aspects of process executions. In addition, the applied strategy also evaluates the proposed handling of unexpected execution behavior. This is, because the test data (in its mutated but also non-mutated form) contains behavior that is not represented in the training data (e.g., manual ad-hoc changes). The following results consist of the average of 100 evaluation runs – individually for each log file – to even out the randomness in the data separation and mutation.

Metrics and Evaluation. The feasibility of the presented anomaly detection approach is analyzed. For this, the training data is utilized to construct signatures which are applied on the test data to differentiate between known randomly mutated (i.e., anomalous) the known non-mutated (i.e., non-anomalous) test data entries. This enables to collect four key performance indicators: *True Positive* (TP) and *True Negative* (TN) represent data entries that were correctly identified as anomalous (TP) or non-anomalous (TN). In comparison, *False Positive* (FP) and *False Negative* (FN) represent data entries which were incorrectly identified as anomalous (FP) or non-anomalous (FN). For example, FP counts non-anomalous test data entries which were incorrectly identified as being anomalous. Based on this performance indicators three standard metrics are calculated for each log file (i.e., BPIC5_1-5, BPIC7, and HEP_1-3):

(*a*) *Precision* $P = TP/(TP + FP)$ – indicates if the identified anomalous test data entries were in fact anomalies; and (*b*) *Recall* $R = TP/(TP + FN)$ – indicates if anomalies were "missed", i.e., not identified; and (*c*) *Accuracy* $A = (TP+TN)/(TP+TN+FP+FN)$ – provides a general anomaly detection performance overview; $TP, TN, FP, FN \in \mathbb{N}_{>0}$; $P, R, A \in [0; 1]$.

For this paper we assume that the number of False Positives (FP) or Negatives (FN) should be low while the number of True Positives (TP) or Negatives (TN) should be high, i.e., the accuracy becomes close to one. In addition the F_β-measure, Eq. 1, is applied because it provides a configurable harmonic mean between Precision (P) and Recall (R), cf. [8]. Hereby, β controls the balance between P and R. So, if $\beta = 1$ then a harmonic mean between P and R is calculated. In comparison a $\beta < 1$ results in a precision and a $\beta > 1$ in a recall-oriented result. $F_{0.5}, F_1, F_{1.5}$-measures were used to present the evaluation results.

$$F_\beta = \frac{(\beta^2 + 1) \cdot P \cdot R}{\beta^2 \cdot P + R} \tag{1}$$

Results. The results were generated based on BPIC 2015/2017 and HEP process execution logs and a publicly available proof-of-concept implementation of the presented approach: https://github.com/KristofGit/MultiInstanceAnomaly.

The implementation calculated a signature in minutes (i.e., about 2 min on average) and required only seconds (i.e., below 3 s on average) to identify a test data entry as anomalous or non-anomalous on a standard 2.6 Ghz Intel Q6300 CPU with 8 GB of RAM. Of course, the signatures can be reused, i.e., calculated once and subsequently applied in order to analyze multiple following process executions. Moreover, the presented approach can be concurrently applied to analyze multiple instances in parallel. This suggests an applicability even on larger process repositories and execution logs.

Primary tests were applied to identify appropriate configuration values for the presented approach. The *punishment factors* for unexpected behavior were set to $pNDC = 0.60$ (known behavior but unexpected order), $pDP = 0.40$ (known behavior but unexpected parallelism), $pOS = 0.30$ (unknown behavior). A lower punishment factor results in a stronger punishment. So for example, pDP is higher than pOS because the latter is only utilized if completely unknown execution behavior is observed. In comparisons the former is applied if "only" an unexpected parallel execution occurred. This is the case, for example, if three parallel executions of process A were observed but only two were expected. As a rule of thumb it can be assumed that a higher punishment improves on the TP/FN side while having a negative impact on the TN/FP performance indicators. A similar conclusion can be drawn for $k = 0.3$, i.e., the percentage of similar time sequences for reference likelihood generation purposes. When k is increased then the proposed approach becomes more relaxed because the reference likelihood typically decreases, i.e., anomalous instances are more likely "overlooked".

A window size ws of 4 (BPIC) and 20 (HEP) was utilized. Hereby, the different ws values compensate that the log sources (e.g., BPIC or HEP) store events with a different density (i.e., the BPIC logs cover more events at the same timespan than the HEP logs). The log dependent ws-value ensures that the generated signatures represent a roughly comparable amount of process execution events for all log sources. Finally, a noise prevention value of $c = 8$ was utilized, i.e., a slot – either original or aggregated – has to cover at least 8 activity executions to not be recognized as volatile or noise and being removed. The chosen values were successfully applied on different processes and domains. Hence, we assume that they can be applied as a valid starting point for future optimizations in scenarios and domains which were not covered by the presented evaluation.

The average evaluation results are shown in Tab. 1. The accuracy metric reached an average result of 78% but also the other metrics show promising results (83% for recall and 77% for precision). Hence, it was found that the proposed approach could successfully identify the constructed anomalies in the analyzed complex multi instance execution evaluation data. It was observed that the proposed approach could more easily identify anomalous behavior for the HEP log based evaluation than during the BPIC based evaluation. This most likely origins from the different complexity of the logs (i.e., the BPIC logs hold substantially more and more complex behavior than the HEP logs). So, it was concluded that the more complex and diverse the signature generation behavior becomes, the harder it is to distinguish correct from anomalous behavior. Nevertheless, even for the challenging BPIC log based evaluation the performance of the presented work achieved an average of 70% anomaly detection accuracy.

Table 1. Anomaly detection performance of the presented approach

	BPIC5_1	BPIC5_2	BPIC5_3	BPIC5_4	BPIC5_5	HEP_1	HEP_2	HEP_3	BPIC7
Accuracy	0.70	0.71	0.73	0.71	0.71	0.94	0.92	0.92	0.66
Precision	0.69	0.68	0.70	0.67	0.67	0.96	0.97	0.96	0.63
Recall	0.78	0.83	0.84	0.84	0.82	0.92	0.88	0.89	0.70
$F_{0.5}$-measure	0.71	0.70	0.72	0.70	0.72	0.95	0.95	0.94	0.64
F_1-measure	0.73	0.75	0.76	0.75	0.75	0.94	0.93	0.92	0.66
$F_{1.5}$-measure	0.75	0.78	0.79	0.78	0.77	0.93	0.91	0.91	0.68

6 Discussion and Outlook

This work applies the common assumption that an anomaly is some kind of unlikely behavior that never or hardly occurs during a business process execution, cf. [4,5]. Accordingly the proposed approach compares given multi instance executions with given recorded historic executions in L to calculate their likelihood in relation to the behavior in L, such that, unlikely process instance executions are identified as anomalous. Hence, this work applies an unsupervised approach as it neither assumes the historic behavior as anomalous or not. It is hard to propose a rule of thumb for predicting the required training data size (i.e., the amount of historic behavior in L). This is because the size of the required training data heavily depends on the amount of execution variety that can be observed for the analyzed instances – which is unique for each organization.

The evaluation showed an average anomaly detection accuracy of 78%, which suggests an applicability in additional scenarios. In addition a detailed analysis of the evaluation results revealed that the proposed behavior likelihood assessment based anomaly detection approach substantially improved the detection results for the analyzed complex real life execution behavior. This is, because the utilized real life evaluation data showed a substantial amount of behavior drift in the analyzed multi instance behavior data, caused, e.g., by fluctuating instance durations or varying parallel instance execution behavior. Hence, not taking these dynamics, by design, into account would have resulted in substantially worse evaluation results, e.g., by causing a high number of false positives. Note, a large amount of false positives could harm an organization's performance, e.g., through process executions which are unnecessarily halted or terminated.

The presented approach determines process executions as anomalous based on their relations to other preceding, succeeding, or simultaneous instances. In comparison to existing work this is a rather big picture focused approach which, by purpose, ignores more fine granular details (e.g., which resource has executed an activity or what data was exchanged between two activities). Hence, in future work we will strive to combine both worlds. Hereby, multiple views on the instance behavior can be taken into consideration to identify diverse and complex anomalous behavior. We assume this as necessary to identify inside threats that actively hide their malicious intentions. Moreover, we will assess the applicability of the proposed approach to analyze complex dynamic parallel activity executions. Finally, we will strive to integrate correlation features to respect contextual aspects, e.g., by adding support for filters to analyze only process instances that meet specific conditions (e.g., based on the involved resources).

References

1. Allen, J.F.: Maintaining knowledge about temporal intervals. ACM **26**(11), 832–843 (1983)
2. Atallah, M., Szpankowski, W., Gwadera, R.: Detection of significant sets of episodes in event sequences. In: Data Mining, pp. 3–10. IEEE (2004)
3. Bezerra, F., Wainer, J., Aalst, W.M.P.: Anomaly detection using process mining. In: Halpin, T., Krogstie, J., Nurcan, S., Proper, E., Schmidt, R., Soffer, P., Ukor, R. (eds.) BPMDS/EMMSAD -2009. LNBIP, vol. 29, pp. 149–161. Springer, Heidelberg (2009). doi:10.1007/978-3-642-01862-6_13
4. Böhmer, K., Rinderle-Ma, S.: Automatic signature generation for anomaly detection in business process instance data. In: Schmidt, R., Guédria, W., Bider, I., Guerreiro, S. (eds.) BPMDS/EMMSAD -2016. LNBIP, vol. 248, pp. 196–211. Springer, Cham (2016). doi:10.1007/978-3-319-39429-9_13
5. Böhmer, K., Rinderle-Ma, S.: Multi-perspective anomaly detection in business process execution events. In: Debruyne, C., et al. (eds.) OTM 2016. LNCS, vol. 10033, pp. 80–98. Springer, Cham (2016). doi:10.1007/978-3-319-48472-3_5
6. Böhmer, K., Rinderle-Ma, S.: Anomaly detection in business process runtime behavior - challenges and limitations. arXiv (2017)
7. Chaoji, V., Rastogi, R., Roy, G.: Machine learning in the real world. VLDB Endowment **9**(13), 1597–1600 (2016)
8. Chinchor, N., Sundheim, B.: Muc-5 evaluation metrics. In: Message Understanding, pp. 69–78. Computational Linguistics (1993)
9. Fdhila, W., Rinderle-Ma, S., Knuplesch, D., Reichert, M.: Change and compliance in collaborative processes. In: Services Computing, pp. 162–169. IEEE (2015)
10. Gupta, M., Gao, J., Aggarwal, C.C., Han, J.: Outlier detection for temporal data: a survey. Knowl. Data Eng. **26**(9), 2250–2267 (2014)
11. de Leoni, M., van der Aalst, W.M., Dees, M.: A general process mining framework for correlating, predicting and clustering dynamic behavior based on event logs. Inf. Syst. **56**, 235–257 (2016)
12. Rogge-Solti, A., Kasneci, G.: Temporal anomaly detection in business processes. In: Sadiq, S., Soffer, P., Völzer, H. (eds.) BPM 2014. LNCS, vol. 8659, pp. 234–249. Springer, Cham (2014). doi:10.1007/978-3-319-10172-9_15
13. Vogelgesang, T., et al.: Multidimensional process mining: questions, requirements, and limitations. In: España, S., Ivanović, M., Savić, M. (eds.) CAISE Forum, pp. 169–176. Springer, New York (2016)
14. Wieringa, R.J.: Design Science Methodology for Information Systems and Software Engineering. Springer, Heidelberg (2014)

Path-Colored Flow Diagrams:
Increasing Business Process Insights
by Visualizing Event Logs

Koen Daenen[(✉)]

Nokia Bell Labs, Antwerp, Belgium
koen.daenen@nokia-bell-labs.com

Abstract. Event logs of a self-care troubleshooting portal indicate that most customers do not follow the directions up to a conclusive end. Consequently, customers risk losing confidence in the support channel, which undermines the competitive strength of the business. We present a method for visual analysis of the event logs that employs graph reduction, and the use of path classification to create a novel type of flow diagram. These diagrams help to discover and communicate new insights, such as important trends about the way the customer traverses through the underlying business process.

Keywords: Workflow analysis · Graph visualization · Business process insights · Troubleshooting process

1 Introduction

Visual analysis is a good communicator of insights to the person or team you need to inform or convince [9]. We introduce the Path-Colored Flow (PCF) diagram to visualize event logs of a business process. The diagram aims to derive new insights, and to communicate these to the process design team. The paper describes three techniques to create this novel diagram. We illustrate the effectiveness based on event logs of a real-life workflow implementation.

A mobile service operator introduced a self-care portal that assists consumers with troubleshooting. The aim of the new care channel was to reduce the number of calls for technical help. The portal asks multiple choice questions, and proposes remediation actions. A workflow specification drives the order of interactions (Fig. 1). When the problem remains, the portal invites the consumer to call an assistant. The event logs revealed that most consumers did not follow the directions up to a conclusive end. This indicated a risk of customers losing their confidence in the support channel, leading to churn, and negatively effecting the operator's business.

The Customer Care team could not pinpoint a clarification for the issue. The design team of the workflow got the mandate to improve the workflow model. But how? Could a re-ordering of the questions or proposed actions reduce the abandon rate? Is there a critical point where most customers get lost? A histogram

© Springer International Publishing AG 2017
J. Carmona et al. (Eds.): BPM 2017, LNCS 10445, pp. 94–109, 2017.
DOI: 10.1007/978-3-319-65000-5_6

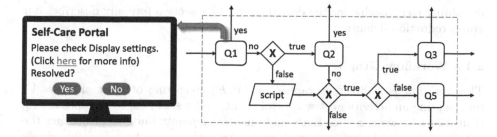

Fig. 1. Snippet of the self-care workflow

of abandons per question or a heat-map confirmed the size of the problem, and indicated that it was widely spread throughout the process. Abandons occurred both on popular as well as rarely consulted pages. The design team still had no grounds to start modifying the workflow design.

We applied the PCF diagram to gain more insights and to communicate these to the design team of the workflow. The goal is to create new insights on where and why the consumers abandoned. The diagram should visually reflect the workflow specification. This drove our choice to use a diagram in which the original flowchart is recognizable.

The self-care workflow is complex and consists of many nodes (e.g. 2324) and edges (e.g. 3185). The possibility to step-back during a process makes the flow even more complex. To mitigate this complexity, we created diagrams based on three contributions:

1. Section 2 defines a graph reduction algorithm that preserves the labels of the outgoing edges of a selected set of nodes. We use it to reduce the workflow graph to the pages presented to the consumer.
2. Section 3 discusses why a customized Sankey diagram is appealing for our analysis. We introduce the Leaky Sankey diagram by extending the diagram with half-edges. We define a formal graph to model the step-back transitions and abandons in our diagram.
3. We enrich the visualization with context data, and introduce *Path-Colored Flows*. Section 4 first describes the technique in its generic form on a graph. Then, we define the *Path-Colored Flow* (PCF) diagram.

Section 5 combines the three techniques and applies them to generate PCF diagrams for the self-care event logs.

2   Reducing the Graph

The original workflow is manually designed to specify the process execution of the self-care portal. It determines the control flow through the questions, but also contains many computational actions. We reduce the workflow graph to create a clear overview of the paths followed. We remove all the nodes and edges that represent automated data-driven logic. The reduction preserves all questions to the user, and all possible answers. This reduces the complexity, the model remaining recognizable for the design team. This section formally describes our graph reduction definition.

2.1 Workflow Graph

The workflow is a directed graph $G = (V, E)$ consisting of a set of nodes V, representing an activity or flow control event, and a set of edges E, representing transitions between nodes. Each node has an *type* attribute, representing the specific type of activity or control event. An activity can be a multiple-choice question (Q1, Q2, etc. in Fig. 1), an information message to be presented to the customer, or a pure computational activity called a *script*. Each activity that requires input from the consumer corresponds with a web page from the self-care portal. We call this a *display* node. Control events are *start-event*, *end-event* or a data-driven decision point (depicted as an X in diamond shape). The latter is an exclusive OR-split. The outcome is determined by the user's profile or a process control variable. A node representing a multiple-choice question, is also an exclusive OR-split. Its outgoing edges represent the possible answers to the question. All joins are OR-joins. The workflows describe a sequential process in which no parallel or inclusive gateways occur.

The edges have a label $a \in \mathscr{L}$, which represents the choice or outcome of the source node; $\mathscr{L}$ is the set of all labels. An edge is defined as the triple (*source*-node u, label a, *target*-node v), which we denote in this paper as $u \overset{a}{\frown} v$.

2.2 Reduction Rules

In this section, we present three reduction rules for sequential workflows. The individual rules are based on the existing reduction rules for process models [14]. Sadiq et al. introduced these rules to analyze the control structure of a process model. A DAG with two types of nodes represents both *or*-split/join and *and*-split/join. The reduction rules are defined to preserve the properties of the graph in terms of concurrency behavior: deadlocks and correct synchronization. The reduction rules presented in this paper reduce a sequential workflow, with only *or*-join/split, represented by a labeled directed (cyclic) graph. Such a workflow graph can be seen as a control program graph. The reduction rule to remove subgraphs with a unique entry and exit node, as described by McCabe [11], does not meet our goal, as this can remove *display* nodes.

Our contribution lies in the definition of criteria to preserve a set of selected nodes and their transitions, and using this to reduce a labeled directed (cyclic) graph to a partially labeled graph. The reduction rules are based on a node selection criterion, e.g. all *display*-nodes. We denote the set of selected nodes as V_S. The reduced graph is represented by $G_R = (V_R, E_R)$ and must meet following **criteria**:

1. V_R must at least contain all selected nodes. $V_R \supset V_S$
2. G_R must hold the same transitions for all selected nodes. We say that two edges represent the same transition when they have the same source node and edge label: $\forall u \overset{a}{\frown} v \in E(u \in V_S \implies \exists! w \in V_R : u \overset{a}{\frown} w \in E_R)$

3. For each path through G that starts and ends in a selected node, a path through G_R must exist that traverses through the same transitions of selected nodes in the same order.

We introduce two variants of the *Adjacent Reduction Rule* [14] (ARR). The main differences are (1) that the condition on the node type is expressed in terms of the set of selected nodes V_S, and (2) that our rules target graphs with labeled edges. The *Transition-safe* ARR preserves the edge labels while the *Unlabeling* ARR removes the labels. Note that the latter rule has no conditions on the control structure. The third reduction rule that we apply, is an exact re-use of the *Closed Reduction Rule* [14]. Figure 2 shows an example of each rule.

(a) Transition-safe ARR. A node $v \notin V_S$ with a single direct successor w is removed, and its connected edges. Each incoming edge $u \overset{a}{\frown} v$ is replaced by an edge $u \overset{a}{\frown} w$ with a the label of the original incoming edge.

(b) Unlabeling ARR. A node $v \notin V_S$ that is not a direct successor of a selected node, is removed. When v is removed, all the connected edges (with or without label) are removed, and replaced by unlabeled edges connecting each of v's direct predecessors with each of v's direct successors. A direct loop $v \overset{a}{\frown} v$ is removed entirely.

(c) Closed Reduction Rule. The application of the adjacent reduction rule generally introduces closed components in the workflow graph. Two or more unlabeled edges with the same source and target node are reduced to a single edge.

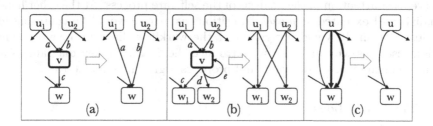

Fig. 2. Examples of reduction rules: $v \notin V_S$, (b) $u_1, u_2 \notin V_S$

To construct G_R, we first apply reduction rules (b) and (c). We visit all nodes of the graph G_i (with $G_0 = G$) and check if we can apply the Unlabeling ARR. Then we apply the Closed Reduction Rule to remove all double edges and create G_{i+1}. We repeat this recursively until G_n can't be reduced any further. Finally we visit all nodes of the graph G_n and check if we can apply the Transition-safe ARR to create G_R.

Since no rules remove a node of V_S, the first criterion is met. Only the Transition-safe ARR removes outgoing edges of selected nodes, but preserves the label, and as such meets the second criterion. Each reduction rule preserves all possible flows between the remaining nodes, and does not introduce additional flows. This ensures that G_R meets the third criterion.

Figure 3 shows an example of a graph reduction with $V_S = \{v_0, ..., v_5\}$. We call the set of nodes that are not selected but not removed, the set of *supporting* nodes $V_T = V_R \setminus V_S$. V_T, consists of nodes that are direct successors of selected nodes and have an outgoing degree greater than 1.

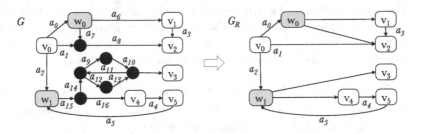

Fig. 3. Reduction of example graph G with $V_S = \{v_0, ..., v_5\}$; $V_T = \{w_0, w_1\}$

3   Leaky Sankey Diagram

The self-care portal allows a consumer to step-back to an earlier question. Via the **back** button of the browser, or the use of breadcrumbs, he or she can rewind the process. These transitions are not modeled by the workflow graph. Nor is the *abandon* event modeled. As such, the original (or reduced) graph model is not suited to represent the event logs. The transitions *step-back* and *abandon* deserve most attention in the analysis of the self-care process. At the other hand, visualizing all executed *step-back* transitions as arrows would be confusing for the design team. It would not represent the workflow diagram anymore. This section discusses why and how we introduce the *Leaky* Sankey diagram as an extension of the Sankey diagram, and proposes a graph to model it.

3.1   The Event Logs

The event logs consist of an ordered set of step records, one record per transition taken during a process execution. Each process instance has a unique report identifier, logged in each step record. A step report r is the sequence of all step records containing a given report identifier. Each record consist further out of a time stamp, an account name, a process identifier, a node identifier, the transition name, and a transition type flag. The *normal* transition type refers to the edge $v \overset{a}{\frown} w$ with v the node with the identifier of the given record, a the transition name, and w the node identified in the next step record of the same step report. The other transition types are *step-back*, *abandon* and *end*.

Before a consumer consults the actual process, he must register and provide some information such as his name, telephone number, and mobile device type. This information is stored in an account profile and can be correlated with a step report based on the account name. In the remainder of this section, and next section, we use a simple workflow with an example data set. Section 5 describes the results based on real customer data.

3.2 Diagram

Schmidt describes Sankey diagrams as *"the visual language of industrial ecology"* — *"they are ideal for interpreting complicated sets of resource flows"* [16]. For a service provider, the customers are its source of income. It is important to follow their journey during a troubleshooting session and visualize where they disconnect.

Our diagram represents the flow of process executions (see example in Fig. 4). We represent the *display* nodes as boxes, interconnected with *pipes* corresponding to the edges of the workflow graph. We represent the event logs as a flow through the machinery of nodes and pipes. We draw the width of the flow proportional to the number of sessions that ran through an edge or node. The start node and end nodes act respectively as the source and sinks of the flow. At the *display* nodes, we preserve the conservation of flow by drawing the flow of the abandoned sessions literally through the node. To visualize the *step-back* transitions without interconnecting the nodes with additional arrows, we also use the metaphor of leaking out and in. The flow leaks out at the node where the back button is pressed, and leaks in at the node to which the process returns. To mark this transition visually, we adopt the use of reference labels (A, B, ...) from engineering drawing practice.

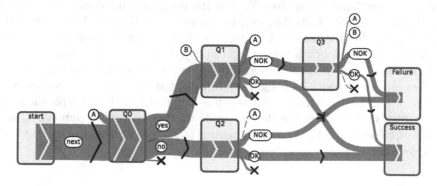

Fig. 4. A leaky Sankey diagram clearly shows *step-back* transitions and abandons.

To describe this as a formal graph model, we first define the set of *step-back* edges. A *step-back* transition is possible in any *display* node returning to a previous *display* node depending on the path traversed so far. In the event logs, these transitions are marked with a *step-back* flag. We define

$$E_B := \{x \overset{\text{back}}{\frown} y | (x, y \in V_S) \wedge (\exists x \overset{p}{\rightsquigarrow} y \in \mathscr{P}_{G_R})\} \tag{1}$$

in which V_S is the set of *display*-nodes, and $\mathscr{P}_{G_R}$ is the set of all paths through G_R. We don't add E_B to the graph definition as it would not represent the workflow diagram anymore: all the *step-back* edges are conditional in the sense that these transitions can only be taken depending on the flow followed so far. Instead, we add *abandon* and *step-back* half-edges to the workflow graph to construct accurate flow diagrams.

To model the *step-back* transitions in our visual analysis, we define $\mathscr{L}_{V_S}$ as a set of unique reference labels for each *display* node, and $\rho : \mathscr{L}_{V_S} \rightarrow V_S$ a bijection that maps each reference label to the corresponding *display* node. We define the label $\mathtt{abandon}$ to explicitly model the *abandon* transitions. Then we define $G_{RH} := (V_R, E_{RH})$, the reduced graph extended with three sets of labeled half-edges $E_{RH} := E_R \cup E_A \cup E_{B'} \cup E_{B''}$, and $G_{RHW} := (G_{RH}, \lambda)$, the graph extended with a weight function; in which

- $E_A := V_S \times \{\mathtt{abandon}\}$, the set of outgoing *abandon* half-edges in each display node;
- $E_{B'} := \{(v, a) \in V_S \times \mathscr{L}_{V_S} | v \overset{\text{back}}{\sim} \rho(a) \in E_B\}$, the set of outgoing *step-back* half-edges in each display node;
- $E_{B''} := \{(a, v) \in \mathscr{L}_{V_S} \times V_S | \rho(a) \overset{\text{back}}{\sim} v \in E_B\}$, the set of incoming *step-back* half-edges in each display node;
- $\lambda : E_{RH} \rightarrow \mathbb{N}$, mapping each (half-)edge onto the number of times the transition appears in the event logs.

Column N_e of Table 1, lists the values of $N_e = \lambda(e)$ for each (half-)edge of the example data set. Figure 4 depicts the Sankey diagram with half-edges. Each arrow is proportional to the flow N_e. On the diagram, the abandoned sessions end with the *abandon* half-edge, clearly showing how the flow leaks out of the graph. Similar for the step-back edges: the flow diagram shows how the flow jumps out and in the graph preserving the flow width. The *abandon* half-edges are represented with a cross pictogram as label; see e.g. the nodes Q1 and Q2 in the example. The *back* half-edges are represented with an arrow, respectively starting or ending on a circle with the reference label. On the example diagram, we can read that parts of the flow at node Q3 return to Q0 (reference A) and parts to Q1 (reference B). Transition names (e.g. yes, no) appear as labels on the edges close to the *source* node.

4   Path-Colored Flows

A flow diagram that reflects the path frequencies, or Sankey diagram, is well suited to highlight the trend of which transitions are dominant in a set of paths. Still, because the *abandon* behavior was spread all over the workflow, this representation falls short to analyze it.

Colors are often used to add information to a visualization. In graphs and flow diagrams, I distinguish three basic techniques:

1. The diagram is composed of distinct traces, each drawn with a different color [4,17]. The colors don't add information, but increase the readability, especially when the traces converge. In case the color of one trace is clearly different from the others, e.g. by saturation, hue, or brightness, it adds one element of global information: the author of the diagram has deliberately chosen to highlight the given trace.
2. A single color is assigned to each edge and node. The colors correspond to a value or item on a color-scale [1,3,6]. This technique adds one dimension of information to the diagram. It requires to limit or reduce the information to a single value per edge and node, e.g. by taking the average [8,20].
3. Riehmann et al. introduced *flow tracing* in an interactive tool for Sankey diagrams [13]. Users may select a node or edge in the diagram and the contributions of all flows are highlighted. This increases the amount of information in the diagram. In the first place, by indicating the selected element, but mainly by revealing the contribution in a partitioning of the width of the edges. The partial width of each edge marked in the highlight color, contributes to the flow that passes the selected item.

We generalize the latter approach. We use a discrete color-scale to add an extra dimension of information as in the second approach. But in stead of flattening per node and edge, we partition the contribution of each edge and each node to each of the color classes. This allows to visually compare to what extent the added dimension has an impact on the flow. This technique allows to generate a single flow diagram of thousands of paths, while highlighting the different classes. This section first describes the classification function. Secondly, we define the formal Path-Colored graph. Finally, we introduce the Path-Colored Flow diagram.

4.1 Classification

As a tool to specify different classifications for process execution logs, we propose a framework of four types classifiers:

- The *duration* classifier is based on the duration of the process execution, measured as the time difference between the last and first node on the execution path. Each class corresponds with a duration range.
- The *result* classifier is based on a field of a step report that describes the outcome of the process execution. This field may be the name of the last executed node, or a dedicated result flag. Each class corresponds with a set of possible outcomes (one or more).
- The *path* classifier is based on the structure of the execution path. The structure of a path is described with the following attributes: length (the number of nodes in the path), is-cyclic and deviates (from the process model). Each class corresponds with a proposition based on these attributes.
- The *context* classifier is based on any contextual information field that may be stored in the step report. Each class corresponds with a set of possible values.

More specialized classifiers can be specified by combining several of the above classifiers, or by using derived fields, i.e. fields that are computed from the original event logs before actually applying the classification. For our example data set, we derive the field mobile device type based on the account name and a lookup in the profile data. We define a *context* classifier to differentiate the mobile device based on the OS: Android, iOS, and Windows.

4.2 Path-Colored Graph

To generalize the different classifiers, we define the classification function f_C : $P_G \to C$, with

- $P_G \subset \mathscr{P}_G$, a set of paths through G, in our example case the set that corresponds with the set of step reports.
- C, a set of classes, each represented in a diagram with a distinct color.

We define the Path-Colored graph as $G_C = (G, C, \lambda_C)$ a triple of a graph G, a set of classes C, and a weight function $\lambda_C : E \to C \to \mathbb{N}$ that maps each edge on a count per class. The count represents how many times the edge appears in a set of paths of the related class. Given a classification function f_C, the weight function can be written as:

$$\lambda_C := e \mapsto c \mapsto \sum_{p \in P_G} \left(\mu(f_C(p) = c) \sum_{e_i \in p} \mu(e_i = e) \right) \text{ with } \mu(X) := \begin{cases} 1 & X \\ 0 & \neg X \end{cases}$$

4.3 The PCF Diagram

Based on the definition in the previous section, we define a customized Sankey diagram:

A PCF (*Path-Colored Flow*) diagram is a Sankey diagram, augmented with colors. Each color marks the contribution of all edges and nodes to a class of paths. An arbitrary classification function determines the partitioning of edges. Inside a node, the contributions to each class of all incoming flows are grouped. At the outgoing side, the contributions split again into the outgoing edges. Throughout the whole diagram a fixed color order is applied.

Table 1 lists the weight factors for the *context* classifier applied on the example data set. The total count for an edge $N_e = \sum_{c \in C} \lambda_C(e)(c)$ corresponds to the total flow of all paths through a given edge. The columns Android, iOS, and Win list the flow count per edge and per device type; and the last three columns list the flow percentage per class.

Figure 5 depicts the corresponding PCF diagram. Each edge is represented by a multi-colored arrow. The total width of an arrow corresponds to N_e, and is divided in a section for each class, each proportional to $\frac{\lambda_C(e)(c)}{N_e}$.

We discuss the value of the PCF diagram based on the same example data set used in Sect. 3. We define a *context* classifier to differentiate the mobile

Table 1. Weight function λ_C

(half-)edge	N_e	Andr	iOS	Win	Andr%	iOS%	Win%	(half-)edge	N_e	Andr	iOS	Win	Andr%	iOS%	Win%
start $\overset{next}{\frown}$ Q0	163	50	53	60	31%	33%	37%	Q1 $\overset{abandon}{\frown}$	24	0	0	24	0%	0%	100%
start $\overset{abandon}{\frown}$	0	0	0	0	0%	0%	0%	Q2 $\overset{NOK}{\frown}$ Failure	26	8	10	8	31%	38%	31%
$\overset{A}{\frown}$Q0	28	6	0	22	21%	0%	79%	Q2 $\overset{OK}{\frown}$Success	38	11	15	12	29%	39%	32%
Q0 $\overset{yes}{\frown}$ Q1	109	31	28	50	28%	26%	46%	Q2 $\overset{A}{\frown}$	2	2	0	0	100%	0%	0%
Q0 $\overset{no}{\frown}$ Q2	76	25	25	26	33%	33%	34%	Q2 $\overset{abandon}{\frown}$	10	4	0	6	40%	0%	60%
Q0 $\overset{abandon}{\frown}$	6	0	0	6	0%	0%	100%	Q3 $\overset{NOK}{\frown}$ Failure	21	8	9	4	38%	43%	19%
$\overset{B}{\frown}$Q1	4	0	0	4	0%	0%	100%	Q3 $\overset{OK}{\frown}$Success	10	5	5	0	50%	50%	0%
Q1 $\overset{OK}{\frown}$Success	26	12	14	0	46%	54%	0%	Q3 $\overset{A}{\frown}$	4	0	0	4	0%	0%	100%
Q1 $\overset{NOK}{\frown}$ Q3	41	15	14	12	37%	34%	29%	Q3 $\overset{B}{\frown}$	4	0	0	4	0%	0%	100%
Q1 $\overset{A}{\frown}$	22	4	0	18	18%	0%	82%	Q3 $\overset{abandon}{\frown}$	2	2	0	0	100%	0%	0%

device type based on the OS: iOS (A), Android (B), and Windows (C). Figure 5 represents the resulting PCF diagram. Both Figs. 4 and 5, give a good overview of the flow, the abandons and the *step-back* transitions. What Fig. 5 depicts extra are the abandons for device type C. In addition, at nodes Q1 and Q2, none of the flows of type C take the OK branch, while a significant number perform a step back.

This as such does not lead us to the root cause of this behavior, but we get a strong hint that the first point to investigate is: "What exactly is happening in node Q1, and why is the behavior different for type C?" Based on domain knowledge, the conclusion may be that the question asked at node Q1 is not applicable for devices of type C. Users of such devices get confused by the question. They either abandon immediately, select NOK or step back, because they think they gave a wrong answer in the previous step. For these devices, question Q1 should be either rephrased or skipped. The process definition could be improved with a data-driven check of the device type, to skip this question in case of type C.

The example in this section was based on example data. The next section applies the PCF diagrams on the event logs of a real process execution.

4.4 Implementation

Existing tools to render graphs such as graphviz[1] and the Javascript library *Sankey*[2] are not able to draw the multi-color curved arrows. Hence, we wrote our own Javascript library to draw the PCF diagrams in SVG. Each color partition of an edge is a separate SVG path composed out of lines and arcs. We avoided Bézier curves because they are complex to draw parallel curves. A Cubic Bézier curve can be specified with four control points, but the offset path can not. The offset path of a Bézier curve must be calculated based on an approximation method [5].

---

[1] For information and download we refer to http://graphviz.org.

[2] For information and download we refer to https://bost.ocks.org/mike/sankey/.

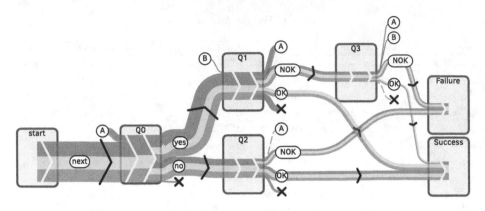

Fig. 5. Context classifier: ■ type A, ▨ type B, ■ type C (Color figure online)

We draw the nodes in the PCF diagrams with fixed height, and scale the widths of the edges such that the maximum flow matches with the node height. We use a minimum stroke width to preserve the readability. When the flow through an edge is lower than this threshold, we use a dashed stroke. When no flow passes through a workflow edge, we use a grey dotted stroke.

5   PCF Diagrams for Self-care Event Logs

This section applies the three techniques introduced in this paper to generate PCF diagrams for the self-care event logs. We describe several classifications applied on the same event log, representing 4475 sessions. The full process specification of the self-care portal encompasses more than 30 interconnected workflows, each handling a certain type of problem. This paper limits the scope to a single workflow handling startup issues. We first reduced the graph by selecting the *display* nodes, as described in Sect. 2. This reduced the set of nodes with 83%, due to the elimination of a lot of script and data-driven decision nodes. Then, we created a Leaky Sankey diagram (see Sect. 3), as basis for the PCF diagrams.

We first discuss the general observations based on the Sankey diagram in Fig. 6, independent from the classification. The diagram is split in two cropped views. At the left (a), we show the start of the workflow with the question "Select issue". At the right (b), the workflow ends in the node Failure or Success. Note that the *supporting* nodes (V_T in Sect. 2.2) are depicted as small rectangles. The incoming edge of the node "Charge and turn... Resolved?" that by-pass the first node of this workflow, may look unexpected. This edge originates from one of the other 29 workflows. Only a fraction of the sessions are successful, and less than half of the sessions arrive at the node Failure. The remaining sessions are abandoned along the way. About 20% of the flows return to the start node at some point.

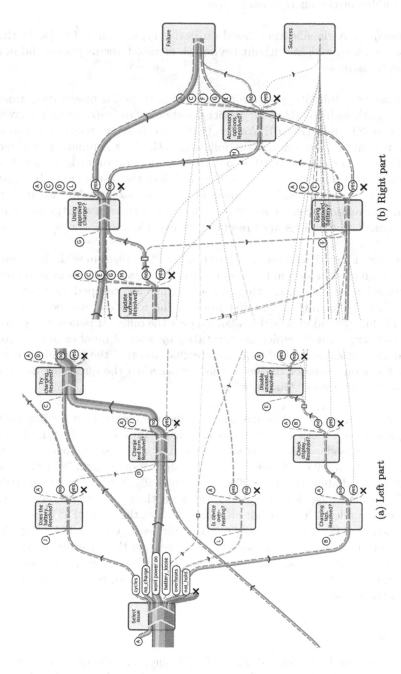

Fig. 6. Duration based classification: ■ <10 s, ■ <5 min, ■ <10 min, ■ <15 min, ■ ≥15 min (Color figure online)

5.1   Classifiers

We applied different classifiers of each type:

Context classifier. A classification based on device type, such as OS (as in the example of Sect. 4.3), with or without keyboard, or tablet versus phone, did not provide useful insights.

Duration classifier. Initially, we didn't expected any special observation from this classifier until we highlighted very short durations with five discrete intervals from <10 s to ≥15 min (Fig. 6). We see that there is no correlation between the total duration of a session and the followed path. This is remarkable given the big difference in duration: <10 s to ≥15 min. A duration of less than 10 s indicates that these consumers don't effectively follow the instructions, but just scan through the question and information pages. They stop reading when they got enough information or can't absorb any more. For those users, the order in which the remediation actions are presented, is less relevant.

Result classifier. Figure 7(a) shows a snippet of the PCF diagram with the classification based on the last step in the step report: `Success`, `Failure`, or *abandon*. This classification has the interesting property that any cropped view reveals the result. Only by looking at the first node of the workflow, we see that the result ratio failure/abandon is fairly balanced over the different issues; except for the edge "`battery loose`", which is only taken by a small number of sessions. Looking at the incoming half-edge `A`, we see that most of the returning flows eventually abandon. None of the users that returned to the question "`Select issue`", eventually reached "`Success`".

Path classifier. We spend extra attention to the possibility of returning to a previous page with a path classifier: no returns, return to the first node, or return to another node. Figure 7(b) shows the first and the last nodes in the PCF diagram. The number of abandons at the first node is high: 80% of these stopped here immediately, the other 20% abandoned after returning. At the node "`Success`", the incoming flows have no returns on their path. This generalizes the observation made with the result classifier. Also at the node "`Failure`", the dominant incoming flows have no returns on their path. We conclude that the possibility to return to a previous page is not increasing the effectiveness of the workflow. The step back events already indicate that the consumer does not find the guidance of the self-care portal useful or well-structured, before he concludes to abandon the process.

5.2   Discussion

So, what makes our Path-Colored Flow (PCF) diagrams effective? The graph reduction decreases the size of the diagram and allows focus. By selecting the *display* nodes, the diagram visualizes the perceived order of the questions shown to

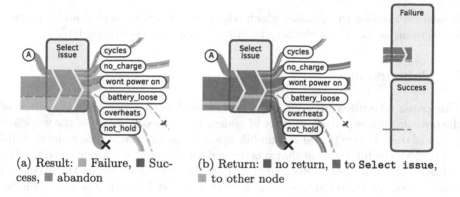

(a) Result: ■ Failure, ■ Success, ■ abandon

(b) Return: ■ no return, ■ to Select issue, ■ to other node

Fig. 7. Result (a) and path (b) classifier

the customer, while preserving all possible paths through the original workflow definition. By using a Sankey diagram, the visualization becomes very transparent on what happens at which point in the flow. The place and proportion of the abandons is immediately visible, as well as the step-back transitions. It is hard to compare large diagrams of different sets of event logs. Path-coloring makes it possible to compare different sets of event logs in a single diagram. The classifications we used, can be used in other statistical methods or graphics. The benefit of the PCF diagram is that it presents the message in the structure of the original workflow design, and is thus very suited to communicate with the design team.

6   Related Work

Reduction rules for process models are presented in several publications [7,14, 15,18]. The work differs in the targeted analysis and semantics of the graphs. Each approach preserves different properties of the process. The Proviado framework [2,12] can generate a personalized view of a large business process model. Elimination and aggregation of process elements can deliver a comprehensive abstraction. Disco[3] and ProM[4], the most popular tools to process and visualize event logs, support features to aggregate and abstract a process model, and use edge width and color codes to reflect path frequencies.

Sankey diagrams are well suited to visualize the patient flow through activities [1,3]. The color of the activity indicates the type of activity. The CareFlow's visual interface assigns colors to elements according to the average outcome of the process [8]. Fill et al. discusses the visual dimensions such as size, color and shape to draw actions in an enterprise model [6]. Outflow [20] visualizes aggregate event progression pathways, using color scales to show the associated statistics. Riehmann et al. introduced *flow tracing* in an interactive tool for Sankey diagrams [13]. Kriglstein et al. presented a tool for process model redesign [10].

---

[3] For information and download we refer to http://www.fluxicon.com/disco/.

[4] For information and download we refer to http://www.promtools.org.

It uses color codes to visualize which edges and nodes are added and removed. Vogelgesang et al. uses a similar visualization for process mining [19].

7   Conclusion

This paper introduced a graph reduction algorithm that preserves the labels of the outgoing edges of a selected set of nodes. We reduced the size of the workflow graph of the self-care portal of a mobile operator significantly. The reduction did not remove the information that was needed for the analysis: i.e. understand the influence of the *display* nodes and their outgoing labeled edges on the execution flows. Secondly, we introduced a new Path-Colored Flow diagram to visualize event logs of a business process. The addition of half-edges to the Sankey diagram made it possible to intuitively differentiate the flows via the explicitly modeled edges, from other transitions allowed during the process execution. A session-based classification was used to visually partition the flow per node and per edge, and as such add an extra dimension of information to the diagram. We gave a framework to specify four types of classifiers and applied them on our use case. We demonstrated how the PCF diagram increased the insights on the event logs of the self-care portal. Most importantly, the visualization was based on the original workflow diagram. The PCF diagrams made the business insights visible and suitable for communication to all stake holders. The diagrams indicated that the customer behavior was not due to an isolated place of bad logic. It raised the need for a general rethinking of the formulation of questions and answers. Our analysis pointed out, that customers that return to a previous page, will not reach success and most probably abandon. Improving the usability of the portal was key to keep the customer's confidence in operator's support channel.

References

1. Basole, R.C., Park, H., Gupta, M., Braunstein, M.L., Chau, D.H., Thompson, M.: A visual analytics approach to understanding care process variation and conformance. In: Proceedings of the 2015 Workshop on Visual Analytics in Healthcare, VAHC 2015, NY, USA, New York, pp. 6:1–6:8. ACM (2015). doi:10.1145/2836034. 2836040
2. Bobrik, R., Bauer, T., Reichert, M.: Proviado – personalized and configurable visualizations of business processes. In: Bauknecht, K., Pröll, B., Werthner, H. (eds.) EC-Web 2006. LNCS, vol. 4082, pp. 61–71. Springer, Heidelberg (2006). doi:10.1007/11823865_7
3. Bos, K., Hasper, W.: Enabling measuring of the patient flow in an orthopaedic clinic. B.S. thesis, University of Twente (2016). http://essay.utwente.nl/70229/
4. Burkhard, R., Meier, M.: Tube map: evaluation of a visual metaphor for interfunctional communication of complex projects. In: Proceedings of I-Know 2004, vol. 4, pp. 449–456 (2004)
5. Elber, G., Lee, I.K., Kim, M.S.: Comparing offset curve approximation methods. IEEE Comput. Graph. Appl. **17**(3), 62–71 (1997). doi:10.1109/38.586019

6. Fill, H.G., Höfferer, P.: Visual enhancements of enterprise models. In: Multikonferenz Wirtschaftsinformatik, pp. 541–550 (2006)
7. Hu, F.H., Jiang, J., Wu, X.N., Ru, F.: Reduction rules of graphical representation in a workflow process model. Adv. Sci. Eng. II Appl. Mech. Mater. **135**, 709–714 (2012). doi:10.4028/www.scientific.net/AMM.135-136.709. Trans Tech Publications
8. Hu, J., Perer, A., Wang, F.: Data driven analytics for personalized healthcare. In: Weaver, C.A., Ball, M.J., Kim, G.R., Kiel, J.M. (eds.) Healthcare Information Management Systems. HI, pp. 529–554. Springer, Cham (2016). doi:10.1007/978-3-319-20765-0_31
9. Keim, D., Andrienko, G., Fekete, J.-D., Görg, C., Kohlhammer, J., Melançon, G.: Visual analytics: definition, process, and challenges. In: Kerren, A., Stasko, J.T., Fekete, J.-D., North, C. (eds.) Information Visualization. LNCS, vol. 4950, pp. 154–175. Springer, Heidelberg (2008). doi:10.1007/978-3-540-70956-5_7
10. Kriglstein, S., Wallner, G., Rinderle-Ma, S.: A visualization approach for difference analysis of process models and instance traffic. In: Daniel, F., Wang, J., Weber, B. (eds.) BPM 2013. LNCS, vol. 8094, pp. 219–226. Springer, Heidelberg (2013). doi:10.1007/978-3-642-40176-3_18
11. McCabe, T.J.: A complexity measure. IEEE Trans. Softw. Eng. SE **2**(4), 308–320 (1976). doi:10.1109/TSE.1976.233837
12. Reichert, M.: Visualizing large business process models: challenges, techniques, applications. In: Rosa, M., Soffer, P. (eds.) BPM 2012. LNBIP, vol. 132, pp. 725–736. Springer, Heidelberg (2013). doi:10.1007/978-3-642-36285-9_73
13. Riehmann, P., Hanfler, M., Froehlich, B.: Interactive sankey diagrams. In: IEEE Symposium on Information Visualization, INFOVIS 2005, pp. 233–240 (2005). doi:10.1109/INFVIS.2005.1532152
14. Sadiq, W., Orlowska, M.E.: Applying graph reduction techniques for identifying structural conflicts in process models. In: Jarke, M., Oberweis, A. (eds.) CAiSE 1999. LNCS, vol. 1626, pp. 195–209. Springer, Heidelberg (1999). doi:10.1007/3-540-48738-7_15
15. Sadiq, W., Orlowska, M.E.: Analyzing process models using graph reduction techniques. Inf. Syst. **25**(2), 117–134 (2000). doi:10.1016/S0306-4379(00)00012-0. http://www.sciencedirect.com/science/article/pii/S0306437900000120
16. Schmidt, M.: The sankey diagram in energy and material flow management. J. Ind. Ecol. **12**(1), 82–94 (2008). doi:10.1111/j.1530-9290.2008.00004.x
17. Tanahashi, Y., Ma, K.L.: Design considerations for optimizing storyline visualizations. IEEE Trans. Vis. Comput. Graph. **18**(12), 2679–2688 (2012). doi:10.1109/TVCG.2012.212
18. Verbeek, H., Wynn, M., van der Aalst, W., ter Hofstede, A.: Reduction rules for reset/inhibitor nets. J. Comput. Syst. Sci. **76**(2), 125–143 (2010). doi:10.1016/j.jcss.2009.06.003
19. Vogelgesang, T., Appelrath, H.-J.: PMCube: a data-warehouse-based approach for multidimensional process mining. In: Reichert, M., Reijers, H.A. (eds.) BPM 2015. LNBIP, vol. 256, pp. 167–178. Springer, Cham (2016). doi:10.1007/978-3-319-42887-1_14
20. Wongsuphasawat, K., Gotz, D.: Exploring flow, factors, and outcomes of temporal event sequences with the outflow visualization. IEEE Trans. Vis. Comput. Graph. **18**(12), 2659–2668 (2012). doi:10.1109/TVCG.2012.225

Assorted BPM Topics

AB-BPM: Performance-Driven Instance Routing for Business Process Improvement

Suhrid Satyal[1,2]([⊠]), Ingo Weber[1,2], Hye-young Paik[2], Claudio Di Ciccio[3], and Jan Mendling[3]

[1] Data61, CSIRO, Sydney, Australia
{suhrid.satyal,ingo.weber}@data61.csiro.au
[2] University of New South Wales, Sydney, Australia
hpaik@cse.unsw.edu.au
[3] Vienna University of Economics and Business, Vienna, Austria
{claudio.di.ciccio,jan.mendling}@wu.ac.at

Abstract. A fundamental assumption of Business Process Management (BPM) is that redesign delivers new and improved versions of business processes. This assumption, however, does not necessarily hold, and required compensatory action may be delayed until a new round in the BPM life-cycle completes. Current approaches to process redesign face this problem in one way or another, which makes rapid process improvement a central research problem of BPM today. In this paper, we address this problem by integrating concepts from process execution with ideas from DevOps. More specifically, we develop a technique called AB-BPM that offers AB testing for process versions with immediate feedback at runtime. We implemented this technique in such a way that two versions (A and B) are operational in parallel and any new process instance is routed to one of them. The routing decision is made at runtime on the basis of the achieved results for the registered performance metrics of each version. AB-BPM provides for ultimate convergence towards the best performing version, no matter if it is the old or the new version. We demonstrate the efficacy of our technique by conducting an extensive evaluation based on both synthetic and real-life data.

Keywords: Business Process Management · DevOps · AB testing · Process performance indicators

1 Introduction

Various lifecycle approaches to Business Process Management (BPM) have a common assumption that a process is incrementally improved in the redesign phase [9, Chap. 1]. While this assumption is hardly questioned in BPM research, there is evidence from the field of AB testing that improvement concepts often do *not* lead to actual improvements. For instance, work on business improvement ideas found that 75% did not lead to improvement: half of them had no impact while approximately a quarter turned out to be even harmful [12]. The results

© Springer International Publishing AG 2017
J. Carmona et al. (Eds.): BPM 2017, LNCS 10445, pp. 113–129, 2017.
DOI: 10.1007/978-3-319-65000-5_7

are comparable to a study of the Microsoft website, in which only one third of the ideas observed had a positive impact, while the remaining had no or negative impact [15]. The same study also observed that customer preferences were difficult to anticipate before deployment, and that customer research did not predict customer behaviour accurately.

If incremental process improvement can only be achieved in a fraction of the cases, there is a need to rapidly validate the assumed benefits. Unfortunately, there are currently two major challenges for such an immediate validation. The first one is *methodological*. Classical BPM lifecycle approaches build on a labour-intensive analysis of the current process, which leads to the deployment of a redesigned version. This new version is monitored in operation, and if it does not meet performance objectives, it is made subject to analysis again. All this takes time. The second challenge is *architectural*. Contemporary Business Process Management Systems (BPMSs) enable quick deployment of process improvements, but they do not offer support for validating improvement assumptions. A performance comparison between the old and the new version may be biased since contextual factors might have changed at the same time. How a rapid validation of improvement assumptions can be integrated in the BPM lifecycle and in BPMSs is an open research question.

In this paper, we address this question by integrating business process execution concepts with ideas from DevOps. More specifically, we develop a technique called AB-BPM that offers AB testing for redesigned processes with immediate feedback at runtime. AB testing in DevOps compares two versions of a deployed product (e.g., a Web page) by observing users' responses to versions A/B, and determines which one performs better [8]. We implemented this technique in such a way that two versions (A and B) of a process are operational in parallel and any new process instance is routed to one of them. Through a series of experiments and observations, we have developed an instance routing algorithm, *LTAvgR*, which is adapted to the context of executing business processes. The routing decision is guided by the observed results for registered performance metrics of each version at runtime. The technique has been evaluated extensively on both synthetic and real-life data. The results showed that AB-BPM provides for ultimate convergence towards the best performing version.

The remainder of this paper starts with a discussion of the background and related work in Sect. 2. Section 3 describes the framework and algorithms for performing AB tests. In Sect. 4, we evaluate our approach on two use cases. In Sect. 5, we discuss the results, limitations, and validity of our approach, and finally draw conclusions in Sect. 6.

2   Background and Related Work

Business Process Management Systems (BPMSs) allow for a rapid deployment of process improvements into operation. However, there is currently no support to test the often implicit assumption that a modification of a process actually represents an improvement. One anecdote of a leading European bank (EB) illustrates this problem. The EB improved their loan approval process by cutting its turnaround time down from one week to a few hours. What happened though was a steep decline in customer satisfaction: customers with a negative notice would complain that their application might have been declined unjustifiably; customers with a positive notice would inquire whether their application had been checked with due diligence. This anecdote emphasizes the need to carefully test improvement hypotheses in practice since customers and process participants might not act in a way that can be predicted deterministically.

Given the current architecture of BPMSs, it is not possible to conduct a fair comparison between the old and the new version of a process since they are not operational at the same point of time. That means, doing a post-hoc analysis of data generated from the old process being operational in time interval $[t(n-1), t(n)]$ and the new process running from $[t(n), t(n+1)]$ is biased towards the respective conditions of each time interval.

The need for continuous improvement is rarely disputed, but it should be complemented with the motto "test fairly" and "fail fast". This motto entails reducing the time between inception and deployment of new versions, and managing business risks brought forth by deployment of the new process versions.

From the above analysis, we derive the following three requirements:

R1 Rapidly validate the improvement assumption: a proposed improvement should be tested within a short time frame after its introduction.
R2 Ensure a fair comparison: the environment in which a comparison is conducted should minimize bias, and avoid the time bias discussed above.
R3 Enable rapid adjustments on process model level: the benefits of a solution should be suited to process models and their specific characteristics.

Concepts from DevOps [3], which aim to bring software development (Dev) and operations (Ops) closer together, may help to address this problem. One DevOps practice is live testing, where new versions of the software are tested in production with actual users of the system. The most popular form of live testing is *AB testing*, where two versions (A and B) are deployed side by side and both receive a share of the production workload while being monitored closely. The monitoring data is then used to draw conclusions about the effectiveness of one version over the other, for instance in the form of increased revenue from higher click-through rates.

So far, AB testing has been used for micro changes of websites, like changing the color of a button [8,15]. The effectiveness of this technique is surveyed by Kohavi et al. for the user interfaces of web applications [15,16]. In this paper, we adopt the idea of AB testing on the process level in order to address R1–R3. Our technique is called AB-BPM. In the following, we discuss in how far previous BPM research is related to R1–R3. We distinguish methodological and technical approaches to process improvement.

Methodological approaches include business process re-engineering and the BPM lifecycle. Business Process Re-Engineering (BPR) offers a methodology for selecting and redesigning a process to improve efficiency, often exploiting IT to support the changes [11]. BPR promotes radical changes to the processes. An explicit perspective for testing re-engineered processes is missing.

Kettinger et al. summarize methods for process improvement project [14]. Lifecycle models like the one described by Dumas et al. [9] propose a more incremental improvement of processes with periodic controlling and revisions. Our research complements this stream of research by providing techniques to experiment with process improvement hypotheses and perform statistical evaluation on them. We assume that a redesigned process is made available, so that such experimentation and analysis can be done. We envision that our approach can be used in conjunction with works that automatically generate process versions just as [4]. Other techniques include root-cause analysis [9,19], e.g. by the help of cause-effect diagrams [9,10], and the consideration of best-practises [2,20]. However, evaluation of effectiveness requires the involvement of process analysts.

Technical approaches focus on monitoring processes at runtime with a focus on specific performance metrics. Concepts based on Statistical Process Control (SPC) [13], Complex Event Processing (CEP) [24,25], and predictive analytics [6] have been proposed and adapted for monitoring business processes. However, these monitoring techniques have not been used to carry out controlled experiments. In our work, the monitoring is performed by the instance router by observing a Process Performance Indicator (PPI) like satisfaction ratings obtained from end users. Based on the chosen PPI, the instance router dynamically adjusts the request distribution rates (Table 1).

Table 1. Mapping existing works to the requirements

Approach	R1	R2	R3
Process re-engineering (Hammer/Champy) [11]	−	−	−
Process improvement [14]	−	−	−
Process lifecycle [9]	+/−	−	−
Statistical Process Control (SPC) [13]	+	−	−
Complex Event Processing (CEP) [24,25]	−	−	+
AB testing, see e.g., [3,15]	+	+	−
AB-BPM (this work)	+	+	+

Our research addresses the gap of an explicit testing of improvement hypotheses in BPM-related research and the lack of an explicit consideration of business processes in the works on AB testing. In the following, we devise our AB-BPM approach so that it meets requirements R1–R3.

3   Approach and Architecture

In this section, we present our approach, starting with mapping the instance routing problem to algorithms from the literature. Based on a small experiment, we choose one algorithm and adapt it to the context of business processes. Then we present our high-level framework, architecture, and implementation.

3.1 Instance Routing – A Multi-armed Bandit Problem

In order to integrate concepts of process execution with AB testing, we have to discuss how new instances are assigned to a specific version of the process.

Therefore, we need an instance router that distributes requests to versions in such a way that any relevant Process Performance Indicator (PPI) is maximized. The instance router also needs to deal effectively with the issue that processes can be long-running, and that PPI measurements can be delayed.

The PPI maximization can be mapped to the so-called *multi-armed bandit problem* [1,5]. The multi-armed bandit problem models a hypothetical experiment where, given some slot machines with different payoff probability distributions, a gambler has to decide on which machines to play. The objective of the gambler is to maximize the total payoff during a sequence of plays. Since the gamblers are unaware of the payoff distribution, they can approach the plays with two strategies: *exploring* the payoffs by pulling different arms on the machines or *exploiting* the current knowledge by pulling arms that are known to give good payoffs. The exploration strategy builds knowledge about the payoffs, and the exploitation strategy accumulates the payoffs. Multi-armed bandit algorithms aim to find a balance between these strategies. If the performance is affected by some context, this can be seen as the so-called *contextual multi-armed bandit problem*, where the gambler sees context (typically represented as a multi-dimensional feature vector) associated with the current iteration before making the choice.

We model the routing algorithm as a multi-armed bandit problem by representing the process versions as the "arms", and the PPI as "payoffs/rewards". The objective of the instance router is to find a good tradeoff between exploration and exploitation, possibly based on the context. To learn the performance of a version in exploration, it sends some of the process instantiation requests to either version. Based on the instance router's experience, it can exploit its knowledge to send more or even all request to the better-performing version. The reward for routing algorithm can be designed to use a PPI like user satisfaction.

3.2 Instance Routing Algorithms and Selection

The multi-armed bandit problem has been explored in related literature. LinUCB [17] is a contextual multi-armed bandit algorithm that has been employed to serve news articles to users with the objective to maximize the total number of clicks. Tompson sampling [23] is one of the simplest approaches to address multi-armed bandits. It is based on a Bayesian approach where arms are chosen according to their probability of producing optimal rewards [7,23]. Thompson sampling can be used to solve the contextual multi-armed bandit problem with linear rewards [1]. In this paper, we chose these three algorithms as candidates for process instance routing and investigate their effectiveness. We have selected these algorithms based on their demonstrated benefits and simplicity. Other algorithms, such as ϵ-greedy, ϵ-first, UCB, and EXP4 also address multi-armed bandit problems [5].

Since the goal for the routing algorithms is to maximize the cumulative value of the PPI, as preparatory work, we have experimented with different routing algorithms with different configurations to find the best performing algorithm. We have compared variations of Thompson sampling techniques [1,7,23], LinUCB [17] and a baseline algorithm which uniformly distributes requests to process versions regardless of context and rewards. We have found that *LinUCB produced the highest cumulative satisfaction score throughout the experiments.* Therefore we use this algorithm in the following. Our architecture is flexible enough that it can be easily replaced by other algorithms.

3.3   Adapting the Routing Algorithm to Business Processes

As discussed above, we chose LinUCB as our routing algorithm. However, we observed that the algorithm can be derailed by process-specific circumstances, such as the long delays before rewards. Long delays are inherent to long-running processes, and not considered in AB testing solutions for Web applications, where delays are measured in seconds or minutes. In contrast, the real-world data which we use in the evaluation has one process instance with an overall duration of more than 3 years.

This results in the following issue. Oftentimes overly long process completion times correlate with problematic process instances, leading to negative rewards. Thus, instances with short completion times can give a positive impression of a process version early on. If the algorithm receives too many positive rewards from one version during the early stages of the AB test, the algorithm is more likely to see that version as preferable. Such an early determination can introduce a bias in the evaluation. Thus, we need to ensure that the algorithm gets enough samples from both versions.

We solve this issue by adopting the idea of a "warm-up" phase from Reinforcement Learning [21], during which we emphasize exploration over exploitation. We sample the probability of exploration by using an exponential decay function, acting as an injected perturbation that diminishes as the experiment proceeds – the sample determines whether the algorithm follows LinUCB's decision or picks a version at random. We consider the "warm-up" as the experimentation phase: after all instances started during the experimentation phase are completed, no more rewards are collected and the instance router stabilizes.

Finally, the original LinUCB algorithm makes its decision based on the summation of past rewards. We found out during the experiments that this can also deceive the algorithm. Therefore, we have modified the LinUCB algorithm to make its reward estimates on the basis of the average of past rewards rather than their sum. We term our adapted instance routing algorithm *Long-term average router (LTAvgR)*.

3.4   AB-BPM Framework, Architecture, and Implementation

Figure 1 shows the architecture of our AB testing framework. We designed the architecture such that the two versions of the process model are deployed side by side in the same execution engine. The instance router distributes the instance creation requests as per its internal logic. An alternative design would be to run two full copies of the entire application stack, one for each version, and using the instance router to distribute the requests across the two stacks. However, the multi-armed bandit algorithms can identify the superior version during the experimentation and alter the allocation of requests to different versions. When a version is clearly superior to the other, most of the requests are sent to the superior version. In such scenarios, the application stack that hosts the inferior version is underutilized. If we run both versions on the same stack, we can keep utilization of the system high, no matter which version is superior.

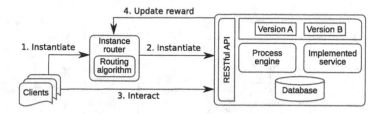

Fig. 1. Application architecture

Given this design choice, the process definitions, implementation, PPI collection, and the shared process execution engine are wrapped by a web application. Process execution data are stored in a shared database. Process instantiation, metrics, and other operations are exposed using RESTful APIs. Upon receiving a request, the instance router instantiates a particular version and receives an identifier. Identifiers of process instances for which rewards have not been observed are stored in a queue. The instance router uses a polling mechanism in parallel to retrieve PPI metrics from the server and update the rewards.

We implemented the architecture prototypically in Java and Python, in part based on the Activiti BPMS. As outlined earlier, our framework is flexible in the choice of the instance routing algorithm: we implemented and tested all five variants discussed above, i.e., LTAvgR, LinUCB, Thompson-sampling with and without linear rewards, and random uniform distribution. The experiments reported in the following section are run with the presented implementation with LTAvgR. We simulate the requests from users by replaying process logs.

4   Evaluation

In this section, we present the methodology and outcomes of our evaluation of the proposed approach. We assess the AB-BPM framework and LTAvgR algorithm first on synthetic data, where we have full control over the environment and parameters. Then we test the approach on real-world data, taken from the building permit process logs from five Dutch municipalities.

4.1   Evaluation on Synthetic Data

In this section, we demonstrate our approach using two example process versions stemming from the domain of helicopter pilot licensing. Version A of the process sequentially schedules the activities based on the cost of performing them. Based on the result (pass/fail), the process either schedules the next activity or terminates the process. In this version, we expect that successful candidates will pay more because of multiple scheduling costs. In contrast, version B of the process schedules all such activities at the beginning, thus reducing the scheduling costs. The processes are illustrated in Fig. 2. These processes have as a result the final status of the license: either *approved* or *rejected*.

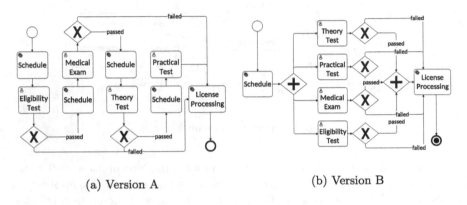

(a) Version A   (b) Version B

Fig. 2. Process versions for the AB testing experiment

As PPI we simulate the user satisfaction, here calculated as a combination of cost, completion time, and result of the process execution. Costs and processing times of each task were derived from the Australian Civil Aviation Safety Authority (CASA)[1] and helicopter hiring rates from an Australian flight school.

Table 2. Cost model of the activities

Activity	Cost	Min. processing time	Max. processing time
Schedule	25	1 day	1 day
Eligibility test	190	1 day	3 days
Medical exam	75	1 day	3 days
Theory test	455	2 weeks	5 weeks
Practical test	1145	1 week	2 weeks
License processing	0 if rejected, 100 if approved	Immediate	Immediate

---

[1] https://www.legislation.gov.au/Details/F2016C00882, Accessed: 03-01-2017.

Table 3. User satisfaction model

Outcome	Cost	Duration	Satisfaction
Approved	[0, 1990]	≤5 weeks	5
	(1990, ∞]	≤5 weeks	4
	[0, ∞)	>5 weeks	3
Rejected	[0, 1890]	≤5 weeks	3
	[0, 1890]	>5 weeks	2
	(1890, ∞]	≤5 weeks	2
	(1890, ∞)	>5 weeks	1

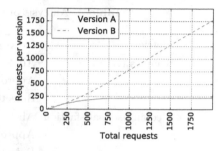

Fig. 3. Requests routing in AB tests

Table 2 shows the costs and processing times for both process versions. Table 3 shows how user satisfaction scores from 1 (lowest) to 5 (highest) are derived. The basic rationale is, the shorter and cheaper, the better. The score ranges from 1 to 3 if the outcome is negative, and from 3 to 5 if the outcome is positive.

Experiment design. We have designed the AB testing experiments such that the instance router receives requests with embedded contexts at the rate of 1 request per second. For the execution, we scale each day to 1 s. In this experiment, we introduce non-determinism in two ways: *(i)* adjusting success rates of an activity based on contextual information, and *(ii)* sampling processing times for each activity using a probability distribution function.

Results. Figure 3 shows the request distribution throughout the AB Tests. When the experimentation or "warm-up" phase ends at approximately 1000 requests, the router stops updating the reward estimates and chooses version B decisively.

A post-hoc analysis shows that the median user satisfaction was similar for both versions. The distributions of user satisfaction scores differed significantly (a Mann-Whitney test [18] resulted in $U = 54072$, p-value $<10^{-6}$ two-tailed, $n_A = 222$, $n_B = 778$). The median delay of the reward was 22.3 s. Table 4 shows the differences between the two versions. Version B produces a better user satisfaction in those cases where an application is approved. It is also faster in all cases. However, the median cost of version A is lower than that of version B when the applications are rejected.

We used an evaluation based on synthetic log data in order to investigate the convergence behaviour of the implementation. We observe that our approach leads to a rapid identification of the more rewarding process version, which receives an increasing share of traffic. This observation is instrumental with respect to the requirements R1–R3, which demand rapid validation, fair comparison and rapid adjustment on the process level.

Table 4. Analysis of versions A and B by cases

Metric	Outcome	Version A	Version B	Overall
Samples (N)	All	222	778	1000
	Approved	72	275	347
	Rejected	150	503	653
Median user satisfaction	All	3	3	3
	Approved	3	5	5
	Rejected	3	3	3
Median cost	All	795	1890	1890
	Approved	2065	1990	1990
	Rejected	795	1435	1435
Median duration	All	28.6 s	17.4 s	19.6 s
	Approved	35.5 s	21.8 s	22.4 s
	Rejected	24.5 s	8.7 s	8.9 s

4.2 Evaluation on Real-World Data

To assess the applicability of our approach over real-world data, we have analysed the data stemming from the five logs in the BPI Challenge 2015 [2], herein identified as $L^1, \ldots, L^5$. Those logs contain the execution data of the building permit issuing processes in five Dutch municipalities. The processes behind each log reportedly contain variations, which allow us to consider them as different versions of the same main process. In this experiment, we simulate the situation where one version is in use, when a new version is suggested and AB-tested in competition with the previous one. Better performance here is equated to shorter time to complete a process instance. Subsequently, the version that won the first round competes against the next version, and so on, until all versions have competed.

Based on the insights from [22], we filtered the events to retain only those activity instances that belong to a so-called "phase", namely constituting the core part of the process. Using the Inductive Miner, we discovered five process models $P^1, \ldots, P^5$ from $L^1, \ldots, L^5$, respectively. We mimicked the execution of the processes by replaying the logs on the process versions. The instance router decided to which alternative version to route each request to create a new instance. In the following, we describe how the execution times were derived, define the reward function, clarify how the competition was organized, and finally report on the achieved results.

Execution Time Simulation. For fairness, in this experiment we replay only the logs that did *not* stem from the original processes. Say, we are AB-testing

---

[2] BPI Challenge 2015, including logs, reports, and process models: https://www.win.tue.nl/bpi/doku.php?id=2015:challenge, Accessed 20-03-2017.

P^i vs. P^j; then we use the logs from $\mathcal{L}_{test} = \{L^1, \ldots, L^5\} \setminus \{L^i, L^j\}$ with $1 \leqslant i, j \leqslant 5$. However, we want to test how the event traces from $\mathcal{L}_{test}$ behave on P^i and P^j in terms of timing. To this end, we assign an activity duration to the execution of every replayed activity from the process version the activity was routed to. Say this is P^i, and the current activity is a; then, we randomly sample a duration value from the durations of a among all the traces of L^i sharing the same execution history (prefix) as the replayed trace.

For instance, consider the execution of activity *phase concept draft decision ready* for process P^1 after the sub-trace [*phase application received, phase application receptive, phase advice known*]. This activity has been assigned with a random sample among the registered execution times of *phase concept draft decision ready*-events following [*phase application received, phase application receptive, phase advice known*] in the 593 traces of log L^1 sharing that prefix.

To that extent, we have folded the traces of every log into a dedicated poly-tree auxiliary data structure, collapsing the traces that share the same prefix on common paths. Every node keeps the activity name and the list of registered execution times of the related events. In addition, event transitions in the log are stored in a table structure along with the list of transition times. When the traces cannot be followed in the poly-tree structure, we perform a lookup on the table structure and derive execution times. If the current transition has not been observed in L^i, we discard the trace as non-conforming and disregard it in the reward calculation.

The events in the logs signal the completion of an activity, and bear eight timestamp fields. However, most of those attributes were missing or unreliable. Therefore, we followed the approach of [22], and used solely the completion *time:timestamp* attribute for each event. We computed the duration of every activity as the difference between the timestamp of its completion and the preceding completion timestamp. We thus included in the activities' duration estimation both the execution time and the waiting time before starting.

Reward Strategy. The filtered BPIC 2015 dataset contains numerous outliers: while the median duration for processes are 39–46 days, outliers can take up to 1158 days, i.e., 3 years and 63 days. To establish a reward function that penalizes very long process completion times, we adopted the following strategy. Given the initial version P^i of a process, we collected all instance execution times reported in its log L^i and computed

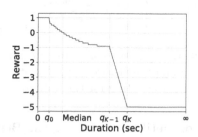

Fig. 4. Reward strategy

$K \geqslant 1$ quantiles $q_1, \ldots, q_K$. We used these quantiles to partition the space of possible execution times into a set of $K + 2$ intervals $I = \{\iota_0, \iota_1, \ldots, \iota_K, \iota_{K+1}\}$ where $\iota_0 = [0, q_0)$, $\iota_{K+1} = [q_K, +\infty)$, and $\iota_k = [q_{k-1}, q_k)$ for $1 \leqslant k \leqslant K$. Those intervals split the range of possible execution times for P^j as follows: ι_0 contains the values below the minimum registered in L^i, ι_{K+1} accounts for any duration

---

Algorithm 1. Strategy for the selection of the best performing version among $\{P^1, P^2, P^3, P^4, P^5\}$.

---

1 $\mathcal{P}_{\text{test}} \leftarrow \{P^1, P^2, P^3, P^4, P^5\}$
2 $P^i \leftarrow$ original process version from $\mathcal{P}_{\text{test}}$
3 $\mathcal{P}_{\text{test}} \leftarrow \mathcal{P}_{\text{test}} \setminus \{P^i\}$
4 **repeat**
5      $P^j \leftarrow$ alternative process version from $\mathcal{P}_{\text{test}}$
6      $\mathcal{P}_{\text{test}} \leftarrow \mathcal{P}_{\text{test}} \setminus \{P^j\}$
7      $\mathcal{L}_{\text{test}} \leftarrow \{L^1, \ldots, L^5\} \setminus \{L^i, L^j\}$
8      $P^i \leftarrow$ best version between P^i and P^j as per the AB test over $\mathcal{L}_{\text{test}}$
9 **until** $\mathcal{P}_{\text{test}} \neq \varnothing$
10 **return** P^i

---

beyond the maximum recorded in L^i, and the K intervals in-between are meant to classify the performance of P^j with respect to their quantile as per L^i. This strategy is illustrated as a step chart in Fig. 4 – every step in the chart represents a quantile.

The idea is to assign a reward of 1 when the duration P^j achieves is lower than any registered execution time of P^i, and decrease it by a step of $2/K$ as long as the measured performance falls into the following intervals. In order to counterbalance the disruptive effect of outliers which take extremely long, we established a penalty value ρ (with $\rho = 4$ in Fig. 4) and defined that the reward linearly decreases from -1 to $(-1 - \rho)$ along ι_K. Finally, any execution time beyond the last quantile is assigned a reward of $(-1 - \rho)$.

Formally, let $\kappa_I(t) : \mathbb{R}^+ \to [0, \ldots, K+1]$ be a mapping function associating a process instance execution time t to the respective interval $\iota_K \in I$ by the index k, e.g. $\kappa_I(t) = 3$ if t is in the range of ι_3. The reward function $r_I : \mathbb{R}^+ \to [-1 - \rho, 1]$ is defined over the set of intervals I as follows:

$$r_I(t) = \begin{cases} 1 - \frac{\kappa_I(t)}{K} \cdot 2 & \text{if } \kappa_I(t) < K \\ -1 - \rho \cdot \frac{t - q_{K-1}}{q_K - q_{K-1}} & \text{if } \kappa_I(t) = K \\ -1 - \rho & \text{if } \kappa_I(t) > K \end{cases} \quad \text{with} \quad \kappa_I(t) = \sum_{k=0}^{K+1} k \cdot \chi_{\iota_k}(t)$$

where χ_{ι_k} is the characteristic function of interval ι_k. The underlying idea was to prefer the process demonstrating a shorter completion time to the slower ones while accounting for the outliers. For our experiment, we set $K = 20$ and $\rho = 4$.

Competition: Selecting the Best Version. To simulate the situation where an organization gradually designs new versions of a process model, we run a competition between the five provided process models. This competition is conducted as a set of pair-wise comparisons between versions, following the schema outlined in Algorithm 1. The idea is to initially consider an original version of the process, P^i, and a new version, P^j. To determine if P^j achieves an actual improvement over P^i while limiting bias as discussed above, the execution of the processes is simulated by replaying the traces in the logs from which P^i and

Table 5. Number of traces   **Table 6.** Ratio of conforming traces

Log	Traces
L^1	1199
L^2	830
L^3	1409
L^4	1051
L^5	1155

Version	L^1	L^2	L^3	L^4	L^5
P^1	1	0.928	0.949	0.974	0.928
P^2	0.913	1	0.928	0.982	0.938
P^3	0.901	0.812	1	0.975	0.886
P^4	0.873	0.731	0.913	1	0.829
P^5	0.897	0.929	0.944	0.979	1

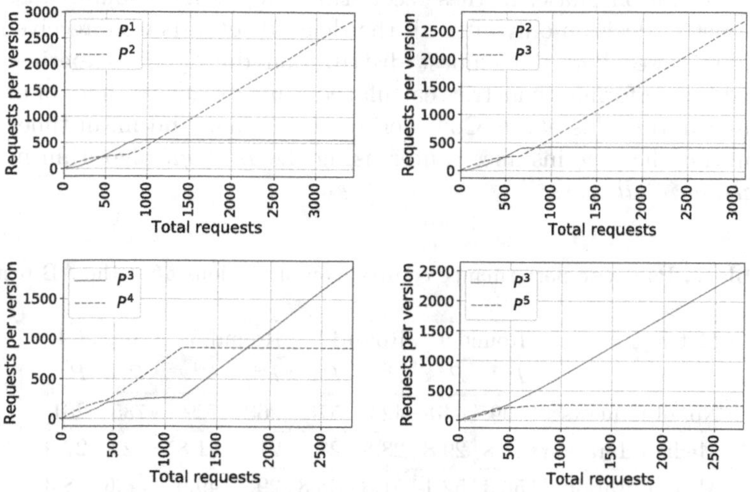

Fig. 5. Request distribution over time

P^j were *not* derived. For instance, P^1 and P^2 are evaluated on the basis of the traces in L^3, L^4, and L^5. If, at the end of a competition round, P^j demonstrated an improvement over P^i, then P^i is replaced with P^j. Otherwise, P^i is maintained. At that stage, another process version is compared to P^i. The selection procedure continues until all process versions have competed. We remark here that the traces which could not be replayed on the process picked by the instance router were discarded. The number of compliant traces still represents the vast majority, because the ratio of conforming traces of all logs over models remained around 0.9, and always above 0.7 as shown in Table 6. Also, the total number of traces per log is shown in Table 5.

Analysis. Without loss of generality, we began the selection considering P^1 as the process currently running on the production system, and progressively entering P^2, P^3, P^4, and P^5 into the competition as described above. We sped up the execution time such that one day in the trace was equated to one second in the experiments.

The sequence of tests was: (1) P^1 vs. P^2, P^2 wins. (2) P^2 vs. P^3, P^3 wins. (3) P^3 vs. P^4, P^3 wins. (4) P^3 vs. P^5, P^3 wins. We can observe that P^3 was the best-performing version. In all tests, the instance router chose the version with lower mean and median execution time.

Figure 5 shows the request distribution throughout the pair-wise tests. The experimentation phase ends roughly after 1000 requests in all cases. We can observe that occasionally the instance router decided to pick another version some time during the post-experimentation phase. In some cases, the decision made during the post-experimentation phase contradicted the decision during the experimentation phase. In these scenarios, the instance router was able to make the better decision only after all the delayed rewards were received.

In Table 7, we show the request distribution during the experimentation phase, and the performance metrics calculated using execution times of processes instantiated during this phase. Considering the median and mean times in this table confirms that the instance router using the *LTAvgR* algorithm made *the right decision in all cases*.

Table 7. Pair-wise performance comparison of versions after the AB tests

Metric	Round 1		Round 2		Round 3		Round 4	
	P^1	P^2	P^2	P^3	P^3	P^4	P^3	P^5
No. of requests	559	440	423	575	263	729	735	261
Median duration	33.8	29.8	28.8	27	21	21.85	22.9	27.9
Mean duration	55.3	52.1	51.8	35.8	29.3	49.9	36.6	38.3

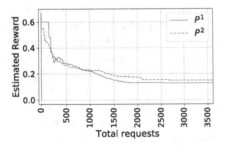

Fig. 6. Estimated rewards during the experiment P^1 vs. P^2

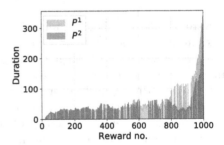

Fig. 7. PPI (duration) during the warm-up phase, smoothed.

For an in-depth view of the reward estimates (the average reward observed by *LTAvgR*) and execution times, we depict in Figs. 6 and 7 how their values changed during the experiment P^1 vs. P^2. The effect that fast completion leads to positive rewards is clearly visible in Fig. 6: shortly after the start of the experiment, the reward estimates for both versions jump to more than 0.6. After some fluctuation, P^1 is preferred approximately from request 280 to request 811. This is also visible in Fig. 5, where we can observe that the change in maximum of the two reward estimates leads to change in the request distribution strategy.

Figure 7 shows the PPIs observed by the instance router in order. Better PPIs, which lead to better rewards, are received early. However, worse PPIs tend to accumulate near the end of the warm-up phase. At request 811, the two estimated rewards are very close to each other – see Fig. 6. At this point, P^2 collects actual rewards from longer durations than P^1 – see Fig. 7. These longer durations result in negative rewards, which cause the reward estimate of P^2 to fall below that of P^1. This development leads to the change in the decision of *LTAvgR*.

5   Discussion

The design of our routing algorithm, *LTAvgR*, was informed by practical observations of the limitations when applying existing algorithms in the process execution context. As we have demonstrated, our approach addresses the key requirements R1-R3. Our evaluation focused on the practical use of AB-BPM and *LTAvgR*; theoretical analyses of the routing algorithm were out of the scope. The in-depth analysis above showed how *business-relevant PPI observations have a direct influence* on the routing decisions taken by *LTAvgR*.

We have used a multi-armed bandit algorithm with rewards derived from a single PPI. In practice, however, multiple PPIs may need to be considered. Furthermore, optimizing routing for one PPI can negatively affect other PPIs. One key challenge in using multiple PPIs is that the reward delay for each PPI can be different. Dealing with such scenarios may require improved collection and reward update mechanisms, which we plan to explore in future work.

One limitation of our evaluation of AB-BPM so far is that they are based on isolated environments with no real user interactions. Factors like effects from the novelty of a process version were not considered. For example, in changing the user interfaces and forms, we may observe that users behave differently when exposed to a new version. As with the case study using real-world logs, we expect to find some patterns unique to business processes when these factors are accounted for. We believe that observations from real-world systems can guide us towards designing a better instance routing algorithm, and identifying best practices for performing AB tests on process versions.

6   Conclusion

Business process improvement ideas do not necessarily manifest in actual improvements. In this paper we proposed the AB-BPM approach which can rapidly validate process improvement efforts, ensure fair comparison, and make process level adjustments in production environments. Our approach uses an instance router that dynamically selects process versions based on their historical performance in terms of the chosen PPI. To this end, we proposed the *LTAvgR* algorithm that can cater for the specifics of business process execution.

We evaluated our approach through synthetic and real-world data. We chose the most effective routing algorithm based on a set of experiments, and evaluated business process versions with synthetic data. Further, we evaluated real-world process versions by performing pair-wise AB tests on them. The evaluation results showed that our instance router dynamically adjusted request distribution to favour the better performing version.

In future work, we aim to integrate our framework with approaches to balance multiple PPIs. In addition, we plan to consider user interaction, and run industrial case studies where we apply our instance router to actual production systems.

Acknowledgements. The work of Claudio Di Ciccio has received funding from the EU H2020 programme under the MSCA-RISE agreement 645751 (RISE_BPM).

References

1. Agrawal, S., Goyal, N.: Thompson sampling for contextual bandits with linear payoffs. In: International Conference on Machine Learning, ICML (2013)
2. Alter, S.: Work system theory: overview of core concepts, extensions, and challenges for the future. J. Assoc. Inf. Syst. **14**, 72 (2013)
3. Bass, L., Weber, I., Zhu, L.: DevOps: A Software Architect's Perspective. Addison-Wesley Professional, New York (2015)
4. Burattin, A.: PLG2: multiperspective processes randomization and simulation for online and offline settings. CoRR abs/1506.08415 (2015)
5. Burtini, G., Loeppky, J., Lawrence, R.: A survey of online experiment design with the stochastic multi-armed bandit. CoRR abs/1510.00757 (2015)
6. Cabanillas, C., Di Ciccio, C., Mendling, J., Baumgrass, A.: Predictive task monitoring for business processes. In: Sadiq, S., Soffer, P., Völzer, H. (eds.) BPM 2014. LNCS, vol. 8659, pp. 424–432. Springer, Cham (2014). doi:10.1007/978-3-319-10172-9_31
7. Chapelle, O., Li, L.: An empirical evaluation of Thompson sampling. In: Neural Information Processing Systems (NIPS) (2011)
8. Crook, T., Frasca, B., Kohavi, R., Longbotham, R.: Seven pitfalls to avoid when running controlled experiments on the web. In: ACM SIGKDD International Conference on Knowledge Discovery and Data Mining, pp. 1105–1114 (2009)
9. Dumas, M., Rosa, M.L., Mendling, J., Reijers, H.A.: Fundamentals of Business Process Management. Springer, Heidelberg (2013)
10. Gregory, F.: Cause, effect, efficiency and soft systems models. J. Oper. Res. Soc. **44**, 333–344 (1993)
11. Hammer, M., Champy, J.: Reengineering the Corporation: A Manifesto for Business Revolution. HarperCollins, New York (1993)
12. Holland, C.W.: Breakthrough Business Results with MVT: A Fast, Cost-Free "Secret Weapon" for Boosting Sales, Cutting Expenses, and Improving Any Business Process. Wiley, Hoboken (2005)

13. Jiang, W., Au, T., Tsui, K.L.: A statistical process control approach to business activity monitoring. IIE Trans. **39**(3), 235–249 (2007)
14. Kettinger, W.J., Teng, J.T.C., Guha, S.: Business process change: a study of methodologies, techniques, and tools. MIS Q. **21**(1), 55–98 (1997). http://dblp.uni-trier.de/rec/bib/journals/misq/KettingerTG97
15. Kohavi, R., Longbotham, R., Sommerfield, D., Henne, R.M.: Controlled experiments on the web: survey and practical guide. Data Min. Knowl. Discov. **18**(1), 140–181 (2009)
16. Kohavi, R., Crook, T., Longbotham, R., Frasca, B., Henne, R., Ferres, J.L., Melamed, T.: Online experimentation at Microsoft. In: Workshop on Data Mining Case Studies (2009)
17. Li, L., Chu, W., Langford, J., Schapire, R.E.: A contextual-bandit approach to personalized news article recommendation. In: International Conference on World Wide Web (2010)
18. Mann, H.B., Whitney, D.R.: On a test of whether one of two random variables is stochastically larger than the other. Ann. Math. Stat. **18**, 50–60 (1947)
19. Ōno, T.: Toyota Production System: Beyond Large-scale Production. Productivity Press, Portland (1988)
20. Poelmans, S., Reijers, H.A., Recker, J.: Investigating the success of operational business process management systems. Inf. Tech. Manage. **14**(4), 295–314 (2013)
21. Sutton, R.S., Barto, A.G.: Introduction to Reinforcement Learning, 1st edn. MIT Press, Cambridge (1998)
22. Teinemaa, I., Leontjeva, A., Masing, K.O.: BPIC 2015: diagnostics of building permit application process in Dutch municipalities. BPI Challenge Report 72 (2015)
23. Thompson, W.R.: On the likelihood that one unknown probability exceeds another in view of the evidence of two samples. Biometrika **25**(3/4), 285–294 (1933)
24. Weidlich, M., Ziekow, H., Gal, A., Mendling, J., Weske, M.: Optimizing event pattern matching using business process models. IEEE Trans. Knowl. Data Eng. **26**(11), 2759–2773 (2014)
25. Weidlich, M., Ziekow, H., Mendling, J., Günther, O., Weske, M., Desai, N.: Event-based monitoring of process execution violations. In: Rinderle-Ma, S., Toumani, F., Wolf, K. (eds.) BPM 2011. LNCS, vol. 6896, pp. 182–198. Springer, Heidelberg (2011). doi:10.1007/978-3-642-23059-2_16

Optimized Execution of Business Processes on Blockchain

Luciano García-Bañuelos[1(✉)], Alexander Ponomarev[2], Marlon Dumas[1],
and Ingo Weber[2,3]

[1] University of Tartu, Tartu, Estonia
{luciano.garcia,marlon.dumas}@ut.ee
[2] Data61, CSIRO, Sydney, Australia
{alex.ponomarev,ingo.weber}@data61.csiro.au
[3] School of Computer Science and Engineering, UNSW, Sydney, Australia

Abstract. Blockchain technology enables the execution of collaborative business processes involving untrusted parties without requiring a central authority. Specifically, a process model comprising tasks performed by multiple parties can be coordinated via smart contracts operating on the blockchain. The consensus mechanism governing the blockchain thereby guarantees that the process model is followed by each party. However, the cost required for blockchain use is highly dependent on the volume of data recorded and the frequency of data updates by smart contracts. This paper proposes an optimized method for executing business processes on top of commodity blockchain technology. Our optimization targets three areas specifically: initialization cost for process instances, task execution cost by means of a space-optimized data structure, and improved runtime components for maximized throughput. The method is empirically compared to a previously proposed baseline by replaying execution logs and measuring resource consumption and throughput.

1 Introduction

Blockchain technology enables an evolving set of parties to maintain a safe, permanent, and tamper-proof ledger of transactions without a central authority [1]. In this technology, transactions are not recorded centrally. Instead, each party maintains a local copy of the ledger. The ledger is a linked list of blocks, each comprising a set of transactions. Transactions are broadcasted and recorded by each participant in the blockchain network. When a new block is proposed, the participants in the network agree upon a single valid copy of this block according to a consensus mechanism. Once a block is collectively accepted, it is practically impossible to change it or remove it. Hence, a blockchain can be seen as a replicated append-only transactional data store, which can replace a centralized register of transactions maintained by a trusted authority. Blockchain platforms such as Ethereum[1] additionally offer the possibility of executing scripts on top of a blockchain. These so-called *smart contracts* allow parties to encode business rules on the blockchain in a way that inherits from its tamper-proofness.

---

[1] https://www.ethereum.org/ – last accessed 4/3/2017.

© Springer International Publishing AG 2017
J. Carmona et al. (Eds.): BPM 2017, LNCS 10445, pp. 130–146, 2017.
DOI: 10.1007/978-3-319-65000-5_8

Blockchain technology opens manifold opportunities to redesign collaborative business processes such as supply chain and logistics processes [2]. Traditionally, such processes are executed by relying on trusted third-party providers such as Electronic Data Interchange (EDI) hubs or escrows. This centralized architecture creates entry barriers and hinders process innovation. Blockchain enables these processes to be executed in a distributed manner without delegating trust to central authorities nor requiring mutual trust between each pair of parties.

Previous work [3] demonstrated the feasibility of executing collaborative processes on a blockchain platform by transforming a collaborative process model into a smart contract serving as a template. From this template, instance-specific smart contracts are spawned to monitor or execute each instance of the process. The evaluation in [3] put into evidence the need to optimize resource usage. Indeed, the cost of using a blockchain platform is highly sensitive to the volume of data recorded and the frequency with which these data are updated by smart contracts. Moreover, the deployment of instance-specific contracts entails a major cost. In order to make blockchain technology a viable medium for executing collaborative processes, we need to minimize the number of contract creations, the code size, the data in the smart contracts, and the frequency of data writes.

This paper proposes an optimized method for executing business processes defined in the standard Business Process Model and Notation (BPMN) on top of commodity blockchain technology. Specifically, the paper presents a method for compiling a BPMN process model into a smart contract defined in the Solidity language – a language supported by Ethereum and other major blockchain platforms. The first idea of the method is to translate the BPMN process model into a minimized Petri net and to compile this Petri net into a Solidity smart contract that encodes the "firing" function of the Petri net using a space-optimized data structure. The second idea is to restrict the number of contract creations to the minimum needed to retain isolation properties. Furthermore, we optimized the runtime components to achieve high throughput rates. The scalability of this method is evaluated and compared to the method proposed in [3] by replaying artificial and real-life business process execution logs of varying sizes and measuring the amount of paid resources (called "gas" in Ethereum) spent to deploy and execute the smart contracts encoding the corresponding process models.

The next section introduces blockchain technology and prior work on blockchain-based process execution. Section 3 presents the translation of BPMN to Petri nets and to Solidity code. Section 4 discusses architectural and implementation optimizations. Section 5 presents the evaluation, and Sect. 6 draws conclusions.

2 Background and Related Work

2.1 Blockchain Technology

The term blockchain refers both to a network and a data structure. As a data structure, a blockchain is a linked list of blocks, each containing a set of transactions. Each block is cryptographically chained to the previous one by including its hash value and a cryptographic signature, in such a way that it is impossible to alter an earlier block without re-creating the entire chain since that block. The data structure is replicated across a network of machines. Each machine holding the entire replica is called a *full node*. In *proof-of-work blockchains,* such as Bitcoin and Ethereum, some full nodes play the role of *miners*: they listen for announcements of new transactions, broadcast them, and try to create new blocks that include previously announced transactions. Block creation requires solving a computationally hard cryptographic puzzle. Miners race to find a block that links to the previous one and solves the puzzle. The winner is rewarded with an amount of new crypto-coins and the transaction fees of all included transactions.

The first generation of blockchains were limited to the above functionality with minor extensions. The second generation added the concept of *smart contracts*: scripts that are executed whenever a certain type of transaction occurs and which read and write from the blockchain. Smart contracts allow parties to enforce that whenever a certain transaction takes place, other transactions also take place. For example, a public registry for land titles can be implemented on a blockchain that records who owns which property at present. By attaching a smart contract to sales transactions, it is possible to enforce that when a sale takes place, the corresponding funds are transferred, the tax is paid, and the land title is transferred, all in a single action.

The *Ethereum* [4] blockchain treats smart contracts as first-class elements. It supports a dedicated language for writing smart contracts, namely Solidity. Solidity code is translated into bytecode to be executed on the so-called *Ethereum Virtual Machine (EVM)*. When a contract is deployed through a designated transaction, the cost depends on the size of the deployed bytecode [5]. A Solidity smart contract offers methods that can be called via transactions. In the above example, the land title registry could offer a method to read current ownership of a title, and another one for transferring a title. When submitting a transaction that calls a smart contract method, the transaction has to be equipped with crypto-coins in the currency *Ether*, in the form of *gas*. This is done by specifying a gas limit (e.g. 2M gas) and gas price (e.g., 10^{-8} Ether/gas), and thus the transaction may use up to gas limit $\times$ price (2M $\times 10^{-8}$ Ether $= 0.02$ Ether). Ethereum's cost model is based on fixed gas consumption per operation [5], e.g., reading a variable costs 50 gas, writing a variable 5–20 K gas, and a comparison statement 3 gas. Data write operations are significantly more expensive than read ones. Hence, when optimizing Solidity code towards cost, it is crucial to minimize data write operations on variables stored on the blockchain. Meanwhile, the size of the bytecode needs to be kept low to minimize deployment costs.

2.2   Related Work

In prior work [3], we proposed a method to translate a BPMN choreography model into a Solidity smart contract, which serves as a factory to create choreography instances. From this factory contract, instance contracts are created by providing the participants' public keys. In the above example, an instance could be created to coordinate a property sale from a vendor to a buyer. Thereon, only they are authorized to execute restricted methods in the instance contract. Upon creation, the initial activity(ies) in the choreography is/are enabled. When an authorized party calls the method corresponding to an enabled activity, the calling transaction is verified. If successful, the method is executed and the instance state is updated, i.e. the executed activity is disabled and subsequent ones are enabled. The set of enabled activities is determined by analyzing the gateways between the activity that has just been completed, and subsequent ones.

The state of the process is captured by a set of Boolean variables, specifically one variable per task and one per incoming edge of each join gateway. In Solidity, Boolean variables are stored as 8-bit unsigned integers, with 0 meaning `false` and 255 meaning `true`.[2] Solidity words are 256 bits long. The Solidity compiler we use has an in-built optimization mechanism that concatenates up to 32 8-bit variables into a 256-bit word, and handles redirection and offsets appropriately. Nevertheless, at most 8 bits in the 256-bit word are actually required to store the information – the remaining are wasted. This waste increases the cost of deployment and write operations. In this paper, we seek to minimize the variables required to capture the process state so as to reduce execution cost (gas).

In a vision paper [6], the authors argue that the data-aware business process modeling paradigm is well suited to model business collaborations over blockchains. The paper advocates the use of the Business Artifact paradigm [7] as the basis for a domain-specific language for business collaborations over blockchains. This vision however is not underpinned by an implementation and does not consider optimization issues. Similarly [8] advocates the use of blockchain to coordinate collaborative business processes based on choreography models, but without considering optimization issues. Another related work [9] proposes a mapping from a domain specific language for "institutions" to Solidity. This work also remains on a high level, and does not indicate a working implementation nor it discusses optimization issues. A Master's thesis [10] proposes to compile smart contracts from the functional programming language Idris to EVM bytecode. According to the authors, the implementation has not been optimized.

3   From Process Models to Smart Contracts

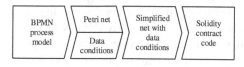

Fig. 1. Chain of transformations

The first and central component of the proposal is a method for transforming a given BPMN process model into a smart contract that can coordinate the execution of one process instance from start to end. Figure 1 shows the main steps of this method. The method takes as input a BPMN process model. The model is first translated into a Petri net. An analysis algorithm is applied to determine, where applicable, the guards that constrain the execution of each task. Next, reduction rules are

---

[2] https://github.com/ethereum/EIPs/issues/93 – last accessed 20/3/2017.

applied to the Petri net to eliminate invisible transitions and spurious places. The transitions in the reduced net are annotated with the guards gathered by the previous analysis. Finally, the reduced net is compiled into Solidity. Below, we discuss each step in turn.

3.1   From BPMN to Petri Nets

The proposed method takes as input a BPMN process model consisting of the following types of nodes: tasks, plain and message events (including start and end events), exclusive decision gateways (both event-based and data-based ones), merge gateways (XOR-joins), parallel gateways (AND-splits), and synchronization gateways (AND-joins). Figure 2 shows a running example of BPMN model. Each node is annotated with a short label (e.g. $A, B, g1 \ldots$) for ease of reference.

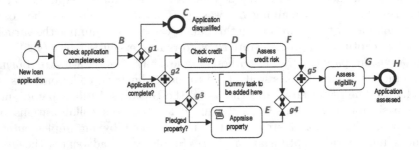

Fig. 2. Loan assessment process in BPMN notation

To simplify subsequent steps, we pre-process the BPMN model to materialize every *skip flow* as a dummy "skip" task. A skip flow is a sequence flow from an XOR-split to a XOR-join gateway such as the one between $g3$ and $g4$ in Fig. 2. Moreover, if the BPMN model has multiple end events, we transform it into an equivalent BPMN model with a single end event using the transformation defined for this purpose in [11]. In the case of the model in Fig. 2, this transformation adds an XOR-join at the end of the process that merges the incoming flows of the two end events, and connects them to a single end event. Conversely, if the process model has multiple start events, we merge them into a single one.

The pre-processed BPMN model is then translated into a Petri net using the transformation defined in [12]. This transformation can turn any BPMN process model (without OR-joins) into a Petri net.[3] The transformation rules in [12] corresponding to the subset of BPMN considered in this paper are presented in Fig. 3. Figure 4 depicts the Petri net derived from the running example. The tasks and events in the BPMN model are encoded as labeled transitions (A, B, ...). Additional transitions without labels (herein called τ transitions) are introduced by the transformation to encode gateways as per the rules in Fig. 3, and to capture the dummy tasks introduced to materialize skip flows.

---

[3] The transformation cannot handle escalation and signal events and non-interrupting boundary events, but these constructs are beyond the scope of this paper.

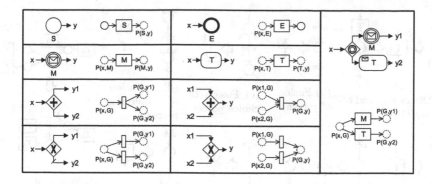

Fig. 3. Mapping of BPMN elements into petri nets

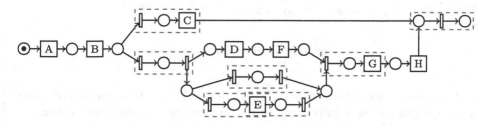

Fig. 4. Petri net derived from the BPMN model in Fig. 2

The transformation in [12] produces so-called *workflow nets*. A workflow net has one source place (start), one sink place (end), and every transition is on a path from the start to the end. Two well-known behavioral correctness properties of workflow nets are (i) *Soundness:* starting from the marking with one token in the start place and no other token elsewhere (the *initial marking*), it is always possible to reach the marking with one token in the end place and no other token elsewhere; and (ii) *Safeness:* starting from the initial marking, it is not possible to reach a marking where a place holds more than one token. These properties can be checked using existing tools [12]. Herein we restrict ourselves to workflow nets fulfilling these correctness properties. The latter property allows us to capture the current marking of the net by associating a boolean to each place (is there a token in this place or not?), thus enabling us to encode a marking as a bit array.

3.2 Petri Net Reduction

The Petri nets produced by the transformation in [12] contain many τ transitions. If we consider each transition as an execution step (and thus a transaction on the blockchain), the number of steps required to execute this Petri net is unnecessarily high. It is well-known that Petri nets with τ transitions can be reduced into smaller equivalent nets [13] under certain notions of equivalence. Here, we use the reduction rules presented in Fig. 5. Rules (a), (b), and (e)–(h)

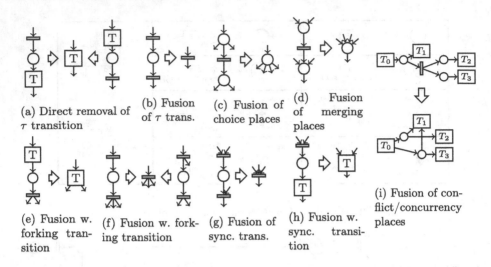

(a) Direct removal of τ transition

(b) Fusion of τ trans.

(c) Fusion of choice places

(d) Fusion of merging places

(e) Fusion w. forking transition

(f) Fusion w. forking transition

(g) Fusion of sync. trans.

(h) Fusion w. sync. transition

(i) Fusion of conflict/concurrency places

Fig. 5. Petri net reduction rules

are fusions of series of transitions, whereas rules (c) and (d) are fusions of series of places. Rule(i) deals with τ transitions created by combinations of decision gateways and AND-splits. It can be proved that each of these reduction rules produces a Petri net that is *weak trace equivalence* to the original one, i.e. it generates the same traces (modulo τ transitions) as the original one.

The red-dashed boxes in Fig. 4 show where the reduction rules can be applied. After these reductions, we get the net shown in Fig. 6a. At this point, we can still apply rule (i), which leads to the Petri net in Fig. 6b.

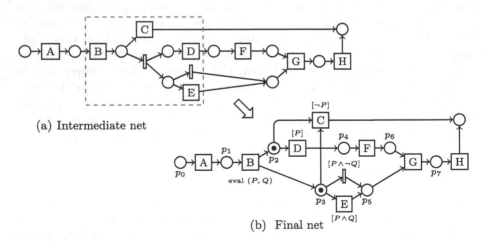

(a) Intermediate net

(b) Final net

Fig. 6. Reduced Petri net corresponding to the BPMN model in Fig. 2

3.3 Data Conditions Collection

Some of the τ transitions generated by the BPMN-to-Petri net transformation correspond to conditions attached to decision gateways in the BPMN model. Since some of these τ transitions are removed by the reduction rules, we need to collect them back from the original model and re-attach them to transitions in the reduced net, so that they are later propagated to the generated code.

Algorithm 1 collects the conditions along every path between two consecutive tasks in a BPMN model, and puts them together into a conjunction. Its output is one conjunctive condition – herein called a *guard* – per task in the original BPMN model. When given the start event as input, the algorithm applies a classical recursive depth-first traversal. It uses two auxiliary functions: (i) SUC-CESSORSOF, which returns the direct successors of a node; and (ii) COND, which returns the condition attached to a flow. Without loss of generality, we assume that every outgoing flow of a decision gateway is labeled with a condition (for a default flow, the condition is the negation of the conjunction of conditions of its sibling flows). We also assume that any other flow in the BPMN model is labeled with condition *true* – these *true* labels can be inserted via pre-processing.

Algorithm 1. Algorithm for collection of data conditions

1: **global guards:** Map⟨Node ↦ Cond⟩ = ∅, visited: Set⟨Node⟩ = ∅
2: **procedure** COLLECTCONDITIONS(curr: Node, predicate: Cond)
3:     guards[curr] ← predicate
4:     visited ← visited ∪ { current }
5:     **for each** succ ∈ SUCCESSORSOF(curr) : succ ∉ visited **do**
6:         **if** curr is a Gateway **then**
7:             COLLECTCONDITIONS(succ, predicate ∧ COND(curr, succ))
8:         **else**
9:             COLLECTCONDITIONS(succ, *true*)

We illustrate the algorithm assuming it traverses the nodes in the model of Fig. 2 in the following order: $[A, B, g1, g2, g3, E, \ldots]$. First, procedure COL-LECTCONDITIONS sets guards = $\{(A, true)\}$ in line 3 and proceeds until it calls itself recursively (line 9) with the only successor node of A, namely B. Note that predicate is reset to *true* in this recursive call. Something similar happens in the second step, where guard is updated to $\{(A, true), (B, true)\}$. Again, the procedure is recursively called in line 9, now with node $g1$. This time guards is updated to $\{(A, true), (B, true), (g1, true)\}$ and since $g1$ is a gateway, the algorithm reaches line 7. There, the procedure is recursively called with succ = $g2$ and predicate = $(true \wedge P)$, or simply P, where P represents the condition "Application complete?". Since the traversal follows the sequence $[A, B, g1, g2, g3, E, \ldots]$, it will eventually reach node E. When that happens, guards will have the value $\{(A, true), (B, true), (g1, true), (g2, P), (g3, P), (E, P \wedge Q)\}$, where Q represents the condition "Pledged property?". Intuitively, the algorithm propagates and combines the conditions P and Q while traversing the path between nodes B

to E. When the algorithm traverses E, the recursive call is done in line 9, where predicate is set to *true*, i.e. the predicate associated with E is not propagated further.

The guards gathered by the algorithm are attached to the corresponding transitions in the reduced net. In Fig. 6b the collected guards are shown as labels above each transition. To avoid cluttering, *true* guards are not shown. The τ transition in the net in Fig. 6b corresponds to the skip task that was inserted in the BPMN model in Fig. 2, hence this τ transition has a guard.

For each transition in the reduced Petri net, we can now determine the conditions that need to be evaluated after it fires. To do so, we first compute the set of transitions that are reachable after traversing a single place from each transition, and then analyze the guards associated to such transitions. In our example, we can reach transitions $\{C, D, E, \tau\}$ by traversing one place starting from transition B. Hence, conditions P and Q need to be evaluated after task B is executed. This is represented by attaching a label $eval(P, Q)$ to transition B.

3.4   From Reduced Petri Net to Solidity

In this step, we generate a Solidity smart contract that simulates the token game of the Petri net. The smart contract uses two integer variables stored on the blockchain: one to encode the current marking and the other to encode the value of the predicates attached to transitions in the reduced net. Variable marking is a bit array with one bit per place. This bit is set to zero when the place does not have a token, or to one otherwise. To minimize space, the marking is encoded as a 256-bits unsigned integer, which is the default word size in the EVM.

Consider the reduced Petri net in Fig. 5b. Let us use the order indicated by the subscripts of the labels associated to the places of the net. The initial marking (i.e. the one with a token in p_0) is encoded as integer 1 (i.e. 2^0). Hence, we initialize variable marking with value 1 when an instance smart contract is created. This marking enables transition A. The firing of A removes the token from p_0 and puts a token in p_1. Token removal is implemented via bitwise operations: marking = marking & uint($\sim$1);. Similarly, the addition of a token in p_1 (i.e. 2^1 hence 2) is implemented via bitwise operations: marking = marking | 2;.

Variable predicates stores the current values of the conditions attached to the Petri net transitions. This variable is also an unsigned integer representing a bit array. As before, we first fix order the set of conditions in the process model, and associate one bit in the array per condition. For safety, particularly in the presence of looping behavior, the evaluation of predicates is reset before storing the new value associated with the conditions that a given transition computes. For instance, transition B first clears the bits associated with conditions P and Q (i.e. 2^0 and 2^1, respectively), and then stores the new values accordingly.

When possible, an additional space optimization is achieved by merging variables marking and predicates into a single unsigned integer variable. The latter is possible if the number of places plus the number of predicates is at most 256. Note that this is not a restriction of our approach: If more space is needed to represent a process model, multiple 256-bit variables can be used.

Algorithm 2 sketches the functions generated for each transition in the reduced Petri net. Item 1 sketches the code for transitions associated to user tasks, while Item 2 does so for transitions associated to script tasks and τ transitions with predicates. For τ transitions without predicates, no function is generated, as these transitions only relay tokens (and this is done by the step function).

In summary, the code generated from the Petri net consists of a contract with the two variables marking and predicates, the functions generated as per Algorithm 2 and the step function. This smart contract offers one public function per user task (i.e. per task that requires external activation). This function calls the internal step function, which fires all enabled transitions until it gets to a point where a new set of user tasks are enabled (or the instance has completed).

Algorithm 2. Sketch of code generated for each transition in the reduced net

1. For each transition associated to a user task, generate a public function with the following code:
   - If task is enabled (i.e. check marking and predicates), then
     (a) Execute the Solidity code associated with the task
     (b) If applicable, compute all predicates associated with this task and store the results in a local bit set, tmpPreds
     (c) Call step function with new marking and tmpPreds, to execute all the internal functions that could become enabled
     (d) Return TRUE to indicate the successful execution of the task
   - Return FALSE to indicate that the task is not enabled
2. For each transition associated with a script task or τ transition that updates predicates, generate an internal function with the following code:
   (a) Execute the Solidity code associated with the task
   (b) If applicable, compute all predicates associated with this task and store the results in a local bit set, tmpPreds
   (c) Return the new marking and tmpPreds (back to the step function)

An excerpt of the smart contract generated for the running example is given in Listing 1.1. The excerpt includes the code corresponding to transitions B, E and the τ transition. Transition B corresponds to task CheckApplication. The corresponding function is shown in lines 4–17 in Listing 1.1. Since this is a user task, the function is called explicitly by an external actor, potentially with some data being passed as input parameters of the call (see line 4). In line 5, the function checks if the marking is such that p_2 holds a token, i.e., if the current call is valid in that it *conforms* to the current state of the process instance. If so, the function executes the script task (line 6 is a placeholder for the script). Then the function evaluates predicates P and Q (lines 8–9). Note that the function does not immediately updates variable predicates but stores the result in a local variable tmpPred, which we initialized in line 7. In this way, we defer updating variable predicates as much as possible (cf. line 39) to save gas (predicates is a contract variable stored in the blockchain and writing to it costs 5000 gas). For the same reason, the new marking is computed in line 11 but the actual update to the respective contract variable marking is deferred (cf. line 39).

Listing 1.1. Excerpt of Solidity contract

```solidity
1  contract BPMNContract {
2    uint marking = 1;
3    uint predicates = 0;
4    function CheckApplication( – input params – ) returns (bool) {
5      if (marking & 2 == 2) { // is there a token in place p1?
6        // Task B's script goes here, e.g. copy value of input params to contract variables
7        uint tmpPreds = 0;
8        if ( – eval P – ) tmpPreds |= 1; // is loan application complete?
9        if ( – eval Q – ) tmpPreds |= 2; // is the property pledged?
10       step(
11         marking & uint(~2) | 12,          // New marking
12       predicates & uint(~3) | tmpPreds // New evaluation for "predicates"
13         );
14       return true;
15     }
16     return false;
17   }
18   function AppraiseProperty(uint tmpMarking) internal returns (uint) {
19     // Task E's script goes here
20     return tmpMarking & uint(~8) | 32;
21   }
22   function step(uint tmpMarking, uint tmpPredicates) internal {
23     if (tmpMarking == 0) { marking = 0; return; } // Reached a process end event!
24     bool done = false;
25     while (!done) {
26       // does p3 have a token and does P ∧ Q hold?
27       if (tmpMarking & 8 == 8 && tmpPredicates & 3 == 3) {
28         tmpMarking = AppraiseProperty(tmpMarking);
29         continue;
30       }
31       // does p3 have a token and does P ∧ ¬Q hold?
32       if (tmpMarking & 8 == 8 && tmpPredicates & 3 == 2) {
33         tmpMarking = tmpMarking & uint(~8) | 32;
34         continue;
35       }
36       ...
37       done = true;
38     }
39     marking = tmpMarking; predicates = tmpPredicates;
40   } ... }
```

After executing B, if condition P holds the execution proceeds with the possibility of executing E or the τ transition. E is a script task and can be executed immediately after B, if condition Q holds, without any further interaction with external actors. For this reason, the Solidity function associated with task E is declared as internal. In the Solidity contracts that we create, all internal functions are tested for enablement, and if positive, executed. Specifically, the last instructions in any public function of the smart contract call a generic step function (cf. lines 22–40 in Listing 1.1). This function iterates over the set of internal functions, and executes the first activated one it finds, if any. For instance, after executing B there are tokens in p_2 and p_3. If $P \wedge Q$ holds, then the step function reaches line 28, where it calls function AppraiseProperty corresponding to transition E. This function executes the task's script in line 19 and updates marking in 20. After this, the control returns to line 29 in the step function, which restarts the while loop. Once all the enabled internal functions are executed, we exit the while loop. In line 39, the step function finally updates the contract variables.

4   Architecture and Implementation Optimization

In this section, we describe the improvements we have made in terms of architecture and implementation, relative to our earlier work on this topic [3].

Architecture Optimization. As introduced in Sect. 2, in [3] we proposed an architecture wherein a process model is mapped to a "factory" smart contract. For each instantiation, this factory contract creates an "instance" smart contract with the code necessary to coordinate the process instance. The instance contract is bound to a set of participants, determined at instantiation time. While this ensures isolation between different groups of participants, it is wasteful if the same group of participants repeatedly executes instances of the same model. In the latter case, the code encapsulating the coordination logic is redeployed for each instance, and contract deployment is particularly expensive.

To avoid this cost, we give the option to combine the factory and instance smart contracts into one, i.e., one smart contract that can handle running multiple instances in parallel. Instead of creating one bitvector per instance, we maintain an extensible array of bitvectors, each encoding the state of a process instance. On deployment, the array is empty. Creating a process instance assigns an instance ID and creates a new bitvector, which is appended to the array. This option is applicable when one group of actors repetitively executes instances of the same process. In situations where the actors differ across process instances, the option with separate factory and instance contracts should be used.

Implementation Optimization. During initial throughput experiments, we discovered that our original trigger implementation was a bottleneck on throughput. Our hypothesis was that we should be able to optimize the performance of the trigger to the point where it is no longer a bottleneck, i.e., a single trigger can handle at least as much throughput as the blockchain itself. To test this hypothesis, we improved our trigger by: (1) switching to asynchronous, non-blocking handling of concurrent requests, to achieve a high degree of parallelism in an environment that lacks full multi-threading. (2) Using the inter-process communications (IPC) channel to communicate with the blockchain software, *geth*, which runs the full blockchain node for a given trigger. During a small experiment we found that IPC can be 25× faster than the previously used HTTP connection. IPC requires the trigger and *geth* to run on the same machine, HTTP allows more flexible deployment architectures – but the performance advantage was too significant to ignore it during performance optimization. (3) Switching to asynchronous interaction with *geth*, which is a prerequisite for using IPC. The above changes required an almost complete rewrite of the code. The resulting throughput performance results are presented in the next section.

5   Evaluation

The goal of the proposed method is to lower the cost, measured in *gas*, for executing collaborative business processes when executed as smart contracts on the Ethereum blockchain. Thus, we evaluate costs with our improvements comparatively against the previous version. The second question we investigate is that of throughput: is the approach sufficiently scalable to handle real workloads?

5.1 Datasets

We draw on four datasets (i.e., logs and process models) described in Table 1. Three datasets are taken from our earlier work [3], the *supply chain*, *incident management*, and *insurance claim* processes, for which we obtained process models from the literature and generated the set of conforming traces. Through random manipulation, we generated sets of non-conforming traces from the conforming ones. The fourth dataset is stemming from a real-world invoicing process, which we received in the form of an event log with 65,905 events. This log was provided to us by the Minit process mining platform[4]. Given this log, we discovered a business process model using the Structured BPMN Miner [14], which showed a high level of conformance ($> 99\%$). After filtering out non-conforming traces, we ended up with dataset that contains 5,316 traces, out of which 49 traces are distinct. The traces are based on 21 distinct event types, including one for instance creation, and have an average length of 11.6 events.

5.2 Methodology and Setup

We translated the process models into Solidity code, using the previous version of the translator from [3] – referred to as *default* – and the newly implemented translator proposed in this paper – referred to as *optimized*. For the optimized version, we distinguish between the two architectures, i.e., the

Table 1. Datasets used in the evaluation.

Process	Tasks	Gateways	Trace type	Traces
Invoicing	40	18	Conforming	5,316
Supply chain	10	2	Conforming	5
			Not conforming	57
Incident mgmt.	9	6	Conforming	4
			Not conforming	120
Insurance claim	13	8	Conforming	17
			Not conforming	262

previous architecture that deploys a new contract for each instance, and the architecture that runs all process instances in a single contract. We refer to these options at *Opt-CF* ("CF" for control flow) and *Opt-Full*, respectively. Then we compiled the Solidity code for these smart contracts into EVM bytecode and deployed them on a private Ethereum blockchain.

To assess gas cost and correctness on conformance checking, we replayed the log traces against all three versions of the contracts and recorded the results. We hereby relied on the log replayer and trigger components from [3], with the trigger improvements discussed in Sect. 4. The replayer iterates through the traces in a log and sends the events, one by one, via a RESTful Web service call to the trigger. The trigger accepts the service call, packages the content into a blockchain transaction and submits it. Once it observes a block that includes the

---

[4] http://www.minitlabs.com/ – last accessed 13/3/2017.

transaction, it replies to the replayer with meta-data that includes block number, consumed gas, transaction outcome (accepted or failed, i.e., non-conforming), and whether the transaction completed this process instance successfully. The replayer has been modified to cater for concurrent replay of thousands of traces.

Experiments were run using a desktop PC with an Intel i5-4570 quadcore CPU without hyperthreading. Ethereum mining for our private blockchain was set to use one core. The log replayer and the trigger ran on the same machine, interacting via the network interface with one another. For comparability, we used the same software versions as in the experiments reported in [3], and a similar blockchain state as when they were run in February–March 2016. For Ethereum mining we used the open-source software geth[5], version v1.5.4-stable.

5.3   Gas Costs and Correctness of Conformance Checking

For each trace, we recorded the gas required for initialization of a new process instance (deploying an instance contract or creating a new bitvector, depending on the architecture), the sum of the gas required to perform all the required contract function invocations, the number of rejected transactions due to non-conformance and the successful completion of the process instance.

The results of this experiment are shown in Table 2. The base requirement was to maintain 100% conformance checking correctness with the new translator, which we achieved. Our hypothesis was that the optimized translator leads to *strictly monotonic improvements* in cost on the process instance level. We tested this hypothesis by pairwise comparison of the gas consumption per trace, and confirmed it: all traces for all models incurred less cost in *Opt-CF*. In addition to these statistics, we report the absolute costs as averages.

Table 2. Gas cost experiment results

Process	Tested traces	Variant	Avg. Cost		Savings (%)
			Instant.	Exec.	
Invoicing	5316	Default	1,089,000	33,619	–
		Opt-CF	807,123	26,093	−24.97
		Opt-Full	54,639	26,904	−75.46
Supply chain	62	Default	304,084	25,564	–
		Opt-CF	298,564	24,744	−2.48
		Opt-Full	54,248	25,409	−42.98
Incident mgmt.	124	Default	365,207	26,961	–
		Opt-CF	345,743	24,153	−7.04
		Opt-Full	54,499	25,711	−57.96
Insurance claim	279	Default	439,143	27,310	–
		Opt-CF	391,510	25,453	−8.59
		Opt-Full	54,395	26,169	−41.14

As can be seen from the table, the savings for *Opt-CF* over *default* are small for the simple *supply chain* process with only two gateways – cf. Table 1 – whereas they are considerably larger for the complex *invoicing* process with 18 gateways. Considering the reduction rules we applied, this can be expected. The other

---

[5] https://github.com/ethereum/go-ethereum/wiki/geth – last accessed 20/3/2017.

major observation is that *Opt-Full* yields massive savings over *Opt-CF*. When considering the absolute cost of deploying a contract vs. the average cost for executing a single transaction and the resulting relative savings, it is clear that the improved initialization is preferable whenever the respective architecture is applicable. As discussed in Sect. 4, this cost reduction also results in a loss of flexibility, and thus the choice requires a careful, case-specific tradeoff.

5.4   Throughput Experiment

To comparatively test scalability of the approach, we analyze the throughput using the three variants of contracts, *default, Opt-CF,* and *Opt-Full.* To this end, we used the largest of the four datasets, *invoicing,* where we ordered all the events in this log chronologically and replayed all 5,316 traces at a high frequency. The three variants were tested in separate campaigns. To ensure conformance, the events within a single trace were replayed sequentially. Ethereum's miners keeps a transaction pool, where pending transactions wait to be processed.

One major limiting factor for throughput is the gas limit per block: the sum of consumed gas by all transactions in a block cannot exceed this limit, which is set through a voting mechanism by the miners in the network. To be consistent with the rest of the experimental setup, we used the block gas limit from March 2016 at approx. 4.7M gas, although the miner in its default setting has the option to increase that limit slowly by small increments. Given the absolute gas cost in Table 2, it becomes clear that this is fairly limiting: for *Opt-CF,* instance contract creation for the invoicing dataset costs approx. 807K gas, and thus no more than 5 instances can be created within a single block; for *default,* this number drops to 4. Regular message calls cost on average 26.1 K/33.6 K gas, respectively for *optimized/default,* and thus a single block can contain around 180/140 such transactions at most. These numbers would decrease further when using a public blockchain where we are not the only user of the network.

Block limit is a major consideration. However, block frequency can vary: on the public Ethereum blockchain, mining difficulty is controlled by a formula that aims at a median inter-block time of 13–14 s. As we have demonstrated in [3], for a private blockchain we can increase block frequency to as little as a second. Therefore, when reporting results below *we use blocks as a unit of relative time.*

Figure 7 shows the process instance backlog and transactions per block. Note that each datapoint in the right figure is averaged over 20 blocks for smoothing. The main observation is that *Opt-Full* completed all 5,316 instances after 403 blocks, *Opt-CF* needed 1,053 blocks, and for *default* it took 1,362 blocks. This underlines the cost results above: due to the network-controlled gas limit per block, the reduced cost results in significant increases in throughput.

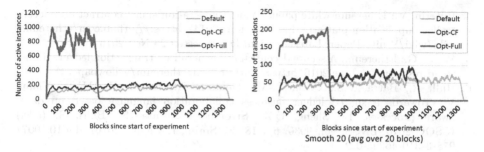

Fig. 7. Throughput results. Left: # of active instances. Right: # of transactions per block, smoothed over a 20-block time window.

6   Conclusion

This paper presented a method to compile a BPMN process model into a Solidity smart contract, which can be deployed on the Ethereum platform and used to enforce the correct execution of process instances. The method minimizes gas consumption by encoding the current state of the process model as a space-optimized data structure (i.e. a bit array with a minimized number of bits), reducing the number of operations required to execute a process step, and reducing initialization cost where possible. The experimental evaluation showed that the method significantly reduces gas consumption and achieves considerably higher throughput relative to a previous baseline.

The presented method is a building block towards a blockchain-based collaborative business process execution engine. However, it has several limitations, including: (i) it focuses on encoding control-flow relations and data condition evaluation, leaving aside issues such as how parties in a collaboration are bound to a process instance and access control issues; (ii) it focuses on a "core subset" of the BPMN notation, excluding timer events, subprocesses and boundary events for example. Addressing these limitations is a direction for future work.

Acknowledgements. This research was started at the Dagstuhl seminar #16191 – *Fresh Approaches to Business Process Modeling*. The research is partly supported by the Estonian Research Council (grant IUT20-55).

References

1. UK Government Chief Scientific Adviser: Distributed ledger technology: Beyond block chain. Technical report, UK Government Office of Science (2016)
2. Milani, F., García-Bañuelos, L., Dumas, M.: Blockchain and business process improvement. BPTrends newsletter, October 2016
3. Weber, I., Xu, X., Riveret, R., Governatori, G., Ponomarev, A., Mendling, J.: Untrusted business process monitoring and execution using blockchain. In: La Rosa, M., Loos, P., Pastor, O. (eds.) BPM 2016. LNCS, vol. 9850, pp. 329–347. Springer, Cham (2016). doi:10.1007/978-3-319-45348-4_19

4. Buterin, V.: Ethereum white paper: A next-generation smart contract and decentralized application platform. First version (2014). https://github.com/ethereum/wiki/wiki/White-Paper. Latest version: last accessed 29 Nov 2016

5. Wood, G.: Ethereum: A secure decentralised generalised transaction ledger. Homestead revision, 23 June 2016. https://github.com/ethereum/yellowpaper

6. Hull, R., Batra, V.S., Chen, Y.-M., Deutsch, A., Heath III, F.F.T., Vianu, V.: Towards a shared ledger business collaboration language based on data-aware processes. In: Sheng, Q.Z., Stroulia, E., Tata, S., Bhiri, S. (eds.) ICSOC 2016. LNCS, vol. 9936, pp. 18–36. Springer, Cham (2016). doi:10.1007/978-3-319-46295-0_2

7. Nigam, A., Caswell, N.S.: Business artifacts: An approach to operational specification. IBM Syst. J. **42**(3), 428–445 (2003)

8. Norta, A.: Creation of smart-contracting collaborations for decentralized autonomous organizations. In: Matulevičius, R., Dumas, M. (eds.) BIR 2015. LNBIP, vol. 229, pp. 3–17. Springer, Cham (2015). doi:10.1007/978-3-319-21915-8_1

9. Frantz, C.K., Nowostawski, M.: From institutions to code: Towards automated generation of smart contracts. In: Workshop on Engineering Collective Adaptive Systems (eCAS), co-located with SASO, Augsburg (2016)

10. Pettersson, J., Edström, R.: Safer smart contracts through type-driven development. Master's thesis, Department of CS&E, Chalmers University of Technology & University of Gothenburg, Sweden (2015)

11. Kiepuszewski, B., ter Hofstede, A.H.M., van der Aalst, W.M.P.: Fundamentals of control flow in workflows. Acta Inf. **39**(3), 143–209 (2003)

12. Dijkman, R.M., Dumas, M., Ouyang, C.: Semantics and analysis of business process models in BPMN. Inf. Softw. Technol. **50**(12), 1281–1294 (2008)

13. Murata, T.: Petri nets: properties, analysis and applications. Proc. IEEE **77**(4), 541–580 (1989)

14. Augusto, A., Conforti, R., Dumas, M., Rosa, M., Bruno, G.: Automated discovery of structured process models: discover structured vs. discover and structure. In: Comyn-Wattiau, I., Tanaka, K., Song, I.-Y., Yamamoto, S., Saeki, M. (eds.) ER 2016. LNCS, vol. 9974, pp. 313–329. Springer, Cham (2016). doi:10.1007/978-3-319-46397-1_25

Efficient Migration-Aware Algorithms
for Elastic BPMaaS

Guillaume Rosinosky[1,2]([✉]), Samir Youcef[2], and François Charoy[2]

[1] Bonitasoft, Grenoble, France
guillaume.rosinosky@bonitasoft.com
[2] Inria Nancy Grand Est - Université de Lorraine - CNRS, Nancy, France
http://www.bonitasoft.com

Abstract. As for all kind of software, customers expect to find business process execution provided as a service (BPMaaS). They expect it to be provided at the best cost with guaranteed SLA. From the BPMaaS provider point of view it can be done thanks to the provision of an elastic cloud infrastructure. Providers still have to provide the service at the lowest possible cost while meeting customers expectation. We propose a customer-centric service model that link the BP execution requirement to cloud resources, and that optimize the deployment of customer's (or tenants) processes in the cloud to adjust constantly the provision to the needs. However, migrations between cloud configurations can be costly in terms of quality of service and a provider should reduce the number of migrations. We propose a model for BPMaaS cost optimization that take into account a maximum number of migrations for each tenants. We designed a heuristic algorithm and experimented using various customer load configurations based on customer data, and on an actual estimation of the capacity of cloud resources.

Keywords: BPM · Cloud · Elasticity · BPM as a service

1 Introduction

During the last decade, we witnessed a major change in the way companies are delivering software. It is more often distributed as a service, operated by software producers, hosted in public clouds instead of as a package and installed on premises. BPM systems (BPMS) vendors and operators start to propose this kind of distribution. It removes the burden for customers to operate the BPMS and the corresponding infrastructure. They pay for process instances they execute or on a fixed monthly rate per user [1]. The service provider aims at ensuring the required service quality at the lowest possible cost. Thanks to the public cloud and the elasticity it supports, providers can deliver that quality while minimizing resource consumption, and thus the operational cost. Public cloud providers allow to add and remove dynamically computing resources. However, it does not fit well with the deployment stack of a BPMS that include web servers and databases systems to store the process execution data because of its transactional nature that is costly to scale up horizontally.

© Springer International Publishing AG 2017
J. Carmona et al. (Eds.): BPM 2017, LNCS 10445, pp. 147–163, 2017.
DOI: 10.1007/978-3-319-65000-5_9

We propose a method that allows to distribute process execution on a set of cloud computing resources, and to adjust the resources based on the load that the customers require. We call a customer *a tenant*, and we ensure that all the processes of a tenant are executed on a BPMS installation. We consider that a service provider will host hundreds of small to medium size tenants. We consider that resources are paid by discrete time units. We assume that we know for each tenant what will be its maximal resource consumption per time unit. More precisely, we want to take into account the knowledge we can get from the business dimension, i.e. the number of process task execution per hour. In our previous work [2], we proposed to optimize resource consumption from a time slot to the next.

In this paper, we extend the method to optimize resource consumption on an given number of time slots. We also limit tenant migrations from an installation to another to avoid unwanted service disruption. Migrations generally become necessary when the resources on which a tenant is deployed are not sufficient to support its required load for the coming period of time. For each duration, based on the knowledge of the resource consumption profile that we can get from all the tenants, we compute a deployment plan that minimize resource consumption while maintaining the number of migrations for each tenant at an acceptable level. This is the contribution of this paper. We propose a linear optimization model and then a heuristic in two parts. First, we compute the appropriate migration times for each tenant using a time series segmentation method. Then we show how this method can be coupled to a restricted version of the time slot heuristic and provide substantial gain compared to naive ones. We validate it with an experimentation using realistic values regarding the size of customers and the size and price of cloud resources.

In the next section, we present the state of art regarding BPM elasticity in cloud computing. In Sect. 3, we describe the model that we want to optimize. In the following Sect. 4, we explain the condition of the experimentation. Then in Sect. 5, we describe and discuss our results. In the last section, we conclude and present possible extensions to this work.

2   Elasticity in BPM

A lot of work has been done to manage BPM elasticity in the cloud. Schulte et al. [3] made a review on the current status on BPM elasticity, and on the different important criteria. We focus here on the scheduling and resource allocation parts (challenges 1 and 2), with an emphasis on the multi-tenancy, and taking into account data transfer and BPM task throughput as a KPI. Usual autoscaling approaches such as Amazon Auto Scaling Group are not usable in our case, as we do not consider clustered BPMS installations.

Hoenisch et al. [4] proposes an interesting approach: the Service Instance Placement Problem, a cost optimization model concerning the assignment of process instances to VM, scheduling of service invocations, and the provisioning of VM. The authors consider the underlying structure of the BPM processes and propose to optimize the cost while taking into account penalties for violated deadlines. However, they consider only the BPM engine's CPU and RAM capacity in their model - and not the database tier- and do not consider multi-tenancy or migrations. Other previous attempts have the same drawbacks such as [5,6]. Rekik et al. [7] developed a very interesting resource allocation and scheduling linear model for business process deployment in cloud federations based on configurable business processes. It considers CPU, memory, bandwidth, price, security and availability for the cloud resources and corresponding needed capacity for the BPM activities, but without regarding the database tier, multi-tenancy, multiple time slots, or migrations.

Hachicha et al. [8] addresses multi-tenant BPMaaS with the concept of configurable resource assignment operator. It consists of an enrichment of the process with meta information on the required resources for the tasks execution. However, it needs to add informations in the BPM schema, thus requiring to alter BPM engine and the processes of the customers and do not propose a resource allocation and scheduling method. Sellami et al. [9] propose a multi tenant approach based on customizable thresholds, however, it does not take into account migration cost or the database tier.

Our work is an evolution of our previous paper [2] where we proposed an bi-objective optimization model for cost and migrations quantity for all tenants scrambled from a time slot to the next, and a corresponding efficient heuristic. Simply repeating this heuristic on multiple time slots may induce the migration of the same tenants, thus generating interruption of service. We propose here to harness the problem of optimization on multiple consecutive time slots, with a limitation on the number of migrations for each tenant.

3   The BPM Execution Model

In this section we introduce the model for the BPMS execution we want to optimize. Our model relies on a few assumptions regarding the BPM system. First, it must be multi-tenant, i.e. several customers (or tenants) can share the same BPMS installation. Thus we can collate several tenants on the same deployment when their load is small. Second, it is possible to migrate a tenant from one installation to another with minimal disruption. The main issue here is the migration of the process execution data from a database to another [10,11]. Another point is the duration of VM allocation, who can take several minutes: during this time the BPMS cannot be available. We also only consider that IaaS providers bill the computing resources per studied time slot (for instance per hour). Main public cloud providers like AWS, and IBM Bluemix, follow this pattern per hour, while Google Compute Engine or Azure propose also a per minute billing.

The operation of a BPMS requires a complex software stack. It combines a BPM engine, load balancers and relational databases to manage process and business data. They are often deployed on distinct hardware instances or virtual machines from the BPM engine, mostly for performance reasons. We call *"cloud configuration"* the set of resources (e.g. two VM for the engine and for the DBMS) that we use to execute process instances for a group of tenants. Last, we assume that we know the usage requirement for each tenant time slot by time slot in term of the completed BPM task maximum throughput per second for a period of time. This metric is related to process execution. It has the advantage of being representative of the system usage of both the database tier and the application server tier. We will also calibrate different cloud configuration in term of BPM task throughput.

We aim at minimizing the cloud resource cost while ensuring that the throughput for each tenant is at the required level. Migrations count for each tenant should not exceed a fixed number in order to avoid disruptions. We propose here a linear model where we wish to optimize the cost of placement.

Let the following variables:

- $\mathcal{T}$, the set of cloud configuration types, with t its cardinality.
- $\mathcal{I}$, the set of tenants with n its cardinality
- $\mathcal{J}$, is $\mathcal{T} \times \mathcal{I}$ the set of all possible cloud configurations associated with each tenant, its cardinality is $m = t \times n$
- C_j, and W_j, respectively the cost and the capacity for the configuration j, with j in $\mathcal{J}$
- $\mathcal{K}$ defines all the time slots, from 0 to D, where $D + 1$ is the number of time slots.
- $w_i(k)$, the required capacity for the tenant i during time slot k
- $x_j{}^i(k)$, the assignment of tenant i to configuration instance j during time slot k
- $y_j(k)$, the activation of configuration j during time slot k
- M is the defined maximum number of migrations for each tenant

$$\min \sum_{j}^{j \in \mathcal{J}} \sum_{k}^{k \in \mathcal{K}} C_j y_j(k) \tag{1}$$

We have the following constraints:

$$\forall i \in \mathcal{I}, \forall k \in \mathcal{K} \sum_{j}^{j \in \mathcal{J}} x_j{}^i(k) = 1 \tag{2}$$

$$\forall j \in \mathcal{J}, \forall k \in \mathcal{K} \sum_{i}^{i \in \mathcal{I}} w_i(k) x_j{}^i(k) \leq W_j y_j(k) \tag{3}$$

$$\forall i \in \mathcal{I} \sum_{j}^{j \in \mathcal{J}} \sum_{k}^{k \in \mathcal{K} \setminus \{D\}} x_j{}^i(k) x_j{}^i(k+1) \geq |\mathcal{K}| - M \tag{4}$$

$$\forall i \in \mathcal{I}, \forall j \in \mathcal{J}, \forall k \in \mathcal{K}, x_i{}^j(k) \in \{0,1\}, y_j(k) \in \{0,1\} \tag{5}$$

Equation 1 is our optimization objective. We want to minimize the total cost of cloud configurations for all the time slots (a day for instance). The constraint described in Eq. 2 means that, for each time slot, each tenant must be located on one and only cloud configuration. The constraint described in Eq. 3 means that, for each time slot, the sum of required throughput of the tenants co-located on a cloud resource do not exceed the capacity of this resource.

The constraint described in Eq. 4 means that we want to limit the number of migrations per tenant to M. If $x_j{}^i(k)$ and $x_j{}^i(k+1)$ are both equal to 1 for the resource j and the tenant i on two consecutive time slots k and $k+1$, their product will be equal to 1, the tenant did not migrate. In the other cases, the product will be equal to 0. They occur when tenant i migrated from or to another resource from time slot k to time slot $k+1$ or when tenant i remained on a different resource on both time slots. We sum these products resource per resource, on each time slot pair for each tenant. We obtain the number of time slot where a tenant remained on the same configuration. The difference between the total number of time slots and this number is the number of migration. Limiting the number of migrations is then straightforward.

Since we have multiplication between two variables, we obtain a quadratic optimization problem. As it can become is very slow to compute, we linearized the Eq. 4, following the usual method. The result for this linearization is described in Eq. 6:

$$\forall i \in \mathcal{I} \sum_j^{j \in \mathcal{J}} \sum_k^{k \in \mathcal{K}\setminus\{D\}} w_j(k+1) \geq |\mathcal{K}| - M$$

$$\forall i \in \mathcal{I} \sum_j^{j \in \mathcal{J}} \sum_k^{k \in \mathcal{K}\setminus\{D\}} x_j{}^i(k) + x_j{}^i(k+1) - 2w_j(k+1) \leq 1$$

$$\forall i \in \mathcal{I} \sum_j^{j \in \mathcal{J}} \sum_k^{k \in \mathcal{K}\setminus\{D\}} w_j(k+1) \leq x_j{}^i(k)$$

$$\forall i \in \mathcal{I} \sum_j^{j \in \mathcal{J}} \sum_k^{k \in \mathcal{K}\setminus\{D\}} w_j(k+1) \leq x_j{}^i(k+1)$$

$$\forall i \in \mathcal{I}, \forall j \in \mathcal{J}, \forall k \in \mathcal{K}, w_i{}^j(k) \in \{0,1\}$$

(6)

Even with the linearization, the resolution of this problem can be very time consuming. The number of variables will be of $tn(D+1) + tn^2(D+1)$, and the number of constraints will be of $tn(D+1) + tn^2(D+1) + 4n$. For 7 cloud resource types (t), 100 tenants (n), and 24 time slots ($D+1$), it makes 1696800 variables, and 1697200 constraints. It is not reasonable to try to compute the optimal solution when the number of tenants grows, as we will see in the experiment part. In the next section, we propose a heuristic that provide solutions to the problem with reasonable computation time even with large number of tenants.

4   Heuristic Optimization Proposition

4.1   Iterative Time Slot Algorithm

This algorithm is based on our previous time slot heuristic [2]. Its principle is to consider that, regarding an initial distribution, we search the best distribution for the next time slot, knowing that the required capacity of each tenant can change. With the time slot algorithm, we search for a Pareto front of the lowest global number of migrations and resources cost. As the number of migrations is discrete and limited by the number of tenants, we can compute the lowest cost for each number. First, we look at overloading and overloaded tenants as shown in Fig. 1.

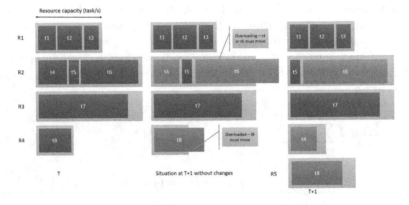

Fig. 1. Example of distribution of tenants on cloud resources at T and T+1 (Color figure online)

It depicts the distribution of tenants on different cloud configurations. The initial state is at time T. At time $T + 1$, the requirement for each tenant changes. Some of them have to migrate (orange and red). The heuristic principle is, that for each possible number of migrations we consider the combination of resources containing the corresponding number of tenants, added to the number of tenants that we must migrate. As we consider the total quantity of tenants, there are several possibilities for each number of migrations. This is a classic subset sum problem, who addresses the following problem: finding a subset of integers in a set who sums to a given integer other. In our case, the subsets are quantity of tenants for each resource, and the sum the number of considered migrations. For instance, in Fig. 1, in order to compute the results for 5 migrations, we could move the tenants $t8$, $t6$, and remove the combinations of resources with 3 remaining tenants ($R1$ or $R2$ and $R3$). Once we have selected the tenants, we first repack the possible tenants in existing resources, and then use a Variable Cost and Size Bin Packing [12] algorithm for the remaining ones. The last step consists in trying to replace the resources with tenants we moved with cheaper resources.

This approach provides good results time slot by time slot but we may have to move some tenant at each time slot, disrupting their service. Thus we must adapt it to limit the number of migrations per tenant described by constraint 4 per time window. In the next section we propose an adaptation of this algorithm to enforce this new constraint.

4.2   A Migration Aware Optimization Strategy

For this new strategy, we add a list of tenants allowed to migrate as an additional parameter to the previous method. If, for a time slot, we allow to migrate every tenant but $T5$ and $T2$ because they have reached their maximum number of migration (M), the new time slot algorithm must ignore them and maintain them on their resources. We cannot delete resources with tenants.

Let a *migration strategy* the set of $h_i(k)$ with $0 \le k \le D - 1$ where each tenant i is allowed to be migrated. $h_i(k)$ is equal to 0 if the tenant is not allowed to move between time slot k and $k + 1$, and equal to 1, if it is allowed. The Eq. 7 describe the maximum number of migrations.

$$\forall i \in \mathcal{I} \sum_{k}^{k \in \mathcal{K} \backslash \{D\}} h_i(k) = M \tag{7}$$

We want to find the best values for each $h_i(k)$ respecting the maximum number of migrations to obtain the best cost.

Once we have determined the different migrations time slots, we choose the required capacity level. As our algorithm does not consider multiple time slots simultaneously, we assign a fixed capacity for each tenant for each period where it does not migrate. We call $P_i(m)$ the capacity during the period between migrations m_1 and m_2. In order to avoid overloads, we consider the maximum capacity required for the corresponding time slots, as in Eq. 8. An example of a maximum load strategy is presented in Fig. 2. These capacities are used instead of the initial capacities of the tenants in the time slot algorithm.

$$\forall i \in \mathcal{I}, \forall m_1 \le k < m_2, P_i(m_2) = \max_{m_1 \le k < m_2} (w_i(k)) \tag{8}$$

As testing every migration strategy requires too much time, we need an efficient way to evaluate which one we should use. It must give better results than a naive approach, and respect the constraint on the number of migrations for each tenant.

4.3   Time Series Segmentation

We propose a method to identify good migrations strategies. We consider first that the variations in load will produce the need for migrations. If for instance a tenant needs a throughput of 10 tasks per seconds between 12pm and 6am, and then a throughput of 50 tasks per second between 6am and 12am, the best

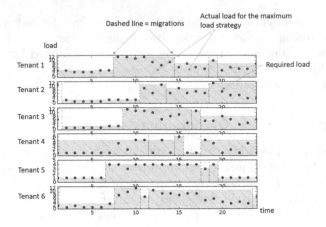

Fig. 2. Migration strategy of 6 tenants on 24 h

time to migrate is at 6am. Our approach here is, for k migrations, to find a way to fragment the load time series in $k + 1$ consecutive fragments, in a way where they have the minimum load. Time series segmentation techniques address this kind of problems.

As Lovric [13] explains, we can see time series segmentation as a processing step and core task for variety of data mining tasks, as a trend analysis technique, as a discretization problem in function of dimensionality reduction, etc. The latter point interests us as we want to find a way to discretize the load time series, with discrete periods of remaining tenants, separated by migrations. The main common algorithms based on Piecewise Linear Representation are originally reported by Keogh et al. [14]: top-down, bottom-up and sliding window. As our approach is offline (we know the future load), we focus only top-down and bottom-up.

The principle is to iteratively separate (top-down) or merge (bottom-up) consecutive sets of observations in the time series, so they keep a minimal error related to real observations. As it can be seen in [13,14], the main version of the algorithm segments the time series using Piecewise Linear Approximation (PLA), fitting each segment with an affine function found by linear regression of the values for each segment. This approach is interesting for our needs, but we can test others. Indeed, as we explained earlier, we want to segment the time series considering the maximum load instead of the mean of segment (we consider for each segment its maximum load as described in Eq. 8). As we will see in the experiment section, we also tested a mean constant piecewise approximation, that consider a fixed mean for the segmentation.

Once we obtain the segments, we compute the corresponding migration strategy. We initialize the matrix to zero, except for the time slots where there is a change of segments, that we initialize to one. We then use this migration strategy with the restricted iterative time slot algorithm as we can see in the next part.

4.4   The Optimization Algorithm

Here is a synthesis of our time slot algorithm:

- First, compute the desired migration strategy using a time series segmentation algorithm.
- Second, initialize the initial time slot (zero) with the initial distribution.
- Then, for each time slot from $k = 1$ to $k = D$:
  - Launch a time slot algorithm, using the distribution of the previous time slot. Only the tenants able to move in the migration strategy for this time slot migrate. Thus we ensure constraint 4 of the model.
  - Keep the less expensive distribution with the least number of migrations. This is the distribution we choose for the current time slot.

This algorithm provides a solution that enforces the constraints. In the next section we describe our experimentation that shows how it provides better results than a naive solution and with a reasonable computation time even for a hundred of tenants.

5   Experimentation

To test our solution we made a few assumptions and relied as much as possible on datasets that gives us realistic foundations for the resolution of the model. As we can see in the model part, we needed to have a good estimation of the customer loads, time slot by time slot on one side, and of the price and capability in BPM task throughput per time slot of a cloud configuration on the other side. We have then compared several segmentation methods on the same datasets of customer loads and cloud configurations to find the best ones. In the next part we describe the datasets.

5.1   Datasets

In order to get meaningful cost and task throughput for our cloud configurations, we have used the data obtained in our test framework experimentation [15]. In this paper, we have set up an experiment on AWS with a BPMN process composed of 20 consecutive automated tasks launching a Fibonacci script. We launched tests on several storage-oriented (r3 family) instance types for the database, and CPU-oriented (c4 family) instance types for the application server. We used the BPM system BonitaBPM 7.3.2[1] in its Open Source version. We compared the resulting BPM task throughput with the price of each cloud configuration. Results are described in Table 1. It provides the number of tasks per second for one \$ for each configuration type. Note that the higher end resources have better absolute performance but the throughput per \$ is lower. This is not in favor of vertical scalability.

---

[1] http://www.bonitasoft.com/.

Table 1. Price, mean BPM task throughput, and mean BPM task throughput by dollar for the given cloud configuration.

DB inst. type	AS inst. type	Price	Task TP	Task TP per $
db.m3.medium	m3.medium	0.177	16.400	92.656
db.m3.medium	c4.large	0.223	23.157	103.845
db.r3.large	c4.large	0.399	55.164	138.255
db.r3.large	c4.xlarge	0.518	58.067	112.100
db.r3.xlarge	c4.large	0.674	65.113	96.607
db.r3.large	c4.2xlarge	0.757	61.474	81.208
db.r3.xlarge	c4.xlarge	0.793	83.236	104.963
db.r3.xlarge	c4.2xlarge	1.032	89.149	86.384
db.r3.2xlarge	c4.2xlarge	1.587	105.794	66.663
db.r3.2xlarge	c4.4xlarge	2.063	107.585	52.150
db.r3.4xlarge	c4.4xlarge	3.173	115.283	36.332
db.r3.4xlarge	c4.8xlarge	4.126	129.279	31.332

Table 2. For each customer, the observed interval in days, the minimum and the maximum task throughput per second for each hour.

Customer	Days	Minimum	Maximum
A	4	1	120
B	1	14	16
C	45	0	120
D	7	1	3
E	45	5	120
F	550	0	4

For the customer load part, we wanted to test multiple tenant quantities (5, 10, 25, 50 and 100), having different throughputs based on real data from BonitaBPM customers. More precisely, we used minimum and maximum throughput per second found in the anonymized execution history tables. The used thresholds are described in Table 2. We have then generated each tenants initial time slot load randomly following an uniform distribution between the two thresholds.

To avoid too much variation between time slots, we also used another parameter we name *tenant gap*, a percentage of the gap between a tenant's minimum and maximum throughput. Using a totally random behaviour makes tenants throughput very chaotic. In general, the required load is relatively stable, as we have noticed in customers data. For each time slot, we compute randomly the percentage of the gap, and we add it to the previous time slot's load. If we obtain a load lower than the minimum, we cap it to the minimum (respectively capped to the maximum for loads superior to the maximum). For instance, for customer E and a gap of 0.25, the change between hours would follow an uniform distribution between $-0.25(120 - 5) = -28.75$ and $0.25(120 - 5) = 28.75$, capped on the minimum and maximum for the customer, here respectively 5 and 120. We have experimented with various values for the gap. A gap percentage of 1 will correspond to a complete random behavior between the minimum and maximum loads. A gap percentage of 0 will correspond to a load that remains the same all the time. For each number of tenants, we tested multiple tenant gaps, 0.25, 0.5, 0.75 and 1, as it shows multiple levels of variability.

We varied the number of tenant and the gap. Nonetheless, given the random nature of the load, we ran the tests multiple times with different random distributions for each pair tenant gap and tenant number. In order to keep repeatable configurations and to test multiple algorithms, we used twenty different random seeds for each parameter pairs. Using the same random seed on a same couple tenant gap/tenant number gives each time the same load distribution.

For each set (the cartesian product between twenty seeds, the 4 tenant gaps, and the 5 different tenant quantities), we tested several migration plan algorithms that we describe in the next part.

5.2   Software and Methods

Our goal here is to determine between the different methods and for different number of migrations, which one gives the better migration strategies i.e. which one between top-down and bottom-up algorithms, grouped and individual strategies, and ConstantMaxPieceWise, ConstantPieceWise and Linear Regression fitters gives the best results.

As our metrics are hourly based on the different configurations and tenant loads, we used hourly time slots, more precisely 48 time slots. The initial setting is the following: for each tenant, we select the least expensive resource that is suited for the maximum required throughput on the study time. We name this method *adapted heuristic*.

For the segmentation part we implemented a fixed and updated version of the Alchemyst library[2]. This library implements top-down and bottom-up algorithms, with mean constant and linear regression fitters. We added max constant fitter as described in Sect. 4.3. We also tested the algorithms for several number of migrations, more precisely 2, 3 and 4 per day, so 4, 6 and 8 for the 48 h studied. Last, we tested two grouping strategies: considering each tenant in a separated manner (*individual strategy*), or executing the segmentation on the sum of the loads, and moving every tenant simultaneously at the obtained migration time slots (*grouped strategy*).

We also enhanced our previous time slot algorithm implementation [2] with the restriction on the tenant list. For performance reasons we don't consider here the subset sum for each number of migrations but only all the tenants.

---

[2] https://github.com/alchemyst/Segmentation developed by Carl Sandrock for his paper [16].

To obtain reference values, we used the *adapted heuristic* to compute the cost. This approach gives us a realistic intuitive reference cost that we can obtain without calculation. We also compared a subset of our test dataset with the solving of the linear model described in Chap. 3 during a limited time, for the lowest number of tenants. For this we used the solver Gurobi [17].

Table 3 summarize the different parameters we used in this experiment. Experiments have been executed on c4.xlarge (CPU optimized) instances on Amazon Web Services. In the next part, we discuss our results.

Table 3. Synthesis of the experiment dataset and algorithm used values.

Group	Variable	Size	Values
Data	Tenant gap %	4	0.25, 0.5, 0.75, 1
Data	Seed number	20	-
Data	Number of days	1	2
Data	Number of tenants	6	5, 10, 25, 50, 100, 200
Data	Number of migrations	3	2, 3, 4
Segmentation	Algorithm	2	Bottom-up, top-down
Segmentation	Fitter	3	Mean constant, max constant, linear regression
Grouping strategy	Tenant load	2	Grouped, individual
Time slot algorithm	Subset sum size	1	1

5.3   Results and Discussion

As we explained in the previous part, we compare our results with the *adapted heuristic* cost we obtained (in Fig. 3). The mean cost is between 370.15 $ for 5 tenants and a tenant gap of 0.25 and 18853.56 $ for 200 tenants and a tenant gap of 1. We can see here that the higher the gap, the higher the cost. Since the variation of the load is less restricted, the maximum load can be higher, and the strategy principle is to consider each tenant's higher load for each cloud resource.

We see in Fig. 4 the results we obtained for a tenant gap of 0.25 and 4 migrations a day with the heuristic. We compute the gain percentage by observing for each experiment run (one seed, one tenant gap percent, on tenant quantity, one number of migrations) the ratio of the difference of the result related to the corresponding adapted cost. We can see that the different strategies give different level of results. Top-down algorithms give better results than bottom-up, except for top-down individual with linear regression fitter. Grouped strategies give almost every time the best results, except for the top-down algorithm on individual strategy with a constant maximum piecewise fitter, who gives results near to the two bests, (top-down grouped constant strategies), and is even more efficient for 200 tenants. For top-down algorithms, constant maximum piecewise is the best algorithm, followed by constant mean piecewise. It is more difficult to compare for bottom-up strategies.

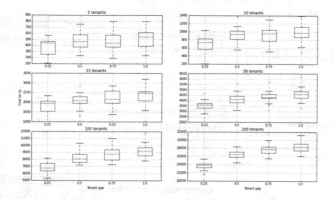

Fig. 3. Distribution of experimentation adapted heuristic costs in dollars

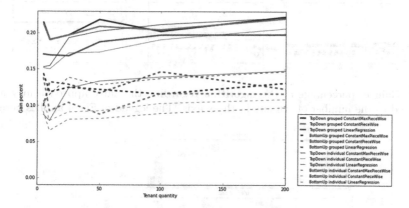

Fig. 4. Gain in percentage of the algorithm regarding the naive approach, compared to the number of tenants, for a tenant gap of 0.25 and 4 migrations per day.

Figure 5 shows a global overview of the results. Most of the time, the top-down algorithm gives better results than the bottom-up. The best global approach are usually more efficient on small number of tenants or with less variation (low tenant gap). The best fitters here are top-down grouped constant max piecewise, top-down individual constant max piecewise and top-down constant piecewise. As expected, we obtain better results when we allow more migrations (from 5 to 10% for 2 migrations to 15 to 25% for 4 migrations). A higher gap percentage generates worse results. The top-down individual Constant Max Piecewise strategy gives the best results for more than 5 tenants almost every time.

Figure 6 show the running time of the iterative time slot algorithm with different parameters. It stays relatively stable, for a defined strategy and number of tenants, mainly for grouped strategies. Individual strategies are always longer, and the duration seems to be multiplied by 3 to 4 each time the number of tenants doubles.

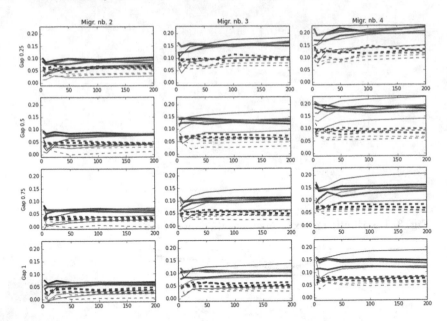

Fig. 5. Gain in percentage of the algorithm regarding the naive approach in y axis, compared to the number of tenants in the x axis. The legend of Fig. 4 applies here.

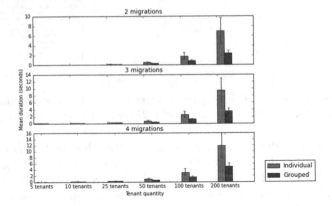

Fig. 6. Mean duration in seconds for individual and grouped approaches, for each studied number of tenants. The standard deviation is represented on each bar, with the mean number of seconds.

We tried to compare our result with the optimal solution but computing the results with the exact model using a solver is very time consuming and does not give very good results for more than 5 tenants and a defined time of 30 min as we can see in the Table 4. For 10 tenants, 24 h of computing time does not give the optimal results: we have obtained a MIP gap of 7% at best with the solver, while with our heuristic we obtained gains of 20% with a mean duration of at most 0.4 s of running time. For 100 tenants, our heuristic duration is 100 s for a grouped strategy.

Table 4. Solver results for the 10 first seeds. Mean MIP gap is the gap between the solution and the inferior bound found.

Tenant qty	Tenant gap	Mean adapted	Nb migr.	Solver duration	Solver gain	Mean MIP gap
5	0.25	419.99	2	1800	41.05%	3.57%
5	0.25	421.91	3	1800	48.28%	3.45%
5	0.25	421.91	4	1800	51.64%	3.91%
10	0.25	685.82	2	1800	3.18%	48.33%
10	0.25	685.82	3	1800	1.28%	52.68%
10	0.25	685.82	4	1800	1.28%	55.33%

We see the correlation between the number of migrations and the gain. The results show us that the top-down algorithm works much better than the bottom-up, and that constant maximum piecewise fitter give almost every time better results than the other fitters. The constant mean piecewise fitter gives also interesting results. Good results for global strategies can be explained by the resource-oriented approach of our algorithm (except for overloading and over-loaded tenants, it considers only resource removal for tenant migration). The very good performance of individual maximum constant piecewise, especially for large number of tenants shows the efficiency of this fitter, but it needs a minimum number of tenants.

We have shown that our heuristic is fast, even when the number of tenants grows, and permits substantial savings for the BPMaaS providers. A gain of 20% on the adapted strategy for 100 tenants corresponds to a mean of 1373.48 $ for two days. Moreover, it is possible to test multiple strategies, as the computing time stays low (for 200 tenants an individual strategy lasts a mean of 11.75 s, and 5.02 s for a grouped one). The longer duration for the individual strategy can be explained by the mechanics of the heuristic. Indeed, in this case we segment every tenant instead of all at once in the grouped strategy. Individual strategies could be more interesting to use in production environment. Indeed, even if we have not considered this constraint, migrating all the tenants together could have some side effects on QoS because all the data of the customers will migrate simultaneously, having negative effects on the available network bandwidth. Of course an individual strategy could give the same results if the tenants have identical workload patterns.

6   Conclusion

In this paper, we have proposed a new linear model for resource allocation and scheduling of BPM execution in the cloud, and a quick, simple and straightforward heuristic giving good results compared to naive approaches and solving of the model. This model relies on assumptions that makes it applicable in an operational setting. First, we consider customers as a whole (tenants) and not a distribution process instance by process instance. This reduces the scope of the calculation and avoids security issues regarding access to the business data. Second we assume that we can migrate a BPMS deployment from one installation to another in a reasonable time. This is not available in current systems but it is possible with very little service interruption.

We have tested it with data from an existing BPMS with a task metric. Of course, we could also use it with other metrics such as the number of processes and even for other services than BPMS. For instance, we could consider web servers and the throughput of HTTP queries as soon as they have a strong data management component. We can also use the algorithm with different temporal dimensions, hours as in the experimentation, but also minutes or seconds. We also plan to use it in an online manner, coupled with a predictive component that computes dynamically the expected load for the following time slots. This is our next step. We also think that can still improve our results using metaheuristics. Even if they are much better than with a naive approach, there is room for improvement as shown by our experiments with a solver for 5 tenants. Last, we want to test this heuristic with customer data on multiple time zones, where the results should be better considering the load patterns we can identify (day/night cycle, working hours, lunch time, etc.).

Acknowledgements. The authors would like to thank Gurobi for the usage of their optimizer, and Amazon Web Services for the EC2 instances credits (this paper is supported by an AWS in Education Research Grant Award). The data and the results are available at: http://doi.org/10.5281/zenodo.401374. The source code of the framework is not free for now, except for the segmentation library, available at https://github.com/guillaumerosinosky/Segmentation/.

References

1. Le, T.M.H., Alfredo, L.A., Choi, H.R., Cho, M.J., Kim, C.S.: A study on BPaaS with TCO model, pp. 249–256. IEEE, December 2014
2. Rosinosky, G., Youcef, S., Charoy, F.: An efficient approach for multi-tenant elastic business processes management in cloud computing environment. In: 2016 IEEE 9th International Conference on Cloud Computing (CLOUD), pp. 311–318. IEEE, June 2016
3. Schulte, S., Janiesch, C., Venugopal, S., Weber, I., Hoenisch, P.: Elastic business process management: state of the art and open challenges for BPM in the cloud. Future Gener. Comput. Syst. **46**, 36–50 (2014)
4. Hoenisch, P., Schuller, D., Schulte, S., Hochreiner, C., Dustdar, S.: Optimization of complex elastic processes. IEEE Trans. Services Comput. **9**(5), 700–713 (2016)
5. Janiesch, C., Weber, I., Kuhlenkamp, J., Menzel, M.: Optimizing the performance of automated business processes executed on virtualized infrastructure. In: 2014 47th Hawaii International Conference on System Sciences (HICSS), pp. 3818–3826. IEEE, January 2014
6. Euting, S., Janiesch, C., Fischer, R., Tai, S., Weber, I.: Scalable business process execution in the cloud. In: 2014 IEEE International Conference on Cloud Engineering (IC2E), pp. 175–184, March 2014

7. Rekik, M., Boukadi, K., Assy, N., Gaaloul, W., Ben-Abdallah, H.: A linear program for optimal configurable business processes deployment into cloud federation. In: 2016 IEEE International Conference on Services Computing (SCC), pp. 34–41. IEEE, June 2016

8. Hachicha, E., Assy, N., Gaaloul, W., Mendling, J.: A configurable resource allocation for multi-tenant process development in the cloud. In: Nurcan, S., Soffer, P., Bajec, M., Eder, J. (eds.) CAiSE 2016. LNCS, vol. 9694, pp. 558–574. Springer, Cham (2016). doi:10.1007/978-3-319-39696-5_34

9. Sellami, W., Kacem, H.H., Kacem, A.H.: Elastic multi-tenant business process based service pattern in cloud computing. In: 2014 IEEE 6th International Conference on Cloud Computing Technology and Science (CloudCom), pp. 154–161. IEEE, December 2014

10. Das, S., Agrawal, D., El Abbadi, A.: ElasTraS: an elastic, scalable, and self-managing transactional database for the cloud. ACM Trans. Database Syst. **38**(1), 1–45 (2013)

11. Barker, S.K., Chi, Y., Hacigümüs, H., Shenoy, P.J., Cecchet, E.: ShuttleDB: database-aware elasticity in the cloud. In: 11th International Conference on Autonomic Computing, ICAC 2014, Philadelphia, PA, USA, 18–20 June 2014, pp. 33–43 (2014)

12. Kang, J., Park, S.: Algorithms for the variable sized bin packing problem. Eur. J. Oper. Res. **147**(2), 365–372 (2003)

13. Lovrić, M., Milanović, M., Stamenković, M.: Algoritmic methods for segmentation of time series: an overview. J. Contemp. Econ. Bus. Issues **1**(1), 31–53 (2014)

14. Keogh, E., Chu, S., Hart, D., Pazzani, M.: An online algorithm for segmenting time series. In: 2001, Proceedings IEEE International Conference on Data Mining, ICDM, pp. 289–296. IEEE (2001)

15. Rosinosky, G., Youcef, S., Charoy, F.: A framework for BPMS performance and cost evaluation on the cloud. In: 2016 IEEE International Conference on Cloud Computing Technology and Science (CloudCom), pp. 653–658. IEEE, December 2016

16. Sandrock, C.: Identification and generation of realistic input sequences for stochastic simulation with Markov processes. In: Cakaj, S. (ed.) Modeling Simulation and Optimization - Tolerance and Optimal Control. InTech, April 2010. doi:10.5772/9035

17. Gurobi Optimization, I.: Gurobi optimizer reference manual (2015)

Uncovering the Hidden Co-evolution
in the Work History of Software Projects

Saimir Bala[1(✉)], Kate Revoredo[2], João Carlos de A.R. Gonçalves[2],
Fernanda Baião[2], Jan Mendling[1], and Flavia Santoro[2]

[1] Vienna University of Economics and Business (WU), Vienna, Austria
{saimir.bala,jan.mendling}@wu.ac.at
[2] Federal University of the State of Rio de Janeiro (UNIRIO), Rio de Janeiro, Brazil
{katerevoredo,joao.goncalves,fernanda.baiao,flavia.santoro}@uniriotec.br

Abstract. The monitoring of project-oriented business processes is difficult because their state is fragmented and represented by the progress of different documents and artifacts being worked on. This observation holds in particular for software development projects in which various developers work on different parts of the software concurrently. Prior contributions in this area have proposed a plethora of techniques to analyze and visualize the current state of the software artifact as a product. It is surprising that these techniques are missing to provide insights into what types of work are conducted at different stages of the project and how they are dependent upon another. In this paper, we address this research gap and present a technique for mining the software process including dependencies between artifacts. Our evaluation of various open-source projects demonstrates the applicability of our technique.

Keywords: Artifact co-evolution · Work history dependencies · Project-oriented business processes · Software project mining

1 Introduction

Project-oriented business processes play an important role in various industries like engineering, health care or software development [2]. Such processes are characterized by the fact that work towards a predefined outcome involves complex tasks executed by different parties. Typically, these processes are not supported by a process engine, but their status is fragmented over different documents and artifacts. This is especially the case for software development processes: the expected outcome is the release of a new software version, but the different project members collaborate with tools like version control systems that are only partially aware of the work process.

This work has been partially funded by the Austrian Research Promotion Agency (FFG) under grant 845638 (SHAPE) and the RISE BPM project (H2020 Marie Curie Program, grant 645751). The second author was partially supported by PROAP/-CAPES and fourth and sixth authors by the National Council for Scientific and Technological Development (CNPq), Brazil.

J. Carmona et al. (Eds.): BPM 2017, LNCS 10445, pp. 164–180, 2017.
DOI: 10.1007/978-3-319-65000-5_10

A key challenge for project-oriented business processes like software development is gaining transparency of the overall project status and work history. Literature has recognized that analyzing the evolution of business process artifacts in projects can help obtaining important clues about the project performance in terms of time [5], cost [22] and quality [13]. This is addressed by functionality of version control systems (VCS) to track versions and changes of informational artifacts like source code and configuration files. While prior research has presented various perspectives for analyzing software artifacts, e.g. [3,14,19,23], there is a notable gap on the discovery of dependencies in the work history. For these reasons, project managers often lack insights into side effects of changes in large software processes.

In this paper, we address this research gap by building on partial solutions from the separate fields of mining software repositories and process mining. More specifically, we develop a technique that uncovers non-hierarchical work dependencies which we call *hidden co-evolution*. This technique extracts the labeled work history from VCS repositories and identifies dependencies beyond simple hierarchical containment. In this way, we help the project manager to spot dependencies in the co-evolution of work histories of different information artifacts. Our technique has been implemented and evaluated using data from a diverse set of open source projects.

The paper is structured as follows. Section 2 describes the research problem along with its requirements and summarizes insights from prior research. Section 3 presents our approach in detail. Section 4 shows a prototypical implementation and evaluates its applicability both in a use case scenario and on real world projects from GitHub. Section 5 concludes the paper.

2 Background

This paper follows the Design Science Research (DSR) paradigm [16]. In this section, we describe the research problem in more detail and define requirements for a solution. Against these requirements, we analyze related work.

2.1 Problem Description

In this paper, we focus on a specific class of project-oriented business processes, namely software development processes. These processes share some common characteristics. First, they involve various resources with different roles. In the simplest case, we can distinguish *project managers* and *project participants*. Project managers are responsible for managing the development process and supervising the work of the project participants, who in turn are responsible for specific work tasks. Second, such processes are usually subject to constraints in terms of cost, time and quality, which is mostly associated with the performance of each of the work tasks. Third, the project participants work on a plethora of artifacts, which are logically organized in a hierarchical structure, with complex interdependencies among them. Given these characteristics, it is the goal of the project manager to organize the software development process in such a way that the work on different files and tasks reflects the complex interdependencies, the constraints and the available participants. Therefore, it is important for the manager to understand the *work history* of the process in order to monitor the progress systematically.

Table 1. An excerpt of a VCS log data

Id	Project Participant	Date	Comment	Diff
1	John	2017-01-31 12:16:30	Create readme file	diff –git a/README.md b/README.md @@ -0,0 +1 @@ +# StoryMiningSoftwareRepositories
2	Mary	2017-02-01 10:13:51	Add a license	diff –git a/README b/README @@ -1,0 +2,3 @@ +The MIT License (MIT) + +Copyright (c) 2015 Mary+
3	Paul	2017-02-02 16:10:22	Updated the requirements.	diff –git a/README.md b/README.md @@ -1,4 +1,5 @@ + # string 1, string 2, string 3 diff –git a/requirements.txt b/requirements.txt @@ -0,0 +1 @@ +The software must solve the problems
4	Paul	2017-02-02 15:00:02	Implement new requirements	diff –git a/model.java b/model.java @@ -1,9 +1,10 @@ +public static methodA(){int newVal=0; @@ -21,10 +23,11 @@ + "1/0",,"0/0", diff –git a/test.java b/test.java @@ -0,0 +1,2 @@ +//test method A +testMethodA()

Software tools like Version Control Systems (VCS) do not provide direct support for monitoring work histories, but they provide a good starting point by continuously collecting event data on successive versions of artifacts. Table 1 shows an excerpt of log data, where the columns, from left to right, indicate the commit identifier, the project participant who committed the changes, the commit date, the comment written by the project participant and the files affected and the change performed[1]. In order to understand the work history and dependencies based upon such data, we identify three major requirements:

R1 (Extract the work history): Discover the process of how artifacts evolve in the project as a *labeled* set of steps. This requirement is difficult because the version changes of a commit in relation to a single file do not directly reveal which type of work has been done. Both commit messages and edit characteristics might inform the labeling.

---

[1] cf. unified diff format https://git-scm.com/docs/git-diff.

R2 (Uncover Work-Related Dependencies): Identify that certain work in one part of the project is connected with work in another part. This requirement is difficult because such dependencies might not only exist between files that reside in the same directory. For example, a change in a source code file might have the side effect of triggering work on a configuration file. We refer to this as *co-evolution* of these files.

R3 (Measure Dependencies): Determine how strong the co-evolution of different artifacts is. This requirement is difficult because measures of *strength* of dependencies and on the *distance* of dependent artifacts have to be devised.

2.2 Related Work

A solution addressing these requirements can partially build upon research in three main areas: *(i)* work on Mining Software Repositories (MSR); *(ii)* Process Mining (PM); *(iii)* and software visualization.

Table 2 shows that these streams of research have mutual strengths, but no contribution covers the full spectrum. In general, methods from MSR have a strength in analyzing dependencies in the structure of the software artifact, but an explicit consideration of the type of work is missing. Contributions in this area focus on the users and the artifacts, mining co-evolution or co-change of project parts [8,24] and network analysis of file dependency graph based on commit distance [1,23,25]. Hidden work dependencies are mentioned as *logical dependencies* [15]. Also techniques for trend analysis [20] and inter-dependencies

Table 2. Requirements addressed by literature and topics covered. Fulfills requirement (✓); Only addresses requirement (⋆)

Main area	Papers	R1	R2	R3	Description
	Zaidman et al. [24]	⋆	✓	✓	Only two labels for processes
	Zimmermann and Nagappan [25]		✓	✓	Only functional dependencies
	Abate et al. [1]		✓	⋆	Only functional dependencies
	D'Ambros et al. [8]		✓	✓	
MSR	Oliva et al. [15]		✓	✓	
	Weicheng et al. [23]		✓	✓	
	Ruohonen et al. [20]		✓	✓	
	Lindberg et al. [13]			⋆	Activity variations
	Kindler et al. [12]	✓			
	Goncalves et al. [9]	✓			
PM	Poncin et al. [17]	✓			
	Beheshti et at. [4]	✓	⋆		
	Mittal and Sureka [14]	⋆			Only bug resolution process
	Bala et al. [2]	⋆		⋆	Unlabelled Gantt chart
	Voinea and Telea. [21]	⋆	✓		Unlabelled processes
Visualization	Ripley et al. [18]	✓		✓	Unlabelled processes
	Greene and Fischer [10]		✓		

between developers [13] are proposed. However, none of these works considers the type of work being done in the process.

In the area of PM, research gives more emphasis to the different tasks of the process. Some works focus on applying process mining for software repositories [2,14,17]. In this context, approaches have been defined that use various queries to extract artifact evolution and resources [4,5]. There is research on identifying the tasks of the process by elicitation from unstructured data of user comments [9]. There are also process mining applications that focus on repetitive steps in software engineering, but not on singular project-oriented processes, such as [12]. All these works only consider the dependencies between work tasks to a limited extend.

There is also work in the area of software visualization. Visualization tools have been proposed in order to allow project managers to have a detailed overview of the software artifact being developed. These tools help to visually inspect artifacts similarities on different levels of granularity [21], observe artifacts evolution or project members contribution [10,18]. In general, they can be characterized as artifact-centric, and largely agnostic to the type of work being done.

In the following, we develop a technique that addresses the three requirements and informs prior research on how to extract work histories and to identify the co-evolution of certain parts of a project-oriented software process.

3   Conceptual Approach

We propose a technique to extract and represent the work history and the dependencies among artifacts of a project-oriented business process. The technique takes as input a VCS log and produces analysis data that describe the evolution of the artifacts, along with metrics about their distance and their similarity in terms of work. The process is depicted in Fig. 1 and consists of three successive steps towards extracting hidden work dependencies from VCS event data. The method works under three main assumptions. First, we assume a *meaningful tree structure*, i.e. the project participants organize the files in a representative hierarchy (e.g., spatially separating documentation from testing into different folders). Second, project participants perform *regular commits* in the VCS. Third, project participants write *descriptive comments* that allow other members to understand the changes.

The first step of the technique is the preprocessing of the VCS log received as input. The main goal of this phase is to generate a set of events and store them into a database. Second, we obtain different views on the stored events. In particular, we are interested in observing *(i)* all the commits that affected the files over time; *(ii)* the amount of change brought by the commits to the files; and *(iii)* the users who issued such commits. The third phase is responsible for considering the different perspectives defined by the project manager and through the generated views extract the necessary knowledge. In the following, we detail the formal concepts and the algorithm of our technique.

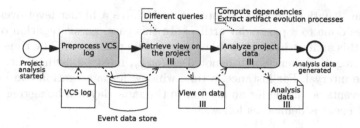

Fig. 1. Approach for generating analysis data from VCS logs

3.1 Preliminaries

As the objective of our technique is to uncover hidden work dependencies, we define the fundamental concepts required to capture them. Work is reflected by *artifacts*, e.g., word documents, spreadsheets, code, etc. Artifacts are leaves in the file tree hierarchy (with directories being special type of non-leaf files). Artifacts evolve over time, while project participants contribute their changes. Each change is an *event* that happens to an artifact in a single point in time. Events can be abstracted into *aggregated events* that allow a coarser grained view on the history. The history of the changes of an artifact over a time interval at a given level of abstraction is referred to as *artifact evolution*. Similar artifact co-evolution establishes a *dependency* between two artifacts.

A software product is subdivided into files and directories. In this work, we consider directories as special type of files which are parents of other files. Formally, let F be the universe of files in a software development project. Files are organized in a file tree. Therefore, each file $f \in F$ has one parent file. The only file without a parent file is the *root* file. We capture this information in the parent relation $Parent : F \times F$. For example, let $f_p \in F$ be the parent of file $f_c \in F$, then $(f_p, f_c) \in Parent$. An *artifact* is a file that is not a parent file, i.e. a file f_a is an artifact if $\forall_{f \in F}(f_a, f) \notin Parent$.

When project participants do a certain amount of work and want to save their current progress, they commit the changes to the VCS. We define changes on artifacts as the *events* of interest on the lowest granularity.

Definition 1 (Event). *Let E be the set of events. An* event $e \in E$ *is a five-tuple* (f, ac, ts, k, u), *where*

- $f \in F$ *is the affected artifact of the event.*
- $ac \in AC = \mathbb{N}$ *is the amount of change done in the artifact.*
- $ts \in TS = \mathbb{N}$ *represents a unix time stamp marking the time of the event occurrence.*
- $k \in \Sigma^*$ *is a comment in natural language text.*
- $u \in U$ *is the project participant responsible for the change.*

For event $e = (f, ac, ts, k, u)$ we overload f, ac, ts, k and u to be used as accessor functions. For example, f is the function $f : E \to F$ mapping an event to its affected artifact.

In some situations, it can be interesting to have a higher level overview of the changes done to a particular artifact. In this case, an aggregation of events related to this artifact in an interval of time can be performed. The time window for the aggregation, henceforth denoted as tw_{agg}, must be defined, i.e. the size of the time interval. For instance, a time window for aggregation can be a day. Thus, all events occurring for an artifact in the same day will be aggregated. An *aggregated event* is defined as follows:

Definition 2 (Aggregated Event). *An* Aggregated Event *for* tw_{agg} $(AE_{tw_{agg}})$ *is a five-tuple* (f, aac, ats, ak, au), *where*

- $f \in F$ *is the affected artifact in the set of events being aggregated.*
- $aac \in AAC = \mathbb{N}$ *is the aggregate amount of change done in the artifact for* tw_{agg}. *It is calculated by summing the amount of changes done in each of the time aggregated.*
- $ats \in ATS = \mathbb{N}$ *represents an aggregate time of the unix time stamp of the events being aggregated.*
- $ak \in \Sigma^*$ *is the concatenation of the comments presented in the events being aggregated.*
- $au \subseteq U$ *are the project participants responsible for the changes in* tw_{agg} *being aggregated.*

The set of aggregated events for a particular artifact defines how this artifact evolves over time. Considering an interval of analysis, henceforth denoted as *ia*, we define artifact evolution as follows.

Definition 3 (Artifact Evolution). *Artifact evolution is the process describing how the file* f *changed over an interval of time ia, i.e., a set of labeled tuples* $A_{evo}(f) = \{(t, a, l) | e \in AE_{ia}, f = f(e), t = ats(e), a = aac, l = ak(e)\}$ *chronologically ordered.*

Note that artifact evolution represents the changes that happened to a file over time. Thus, we can build the time series of a file f as the vectors of changes $X_f = (a_1, ..., a_n)$ in the time window $tw_{agg} = [t_1, t_n]$, with a_i being the sum of the changes of f in of the aggregated intervals t_i of the time window tw_{agg}.

We measure the dependency between two files f_a and f_b in terms of their *degree of co-evolution* as follows.

Definition 4 (Degree of Co-Evolution). *Given two files* f_a *and* f_b, *the degree of co-evolution* $\chi : F \times F \rightarrow [0, 1]$ *is a similarity function of the respective time series.*

In this paper, we fix $\chi(f_a, f_b) = |\sigma(X_{f_a}, X_{f_b})|$, where σ is the correlation function of the two vectors X_{f_a} and X_{f_b}.

The way files are kept in the directory structure establishes an inherent relationship among files being stored close to each other in the hierarchy. For instance, files serving the same purpose are stored close to each other in the file system. Hidden work dependencies are expected to happen between artifacts that are distant in the file structure. We measure this distance as the length of the shortest route connecting two files in the file tree. We adapt the notion of path from [11] to our file tree. Given a file f, the path to the root node can be obtained by navigating the *Parent* relationship up to the root file. The path p from f_a to the root f_r is the set of parent files encountered along such route. i.e. $p(f_1, f_r) = \{(f_1, ..., f_k, f_{k+1}, ..., f_r)\}$ such that for any k, $(f_{k+1}, f_k) \in Parent$. The length of the path is the cardinality $|p|$ of the set. The shortest path between two files f_a, f_b in a tree passes through the Least Common Ancestor (LCA) [6]. This is equivalent to considering the paths from the single files to the root node $p_a = p(f_a, f_r)$ and $p_b = p(f_b, f_r)$ minus their intersection $I_{p_a, p_b} = \{p(f_a, f_r) \cap p(f_b, f_r)\}$. Thus, we define the *file distance* as the length of the shortest path between two files f_a and f_b as follows.

Definition 5 (File Distance). *The distance $d : F \times F \rightarrow \mathbb{N}$ between two files belonging to the same directory structure is defined as the number of nodes in the minimum path connecting the two files in the project file tree:* $d(f_a, f_b) = |p_a| + |p_b| - 2 * (|I_{p_a, p_b}|)$.

3.2   Hidden Dependencies Discovery Algorithm

We are focused on finding interesting hidden work dependencies. These dependencies are typically reflected by changes that happen to couples of allegedly unrelated files during their evolution. This section details the procedure that implements the technique outlined in Fig. 1.

Algorithm 1 presents the steps required to explicate such hidden dependencies. The procedure `PreprocessLog(L)` in line 2 takes as input a VCS log $\mathcal{L}$ structured as in Table 1 and parses out work events at the granularity of line changes. These events are then stored into an event data storage. Events parsed from VCS logs contain rich information about multiple aspects of the work they reflect. In order to represent all these different aspects, we devised the entity-relationship data model. Hence, we are able to store all the information that is possible to obtain after parsing the VCS log. Furthermore, this step allows the user to obtain simple information, such as statistics on the project, already at an early stage of the procedure. The output of the `PreprocessLog(L)` step results in the storage of all the events E into a database.

Next, the iterative call of the procedure `RetrieveView(E, query)` in line 3 performs several querying the data storage containing the set E. For example, a possible query can obtain all the comments associated to each change of a specific file. To obtain information on the evolution of files, we query the database for the changes of all the files within a user defined time interval tw_{agg}. In general several time frames can be chosen, each of them producing a *view* V on the data, i.e., a set of aggregated events chronologically sorted within tw_{agg}. For example, users may be interested in artifact-views aggregated by day, by month, etc. Multiple *views* are possible by defining them in the `queries` parameter. We collect these views into a set $\mathcal{V} = \bigcup_{\text{queries}} V$.

Algorithm 1. Generate project analysis data

Input   : A VCS log $\mathcal{L}$
Output: A set of triples $\{(Dist, Stories, D_{co-evo})\}$, artifact evolutions, and
            dependencies
Data   : E event set, $\mathcal{V}$ views set, $AnalysisData = \{(Dist, Stories, D_{co-evo})\}$,
            degree of co-evolution threshold γ, file distance threshold δ, user
            defined queries queries

1 $Files \leftarrow \emptyset, Stories \leftarrow \emptyset, TimeSeries \leftarrow \emptyset, AnalysisData \leftarrow \emptyset, \mathcal{V} \leftarrow \emptyset,$
   $A_{evo}(f) \leftarrow \emptyset;$
   /* Preprocess VCS log   */
2 $E \leftarrow \texttt{PreprocessLog}(\mathcal{L});$
   /* Retrieve views on the project   */
3 **for** i *from* 1 **to** |queries| **do** $\mathcal{V} \leftarrow \mathcal{V} \cup \texttt{RetrieveView}(E, queries[i]);$
   /* Analyze project data   */
4 **foreach** *view* $V \in \mathcal{V}$ **do**
5   **foreach** *aggreagated event* $ae \in V$ **do**
6   **foreach** $f = f(ae), t = ats(ae), a = aac(ae), l = aak(ae) \in ae$ **do**
             /* Construct the artifact evolution set for the file   */
7              $A_{evo}(f) \leftarrow A_{evo}(f) \cup \{(t, a, l)\};$
             /* Construct the process using story mining   */
8              $Stories \leftarrow Stories \cup (f, \texttt{StoryMining}(l)));$
             /* Collect files and time series   */
9              $Files \leftarrow Files \cup \{f\};$
10             $TimeSeries(f) \leftarrow$ construct time series from $A_{evo}(f);$
11   **end**
12   **end**
13   **foreach** *pair of files* $i, j \in Files$ **do**
          /* Compute degree of co-evolution   */
14         $coEvoDegree \leftarrow \chi(TimesSeries(i), TimeSeries(j));$
          /* Compute file distances   */
15         $distance \leftarrow d(i, j);$
          /* Select based on user defined thresholds   */
16   **if** $coEvoDegree > \gamma$ **then** $D_{co-evo} \leftarrow D_{co-evo} \cup \{coEvoDegree\};$
17   **if** $distance > \delta$ **then** $Dist \leftarrow Dist \cup \{distance\};$
18   **end**
19     $AnalysisData \leftarrow AnalysisData \cup \{Dist, Stories, D_{co-evo}\};$
20 **end**
21 **return** $AnalysisData;$

The step in line 4 starts an iteration over the views set $\mathcal{V}$. Here is where
we collect the analysis data that are returned by the algorithm. For each of the
aggregated artifacts contained in a view V, we retrieve the information necessary
to compute the *degree of co-evolution* between pairs of files and their *file distance*.
First, we construct the artifact evolution of all the artifacts present in $ae \in V$.
Note that an aggregated event $ae \in V$ is a record obtained from a view on the
project which is composed, among other attributes (e.g., file, time, amount of

change), by the comment associated to the specific change. Comments describe multiple changes executed on the file, i.e. they describe a *story* of the artifact. Stories associated to each file are collected and the corresponding labels are chronologically ordered. These file stories are then input to the StoryMining technique [9]. Story Mining was designed to receive as input a story freely written by the participants, describing their work in a particular business process. As an output, the *actors* and the process *activities* executed by them are extracted. Our technique is concerned with the stories of the files. Therefore, they are the actors of the story mining, and the resulting business process consists of the steps describing their evolution process. We collect the resulting processes in the step in line 8. The step in line 10 is concerned with the construction of a time series from the set of artifact evolutions A_{evo} computed in line 7. Specifically, this step gathers the values of the changes of each of the artifact f in A_{evo} and records them in $TimeSeries(f)$.

After all the aggregated events ae have been explored, the algorithm moves on to computing the metrics (lines 13–18). In this loop, the algorithm iterates through all the pairs of files. For each pair, the *degree of co-evolution* and *artifact-distance* metrics are computed according the Definitions 4 and 5, respectively. These two measures are collected only if their values are above the user defined thresholds γ and δ. After the loop is over, the two measurements and the stories mined with the StoryMiner are stored in *AnalysisData*.

Finally, after iterating over all the user defined views, the algorithm returns the *AnalysisData* collection which can now be further inspected and analyzed in more detail, as we show next with an example.

3.3 Example

Let us consider the following example of a software development process. It contains 10 files arranged hierarchically as depicted by the file tree in Fig. 2. At the first level of the file tree there is the README.md file which describes the project. The software product in our case is called *running example* and is contained under the f_3 directory. The product consists of an example for software developers who want to organize their projects according to a predefined structure. The project has 21 commits over 10 days.

An excerpt of the VCS log for this project was illustrated in Table 1 above. The project managers are interested in understanding the work process done by project participants in each of the files and whether there is some hidden work dependency. We show how our technique meets the requirements by applying each step to this project and discussing the outcomes.

Let us suppose we have preprocessed our data and have the events set E already stored in a database. Then $\mathcal{V}$ is obtained by querying the data and aggregating them by day. Then, the *parent* relation is $Parent = \{(f_1, f_2), (f_1, f_3), (f_3, f_4), (f_3, f_6), (f_3, f_{12}), (f_4, f_5), (f_6, f_7), (f_6, f_8), (f_6, f_9), (f_9, f_{10}), (f_{10}, f_{11})\}$. Next, we compute the artifact evolution of for each artifact. For example, the artifact evolution of file REAMDE.md (f_2) limited on the information from

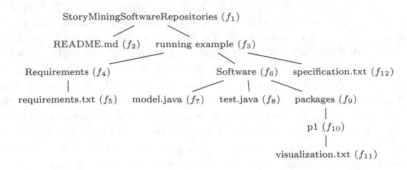

Fig. 2. File tree describing the file structure in our scenario of use.

Fig. 3. Example of business process showing the artifact evolution

Table 1 is $A_{evo} = \{$ (2017-01-31, 1, *Create readme file*), (2017-02-01, 3, *Add a license*), (2017-02-02, 1, *Updated the requirements*)$\}$. The resulting process from the story mining algorithm is shown in Fig. 3.

Next, we calculate the metrics. The dependencies are computed in the steps enclosed in lines 13–18 of Algorithm 1. E.g., the artifacts README.md (f_2) and test.java (f_7) appear in the *TimeSeries* collection as the vectors $\boldsymbol{X_{f_2}} = (1,3,1,0)$ and $\boldsymbol{X_{f_7}} = (0,0,0,2)$. We use the Pearson correlation between the to vectors $\sigma(\boldsymbol{X_{f_2}}, \boldsymbol{X_{f_7}} = -0.66)$ and take its absolute value as degree of co-evolution $\chi = |\sigma|$. Therefore, the *degree of co-evolution* between the considered artifacts is $\chi = 0.66$. The *file distance* is the length of the route from f_2 to f_7, i.e. $d(f_2, f_7) = \{(f_2, f_1), (f_1, f_3), (f_3, f_6), (f_6, f_7)\}$. Therefore, the file distance between README.md and test.java is $d(f_2, f_7) = 4$.

4   Evaluation

In this section, we show the applicability of our technique to project-oriented business processes and its effectiveness in uncovering work dependencies. With respect to the requirements formulated in Sect. 2, we evaluate against requirements R2 and R3 in Sect. 4.1 and against requirement R1 in Sect. 4.2.

We implemented our techniques as a prototype[2] and used it on 10 real world software projects with different sizes. The input of our program is a VCS log and the output is a set of analysis data with information about the evolution of the artifacts and their dependencies. We report the results in Table 3. The results are listed in increasing order of project size. The parameters χ and d are the metrics

---

[2] The source code is available at https://github.com/s41m1r/MiningVCS.

Table 3. Evaluation of real world projects. Respectively the thresholds are: χ^L if $\chi < 0.3$, χ^H if $\chi > 0.7$ low and high degree of co-evolution; d^L if $d \leq 2$, d^H if $d > 2$ respectively low and high distance.

Project	Commits	Files	χ^H	χ^L	(d^L,χ^L)	(d^L,χ^H)	$(d^H,\chi^L,)$	(d^H,χ^H)	$\overline{\|p_f\|}$	$max(\|p_f\|)$	$\|A_{evo}\|$	$\bar{d}$	$max(d)$
mwaligner	21	9	37	7	6	30	1	7	1.11	2	2.40	0.94	3
Biglist	202	15	22	90	31	18	59	4	1.47	3	2.76	1.20	5
camundaRD	11	15	74	26	0	25	26	49	2.18	4	2.05	2.03	7
graphql	256	30	89	357	121	89	236	0	1.40	2	3.18	1.11	4
jgitcookbook	135	89	773	2866	505	289	2361	484	6.93	8	1.33	2.68	14
mysqlpython	749	168	2288	11571	742	591	10829	1697	2.59	7	1.65	2.52	11
gantt	23	228	7006	14343	386	3480	13957	3526	3.30	4	1.71	2.16	7
facebookjavasdk	38	293	16478	26092	2017	16311	24075	167	6.21	8	4.78	5.58	13
caret	864	432	15366	60874	9538	14785	51336	581	3.01	4	3.15	1.60	7
operationcode	1114	1053	84024	444605	2291	5537	442314	78487	4.27	8	2.01	4.85	15

of *degree of co-evolution* and *distance*, respectively. In this example, $\chi > 0.7$ signifies that the co-evolution is high (χ^H) and $\chi < 0.3$ that the co-evolution is low (χ^L). As previously mentioned, this is a user customizable threshold that can be set by the domain expert. Likewise, the distance is considered low (d^L) when $d <= 2$ and high (d^H) when $d > 2$. The parameter $\overline{\|p_f\|}$ and $max(\|p_f\|)$ are respectively the average and the maximum lengths of the path to the root (i.e. average tree depth of the files). The column $\|A_{evo}\|$ shows the average number of activities in the process representing the artifact evolution. Lastly, the columns $\bar{d}$ and $max(d)$ report the average and maximum file distance, respectively. Next, we use these data for a quantitative evaluation of the projects.

4.1 Quantitative Evaluation

Here we address requirements R2 and R3. First, we compute project profiles. These profiles show the distribution of work-related dependencies in a project. Second, we evaluate whether the work on files can be predicted.

Before assessing project profiles, we make the following consideration. Our metrics define four classes: *(i)* low distance low co-evolution; *(ii)* high distance low co-evolution; *(iii)* low distance high co-evolution; *(vi)* high distance high co-evolution. Figure 4b helps clarifying these four classes. In fact, except for values of distance equal to 0, it is possible to see how the density of file pairs is higher when the distance is low. This is a normal situation in project where highly related files are stored closely to each other in the file system. Conversely, the dots on the top right of the plot mark files which are very distant to each other but still highly correlated. These can be, for instance, logical dependencies that can happen because of bad modularization of the project.

Hidden work dependencies belong to the last mentioned case, i.e. files are distant in the file tree but they have similar time series. According to this consideration we computed the project profiles in Fig. 4a. We observe three types of

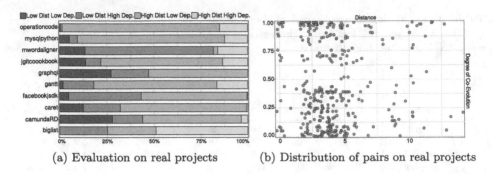

(a) Evaluation on real projects   (b) Distribution of pairs on real projects

Fig. 4. Characterization of the evaluated software projects

processes. First, several projects have hardly any hidden work dependencies. Second, several have a moderate degree between 10% and 20%. Third, the project *Biglist* has a high share of hidden dependencies. This hints at the possibility for better organizing the project according to good modularization best practices. That means, the project can be restructured in a way to reduce the unwanted side-effect the work on one file produces on other files.

Next, we evaluate whether the work on files can be predicted. Zipf's law is typically used in corpus analysis and states that the *frequency* of usage of any word is inversely proportional to its *rank* in the frequency table. This approach has already been applied to software projects for understanding whether the assignment of developers to tasks in a software project could be predicted [7]. Here, we focus on understanding whether the Zipf's law holds true also for work dependencies within a project.

To this end, we selected one big and one small project from Table 3, namely *Biglist* and *Caret*. Biglist is a small project on a list of strings which are known to cause issues when used as user-input data. Caret is a big project consisting in the development of a sublime text editor for Chrome OS. We collected how frequently were the artifacts worked on to generate a ranking. Figure 5 depicts the corresponding charts and the fitted Zipf distribution. We notice that both projects present a similar distribution of values. This holds also for the other projects analyzed. In particular, Zipf's law is valid for the most frequently changed files. Afterwards, the distribution drops because of files not being worked anymore but still being part of the project.

4.2   Qualitative Evaluation

In this section, we address requirement R1 by showing insights on the work history of files that are related. To this end we focused on the project *smsr*, which has 21 commits over a time span of ten days.

Let us consider an example where our technique proves helpful. Our technique finds 6 highly related pairs, as shown in Table 3. We excluded files that have

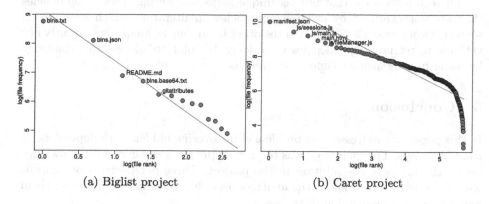

(a) Biglist project   (b) Caret project

Fig. 5. Zipf distribution of the worked files

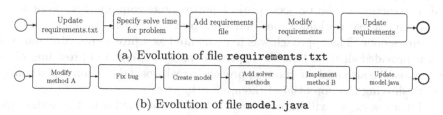

(a) Evolution of file `requirements.txt`

(b) Evolution of file `model.java`

Fig. 6. Processes of two work-dependent files

a functional dependencies, e.g. interface-class relations, where a change in the interface trivially brings change in the class. Thus, we were able to select the files `smsr/running example/Requirements/requirements.txt` and `smsr/running example/Software/model.java`, having $\chi = 0.7$ and $d = 4$. Moreover, by observing the content we verified that they do not have functional dependencies. Therefore, these two files are work dependent. Figure 6 shows the extracted processes after mining their stories. Interestingly, the two processes do not share any activity because they were never changed together in the same commit.

Our technique can fail under some circumstances. Consider the example above. We know that the files `requirements.txt` and `model.java` are work dependent. Let us now assume that the assumption of *regular commits* in the VCS does not hold. Nevertheless, we know that there is the following work pattern: *at irregular times, one change in the requirements produces 2 changes of work that must be implemented in model in the next day.* In a short time window of 4 days, the time series would be $X_{req} = (1, 0, 1, 0)$, $X_{model} = (0, 2, 0, 2)$ and their correlation is $\sigma(f_{req}, f_{model}) = -1$. Hence, they would score a high degree of co-evolution $\chi = 1$. However, if we double the time window and observe only another pattern the correlation would change. We get $X_{req} = (1, 0, 1, 0, 0, 1, 0, 0)$, $X_{model} = (0, 2, 0, 2, 0, 0, 2, 0)$ which score a $\sigma(f_{req}, f_{model}) = -0.66$, $\chi = 0.66$ and therefore not a high value of correlation.

These results show that our technique helps uncovering work dependencies that are not captured by existing approaches in literature which leverage on social network analysis [23, 25]. On the other hand, our technique is currently not yet able to retrieve dependencies with delay. We plan to address this challenge by using moving-average time series models.

5 Conclusion

In this paper, we addressed the problem of uncovering hidden work dependencies from VCS logs. The main goal was to provide project managers with knowledge about the artifacts co-evolution in the project. Three perspectives of analysis were considered, evolution of the artifacts over time, dependencies among them and structural organization of the project.

Our approach works under the assumptions that repositories reflect the hierarchical structure of the project, project participants commit their work regularly during active working times and they provide informative comments for the changes done. The approach was implemented as a prototype. A scenario of use was provided showing how the approach can be applied and providing some discussions. We also evaluated our approach in real-world data from open source projects showing the potential of the approach.

In future work, we will improve our evaluation varying for instance the time window, the dependency threshold and consider a study case with project managers. We plan to investigate other types of dependencies between artifacts. Specifically, we are interested in a semantic analysis of the work performed in both artifacts, considering for instance some similarity measures. We also aim to improve the visualization to consider other knowledge extracted, for instance the type of change performed in the aggregate events could be shown associated to the activities in the artifact process.

References

1. Abate, P., Cosmo, R.D., Boender, J., Zacchiroli, S.: Strong dependencies between software components. In: 3rd International Symposium on Empirical Software Engineering and Measurement ESEM, pp. 89–99 (2009)
2. Bala, S., Cabanillas, C., Mendling, J., Rogge-Solti, A., Polleres, A.: Mining project-oriented business processes. In: Motahari-Nezhad, H.R., Recker, J., Weidlich, M. (eds.) BPM 2015. LNCS, vol. 9253, pp. 425–440. Springer, Cham (2015). doi:10.1007/978-3-319-23063-4_28
3. Bani-Salameh, H., Ahmad, A., Aljammal, A.: Software evolution visualization techniques and methods - a systematic review. In: 2016 7th International Conference on Computer Science and Information Technology (CSIT), pp. 1–6 (2016)
4. Beheshti, S.-M.-R., Benatallah, B., Motahari-Nezhad, H.R.: Enabling the analysis of cross-cutting aspects in ad-hoc processes. In: Salinesi, C., Norrie, M.C., Pastor, Ó. (eds.) CAiSE 2013. LNCS, vol. 7908, pp. 51–67. Springer, Heidelberg (2013). doi:10.1007/978-3-642-38709-8_4

5. Beheshti, S.-M.-R., Benatallah, B., Sakr, S., Grigori, D., Motahari-Nezhad, H.R., Barukh, M.C., Gater, A., Ryu, S.H.: Process Analytics - Concepts and Techniques for Querying and Analyzing Process Data. Springer, Cham (2016)
6. Bender, M.A., Farach-Colton, M.: The LCA problem revisited. In: Gonnet, G.H., Viola, A. (eds.) LATIN 2000. LNCS, vol. 1776, pp. 88–94. Springer, Heidelberg (2000). doi:10.1007/10719839_9
7. Canfora, G., Cerulo, L.: Supporting change request assignment in open source development. In: Proceedings 2006 ACM Symposium on Applied Computing - SAC 2006, p. 1767, April 2016
8. D'Ambros, M., Lanza, M., Lungu, M.: Visualizing co-change information with the evolution radar. IEEE Trans. Softw. Eng. **35**(5), 720–735 (2009)
9. Gonçalves, J., Santoro, F.M., Baião, F.A.: Let me tell you a story - on how to build process models. J. Univers. Comput. Sci. **17**(2), 276–295 (2011)
10. Greene, G.J., Fischer, B.: Interactive tag cloud visualization of software version control repositories. In: 3rd Working Conference on Software Visualization, pp. 56–65 (2015)
11. Gubichev, A., Bedathur, S., Seufert, S., Weikum, G.: Fast and accurate estimation of shortest paths in large graphs. In: 19th ACM International Conference on Information and Knowledge Management, p. 499 (2010)
12. Kindler, E., Rubin, V., Schäfer, W.: Activity mining for discovering software process models. Softw. Eng. **79**, 175–180 (2006)
13. Lindberg, A., Berente, N., Gaskin, J., Lyytinen, K.: Coordinating interdependencies in online communities: a study of an open source software project. Inf. Syst. Res. **27**(4), 751–772 (2016)
14. Mittal, M., Sureka, A.: Process mining software repositories from student projects in an undergraduate software engineering course. In: ISCE Companion, pp. 344–353 (2014)
15. Oliva, G.A., Santana, F.W., Gerosa, M.A., de Souza, C.R.: Towards a classification of logical dependencies origins. In: Proceedings 12th International Workshop 7th Annual ERCIM Workshop on Principles of Software Evolution - IWPSE-EVOL 2011, p. 31 (2011)
16. Peffers, K.E.N., Tuunanen, T., Rothenberger, M., Chatterjee, S.: A design science research methodology for information systems research. J. Manag. Inf. Syst. **24**, 45–77 (2007)
17. Poncin, W., Serebrenik, A., Brand, M.V.D.: Process mining software repositories. In: 2011 15th European Conference Software Maintenance Reengineering, pp. 5–14 (2011)
18. Ripley, R.M., Sarma, A., Van Der Hoek, A.: A visualization for software project awareness and evolution. Visualization 2007 - Proceedings 4th IEEE International Workshop on Visualizing Software for Understanding Analysis, pp. 137–144 (2007)
19. Robles, G., González-Barahona, J.M., Cervigón, C., Capiluppi, A., Izquierdo-Cortázar, D.: Estimating development effort in free/open source software projects by mining software repositories: a case study of openstack. In: 11th Working Conference on Mining Software Repositories, pp. 222–231 (2014)
20. Ruohonen, J., Hyrynsalmi, S., Leppänen, V.: Time series trends in software evolution. J. Soft. Evol. Process **27**(12), 990–1015 (2015)
21. Voinea, L., Telea, A.: CVSgrab: mining the history of large software projects. In: Eurographics/EuroVisualization, pp. 187–194 (2006)
22. Voinea, L., Telea, A.: Visual data mining and analysis of software repositories. Comput. Graph. **31**, 410–428 (2007)

23. Weicheng, Y., Beijun, S., Ben, X.: Mining GitHub: why commit stops - exploring the relationship between developer's commit pattern and file version evolution. In: 20th Asia-Pacific Software Engineering Conference, pp. 165–169 (2013)
24. Zaidman, A., Van Rompaey, B., Demeyer, S., Van Deursen, A.: Mining software repositories to study co-evolution of production & test code. In: 1st International Conference on Software Testing, Verification and Validation, pp. 220–229 (2008)
25. Zimmermann, T., Nagappan, N.: Predicting defects using network analysis on dependency graphs. In: 13th International Conference on Software Engineering, p. 531 (2008)

Decisions and Understanding

Towards a Holistic Discovery of Decisions in Process-Aware Information Systems

Johannes De Smedt[1,2(✉)], Faruk Hasić[1], Seppe K.L.M. vanden Broucke[1], and Jan Vanthienen[1]

[1] Department of Decision Sciences and Information Management, Faculty of Economics and Business, KU Leuven, Leuven, Belgium
{johannes.desmedt,faruk.hasic,seppe.vandenbroucke, jan.vanthienen}@kuleuven.be
[2] Management Science and Business Economics Group, University of Edinburgh Business School, Edinburgh, UK
johannes.desmedt@ed.ac.uk

Abstract. The interest of integrating decision analysis approaches with the automated discovery of processes from data has seen a vast surge over the past few years. Most notably the introduction of the Decision Model and Notation (DMN) standard by the Object Management Group has provided a suitable solution for filling the void of decision representation in business process modeling languages. Process discovery has already embraced DMN for so-called decision mining, however, the efforts are still limited to a control flow point of view, i.e., explaining routing (constructs) or decision points. This work, however, introduces an integrated way of capturing the decisions that are embedded in the process, which is not limited to local characteristics, but provides a decision model in the form of a decision diagram which encompasses the full process execution span. Therefore, a typology is proposed for classifying different activities that contribute to the decision dimension of the process. This enables the possibility for an in-depth analysis of every activity, deciding whether it entails a decision, and what its relation is to other activities. The findings are implemented and illustrated on the 2013 BPI Challenge log, an exemplary dataset originating from a decision-driven process.

Keywords: Decision mining · Decision Model and Notation · Process mining

1 Introduction

The prevalence of new works on decision modeling and mining, as witnessed by the vast amount of new works on Decision Model and Notation [1], shows an increasing interest in documenting, modeling, and analyzing the decision dimension of processes. Many research efforts have pursued the discovery of the decision layer of processes already, including the seminal work on *decision mining* [2], and its extensions and improved versions [3–6]. Nevertheless, this

© Springer International Publishing AG 2017
J. Carmona et al. (Eds.): BPM 2017, LNCS 10445, pp. 183–199, 2017.
DOI: 10.1007/978-3-319-65000-5_11

form of decision mining is focusing on decision point analysis, i.e., the discovery of split-operators in the control flow which are dependent on certain data variables tied to the activities that need to be performed subsequently. Hence, the focus still lies with the extraction of control flow information, rather than decision information. Furthermore, issues arise when dealing with loops and non-local dependencies, i.e., decision variables that are not only affecting its subsequent routing construct(s), but also other variables processed by activities further down the process. This work complements the typical decision point analysis by rather constituting the different types of activities that are present in a process model and establishing how they contribute to the decision layer of the whole model. It proposes a framework for connecting decision variables which can be linked according to any control flow representation and any data mining algorithm by constructing decision requirement diagrams. It consists of a four step approach that decides how activities are influencing variables, classifying whether they form decisions, building a decision requirement diagram, and finally, building the control flow of the process by taking into account the decisions, rather than solely the routing of activities.

This paper is structured as follows. In Sect. 2, an overview of decision modeling and mining is constituted to frame the problem. In Sect. 3, the necessities for an integrated technique are introduced and illustrated, followed by Sect. 4, which introduces the approach for doing so. Section 5 outlines the implementation, as well as the application of the proposed technique to the 2013 BPI Challenge log. Finally, Sect. 6 concludes and discusses future work.

2   Decision Modeling and Mining

This section introduces and situates the concepts used for decision modeling and mining subsequently, i.e., first decision models are elaborated and formalized, next decision mining is discussed in more detail.

2.1   Decision Models and Related Work

The decision modeling approaches present in process management literature often breach the separation of concerns between control and data flow, hence negatively influencing maintenance and reusability. They do this by hard-coding and fixing the decisions in business processes [7]. Consequently, splits and joins in business processes are misused to represent typical decision artifacts such as decision tables. Recently, the separation of processes and decision logic has become an evident trend. Such an approach is supported by the DMN standard [1], since it has the clear intention to be used in conjunction with the Business Process Model and Notation (BPMN) [8]. Decoupling decisions and processes to stimulate flexibility, maintenance, and reusability, yet integrating decision and process models is therefore of paramount importance [9]. The DMN standard allows to model and describe decisions in a declarative way on two levels, the requirements level and decision logic level. For the first level decisions requirement diagrams (DRD) are used to represent the information requirements of the decisions in the model. These diagrams can consist of several types of elements, decisions, input data, business knowledge models, and knowledge sources. Information requirements in the DRDs represent the requirements of decisions in terms of subdecisions and input data, depicted using arrows going from the requirement to the decision. The second level uses the FEEL expression language to describe the decision logic behind every decision. The FEEL language allows to write executable decisions in a declarative language.

Besides DMN, also the Product Data Model (PDM) [10] is a well-known language to capture the dependencies that exist between decisions and their input in workflows. DMN, however, is more driven by the decision and its rationale compared to PDM, which rather focuses on the data and its impact on the workflow.

To support our approach we introduce a formal basis for decisions and requirements in DMN models. We take abstraction from the use of Business Knowledge Models and Knowledge Sources, as defined in the DMN standard. However, all definitions and theorems provided can be readily extended to include the use of these concepts.

2.1.1 Formal Definition

A DMN model can be represented as follows. We adopt the definition of *decisions* and *decision requirement diagrams* from [9].

Definition 1. *A decision requirement diagram DRD is a tuple (D_{dm}, ID, IR) consisting of a finite non-empty set of decision nodes D_{dm}, a finite non-empty set of input data nodes ID, and a finite non-empty set of directed edges IR representing the information requirements such that $IR \subseteq D_{dm} \cup ID \times D_{dm}$, and $(D_{dm} \cup ID, IR)$ is a directed acyclic graph (DAG).*

The DMN specification allows a DRD to be an incomplete or partial representation of the decision requirements in a decision model. The complete set of requirements is derived from the set of all DRDs in the decision model.

Definition 2. *The decision requirements level R_{DM} of a decision model DM is the set of all decisions requirement diagrams in the model.*

The information contained in this set can be combined into a single DRD representing the entire decision requirements level. The DMN standard calls such a DRD a decision requirement graph (DRG). We extend the notion of a DRG, in such a way that a DRG is a DRD which is self-contained, i.e. for every decision in the diagram all its requirements are also represented in the diagram.

Definition 3. *A decision requirement diagram $DRD \in R_{DM}$ is a decision requirement graph DRG if and only if for every decision in the diagram all its modeled requirements, present in at least one diagram in R_{DM}, are also represented in the diagram.*

According to the DMN standard a decision is the logic used to determine an output from a given input. In BPMN a decision is an activity, i.e. the act of using the decision logic. Another common meaning is that a decision is the actual result, which we call the output of a decision. We define a decision using its essential elements.

Definition 4. *A decision* $d \in D_{dm}$ *is a tuple* (I_d, O_d, L), *where* $I_d \subseteq ID$ *is a set of input symbols,* O_d *a set of output symbols and* L *the decision logic defining the relation between symbols in* I_d *and symbols in* O_d.

In case of decision tables, a commonly used reasoning construct in decision models, I_d and O_d contain the names of the input and output elements, respectively, and L is the table itself, i.e. the set of decision rules present in the table. Note that, since a DRD is a DAG, $I_d \cap O_d = \emptyset$.

2.2 Decision Mining and Related Work

In recent business process management literature, decision mining arises as a frequent term. It was first introduced in process literature in the work of [2]. The work derives and describes the routing in so-called decision points in Petri nets [11] through a decision tree algorithm. The main idea is to use the control flow data to determine the overall structure of the process first, and consequently use the instances' attributes to define where the data had an impact on the work flow. Following this seminal work, numerous other studies have been dedicated to refining decision point analysis and assessing variations of the problem [3,5,6,12]. The most holistic outcomes are provided by [4,13]. The former mines for read and write operations on the variables and relates them to the guards of the different activities present in a Petri net, obtaining a data-aware Petri net. The latter incorporates XOR-splits in the decision model which also consists of data attributes, which are either considered inputs or decisions themselves. This way, a combination of attributes and control flow elements is found in the form of a DMN model. A different approach is to mine for the mental actions performed by decision makers [14], captured in a Product Data Model [10].

Contrary to focusing on the control flow, other works exist that rather start from the data perspective while either incorporating control flow for clarification, or by structuring the results. In [15] a general framework for correlating business activity variables and process variables is proposed, and in [12], the resource perspective is mixed with the control flow for recommendations of future executions. In [16], Guard Stage Milestone models [17] are mined by extracting business objects and enriching them with their lifecycle information. Nevertheless, these approaches do not focus on deriving the decision rationale that is present in the process.

In [18], a framework to position all these works was proposed. This framework, depicted in Fig. 1, consists of two dimensions, i.e., the decision control flow dimension and the decision model maturity dimension, to classify each approach into four quadrants. The presence of an elaborate decision control flow dimension is depicted along the vertical axis. Typical data mining approaches belong in Q1, as they do not incorporate dynamic data aspects. On the other hand, Q2 represents an approach where the primary objective is to derive the control flow of activities by fitting process models such as Petri nets [19]. Along the horizontal axis, the decision model maturity dimension is pictured. This dimension evaluates the presence of a decision model. In Q1 and Q2, no such model is available, while a decision model is present in Q3 and Q4. Quadrants Q3 and Q4 differ in prioritisation, as in Q3 the decision model is not orthogonally connected to the process, but rather parts of the decision model are incorporated in segments of the control flow. Hence, a holistic decision model is absent in Q3. On the contrary, Q4 approaches provide a holistic decision model that incorporates all the decisions made and that can be reused throughout the process.

Clearly, the approaches building on [2] display strong abilities to extract the control flow and relating data variables to its routing elements. Other approaches provide a strong decision model output, but do not focus on how the decision was established throughout the process. Hence, there is a gap between strong control flow-driven and decision model-driven approaches, as the challenge is to develop a decision mining approach that is driven by the decision model, rather than by the control flow containing decision points. In [20] both event labels and data attributes are considered, as dependency conditions are discovered using classification and embedded in process discovery. The information on the discovered rules annote the resulting process models. Hence, this method hovers between Q3 and Q4. The approach in [16] constructs artifacts by correlating the data of the events. The control flow over these artifacts is mined as well and the outcome is presented in a holistic model containing both layers. Consequently, this is the only approach to the authors' knowledge that truly belongs in Q4, as it both handles the complexity of the data and the dynamic behavior of its activity generators.

In this paper, we will address the research gap in Q4 by introducing the *Process Mining Integrating Decisions* (**P-MInD**) framework by focusing on constructing DRDs which are compatible with process models that are activity diagram-based (e.g. Petri nets and BPMN).

2.2.1 Event Logs

Process mining and its related techniques employ the notion of the event log to define the structure of data suitable for activity- and case-based discovery.

Definition 5. *An event log is a tuple (E, A, λ, V, var, Val, $\mathcal{L}$), where:*

- *E is a set of events.*
- *A is a set of activities (event types).*
- *$\lambda : E \rightarrow A$ is a labelling function mapping events to activities.*
- *V is a set of variables.*
- *var $: E \rightarrow 2^V$ is a function mapping events to the subset of variables used in this event.*
- *For each $v \in V$ a partial function $val_v : E \rightarrow dom_v$ mapping events to values in the domain of v. We denote the set of these partial functions as Val.*
- *$\mathcal{L} \subseteq \bigcup_{n \in \mathbb{N}} E^n$ the set of event tuples in the log.*

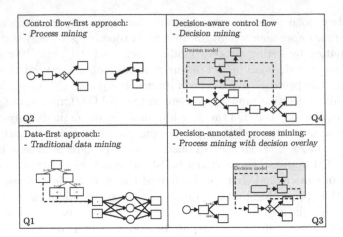

Fig. 1. The decision mining quadrant.

For brevity we use var(a) = {v ∈ var(E)|λ(E) = a}, i.e. var(a) for a ∈ A denotes the variables for the event label with a.

Typically, special variables include the timestamp $(t \in V)$ and resource $(res \in V)$. The timestamp is denoted $T(e) = val_t(e)$.

Consider for example the set of events $E = \{e_1, e_2, e_3, e_4, e_5\}$, a set of variables $V = \{res, time, docid\}$, and a set of activity labels $A = \{register, send, receive\}$. Then $\mathcal{L} = \{(e_1, e_2), (e_3, e_4, e_5)\}$, with $\lambda(e_1) = register$, $\lambda(e_2) = send$, $\lambda(e_3) = receive$, $\lambda(e_4) = send$, $\lambda(e_5) = receive$, $var(e_1) = \{res, time\}$, $var(e_2, e_3) = \{res, time, docid\}$, $var(receive) = \{res, time, docid\}$, and $val_{res}(e_1) = john$, $val_{res}(e_2, e_3) = ann$, $T(e_1) = 1, T(e_2) = 4, T(e_3) = 9$.

3   Business Process Activities and Their Relation to Decisions

In this section, a typology for different activities used for making decisions in processes is proposed, as well as a running example of a decision model intertwined with a process model.

3.1   Business Activities

Decisions do not surface solely as the driver of control flow. Rather, they both encompass the routing of cases, i.e., because of decision outcomes that steer toward a certain activity tailored towards supporting its output, and the changes in the data layer of the process as well. The latter introduces numerous types of activities that are representatives of the *decision* model in the *process* model:

Definition 6. *The input and output data variables of business activities are defined as follows:*

– $I : A \rightarrow V$, *function assigning activities which receive input of a certain variable,*
– $O : A \rightarrow V$, *function assigning activities which deliver output for a certain variable.*

This enables the construction of the following activity types:

1. **Operational activities ((no) inputs, no outputs):** do not have any influence on the process' decision dimension and only act as a performer of a specific action that is tied to that specific place in the control flow. They might serve as the end of a decision. They are provided with the decision inputs needed, which are not used further in the process,
   $A_o = \{a \in A \mid O(a) = \emptyset\}$.
2. **Administrative activities (no inputs, outputs):** have the purpose to introduce decision inputs into the process,
   $A_a = \{a \in A \mid I(a) = \emptyset \wedge O(a) \neq \emptyset\}$.
3. **Decision activities (inputs, outputs):** serve a true autonomous decision purpose as they transform decision inputs into a decision outcome,
   $A_d = \{a \in A \mid I(a) \neq \emptyset \wedge O(a) \neq \emptyset\}$.

It holds that $A_a \cup A_o \cup A_d = A$. Typically, the decision points that are used for decision mining in processes are of the decision activity type, but tailored towards deciding which activity should be performed next based on the event labels. Note that these are not included in V.

We can now make the connection with decisions and decision models.

Definition 7. *A decision in a business process can be defined as follows:*

– *A decision in a process model,* $d^a \in D_{dm}$ *is a tuple* $(I_{d_a}, O_{d_a}, L_{d_a})$, *where* $a \subseteq A_d$, $I_{d_a} \subseteq I(a)$, $O_{d_a} \subseteq O(a)$, *and* $L_{d_a} \subseteq L$.

This last definition connects a decision activity with a decision and it shows than one decision activity can be tied with multiple decisions. The latter implies that, within an event log, the same activity can make different decisions, i.e., changes in variable values, and can be represented as different decision nodes within a decision model, as well as different activity types. This interpretation of how activities are present in process models is the main difference with other decision mining techniques, who keep the one-to-one mapping of activities and decisions.

3.2 Running Example

Consider the example BPMN model in Fig. 2 and the corresponding decision model, which illustrates the different activity types elaborated earlier, in Fig. 3. The process model contains a simple control flow with an AND-split and -join, as well as an XOR-split and -join that gets repeated later on. In the first part of the model, two variables are set, i.e., *Retrieve liability (RL)* and *Retrieve category (RC)* set the liability score and risk category respectively. Since they do not contain any inputs, they are administrative activities. *Determine risk (DR)* uses those variables to set the risk score, hence it is a decision activity with $I_D = \{liability\,score, risk\,category\}$, $O_D = \{risk\,score\}$. The risk score serves as an input for *Evaluate risk level and case (ERLC)*, which also uses the *Case characteristics (CC)* as an input (they are undefined by default). It does not have output variables, however, it decides which activity is performed next. Notice that this is the typical example of a decision point [2]. The label of the subsequent activities is not part of V, hence *ERLC* is not a decision activity, but an operational activity. *Subscribe policy B (SPB)* uses the *CC*, but has no output, hence it is an operational activity. Notice that *Subscribe policy A (SPA)* is the most convoluted activity. It serves both as a decision activity, as it sets the *CC* based on the liability score, as well as an administrative activity when it sets *CC*. Note that this is because there can exist no overlap between the inputs and outputs of decisions, hence there exist two instantiations of the activity and the model is capable of revealing how the activity contributes to the decision-making over the different iterations of the loop. Finally, *Print category (PC)* and *Archive claim (AC)* are two operational activities.

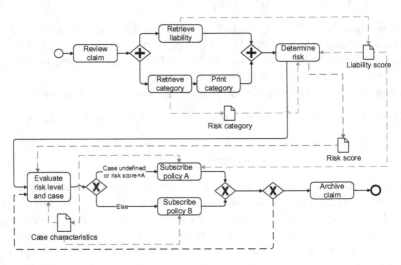

Fig. 2. BPMN model representing a liability claim process based on different decisions throughout.

In Fig. 3 the corresponding DRDs are provided. In DMN, decision activities are the typical nodes present in the model. Only a conjunction of inputs and a rationale to make decisions based on them is incorporated. Nevertheless, in this representation the administrative (indicated with ◇) and operational (indicated with ⊟) activities are added in gray to illustrate how the input data and all the activities in the process are related. Only *DR* and *SPA* fully process their inputs into a *Risk score (RS)*. *ERLC* has inputs, but has the label of the subsequent activities as an output. This is the typical decision point analysis approach.

However, decision point analysis techniques would not be able to discover the long-distance decision dependencies of *Liability score (LS)* and *Risk category (RiC)* with their respective administrative and decision activities, as they only resolve calculations in areas of a process model that introduce XOR-gates.

4 Discovering Decision Models

In this section, an approach is introduced to mine DRDs from event logs. The main driver of the approach is the classification of activities into the different types that were discussed previously, which are matched afterwards with how they influence the different data variables in the log.

4.1 P-MInD Approach

In order to obtain a decision model from an event log, the decisions need to be derived from the activity information in the event log first. The exact steps the approach follows are outlined in Algorithm 1.

4.1.1 Step 1: Evaluate Activity Involvement

Every event is scrutinized and information regarding its variables is stored. In order to get a grasp on the effect a particular event type, i.e., activity, has on a certain variable, it is checked whether the value of a particular variable is changed. Note that the approach assumes that this data is fully and correctly recorded in the event log. This is done by comparing the previous value of the variable in the previous event $(e_{t-1} = l : T(l) < T(e), T(l) > T(f), \forall f, T(f) < T(e))$ and the current event (line 6) in case the event in question is not the start point of the particular execution trace. This is done for every event, and populates the shift metric for a certain activity. This shift is defined as $S : (V, A) \to \mathbb{N}$, i.e., the number of shifts in value of a variable $v \in V$ for an activity $a \in A$. A shift threshold, s_t, is offered to the user to adjust the sensitivity of the algorithm to take a certain variable under scrutiny. The user can also opt to exclude variables, V_u, in order to avoid the inclusion of, e.g., exogenous variables such as time stamps of events.

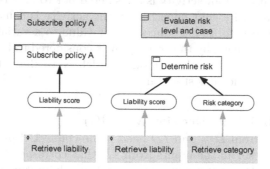

Fig. 3. The corresponding DMN model based on the process in Fig. 2.

Algorithm 1. Mining a DMN model from an event log

1: **procedure** MINE_DMN_MOD($\mathcal{L}, s_t, V_u, minsup$)   ▷ Input: Log and parameters
2:    $D_{dm}, ID \leftarrow \emptyset, S(V, A) \leftarrow 0$ ▷ Initialize variables
3:    **for** $e \in \mathcal{L}$ **do**
4:        **if** $e_{t-1} \neq \emptyset$ **then**   ▷ Skip first event to avoid non-existing e_{t-1}
5:            **for** $v \in var(e) \setminus V_u$ **do**   ▷ V_u excludes user-set variables
6:                **if** $val_v(e_{t-1}) \neq val_v(e)$ **then**
7:                    $S(v, a) + +$ ▷ Raise the variable's shift counter
8:        **for** $a \in A$ **do** $(D_{dm}, ID) \cup$ buildModels($a, s_t, minsup$)
9:    **return** buildDMNmodel(D_{dm}, ID)

4.1.2   Step 2: Classify the Activities

Once the influence of activities over variables is extracted, it can be used to determine whether a rationale between input and output of an activity exists. The shift metric is used to determine whether an activity actually altered the value of the variable, as outlined in Algorithm 2. If this is the case, i.e., the variable a is believed to have changed, then a predictive model is built which takes a as the target variable. Note that any type of predictive model can be used, e.g., decision trees, neural networks, SVMs, and so on. All other variables, i.e., $var(a) \setminus a$, are used as independent variables to determine the value of a. The downside of doing this, however, is that variables which are set together should not serve each other's predictive model as they are completely dependent of each other. This is somewhat remedied in step 3, however, cannot be fully avoided in the current approach. The evaluation of the model L is then considered to justify whether there was a causal link between the newly-set variable (v), and the other ones ($var(a) \setminus v$). For this, the Area Under Receiver Operating Curve (AUROC) value is evaluated. In case the value $L.AUROC$ is high enough, determined by the adjustable parameter $minsup$, the value is considered to be determined by the activity, which gets saved as a decision node $d^a = (v_L, v, L)$ in D_{dm} with v_L the significant independent variables of the predictive model and L the decision logic (e.g. a decision table). Note that only a singleton is considered for output, and the decisions are not multi-objective due to the nature of the predictive models used. If an activity is considered not to be a decision activity, but the event witnessed a shift nonetheless, the activity is considered to be an administrative activity, for it introduces a new value to a variable. It is stored in ID. If no shifts are made by the activity, it is considered to be an operational activity and out of scope for the decision model. Note that at any time, the corresponding decision logic is stored in L.

4.1.3   Step 3: Build the Decision Model(s)

Next, the elements from D_{dm} and ID need to be connected by IR to obtain a DMN model. To do so, all the inputs from ID that correspond with the inputs of the decisions in D_{dm} are connected, as well as the outputs of decisions in D_{dm} that also correspond with the inputs of other decisions. This is shown in

---

Algorithm 2. Constructing relations between variables of an activity

---

1: **procedure** BUILDMODELS($a, s_t, minsup$)   ▷ Input: event log and parameters
2:     **for** $v \in var(a)$ **do**
3:         **if** $S(v, a) > s_t$ **then**
4:             $L \leftarrow buildPredictiveModel(a, var(a) \setminus v)$
5:             **if** $L.AUROC > minsup$ **then**   ▷ Check if the model is explanatory
6:                 $D_{dm} \leftarrow d^a = (v_L, v, L)$ ▷ Save decision as decision node
7:             **else** $ID \leftarrow (a, v)$ ▷ Save variable as input node set by a
8:     **return** (D_{dm}, ID)

Algorithm 3. For every relation between two decisions, it is checked whether the sequence of the relation is correct, i.e., the decision input is indeed decided before the decision is used as an input, as shown as $a_1 < a_2$ on lines 4 and 6. This can be done in numerous ways, according to, e.g., the number of times the decision delivering the input is followed by the decision using the input. This somewhat counters the effect of correlating variables that are set together, as discussed in step 2, because although they are related the check identifies whether there has been a previous value on which the shift might have been based. This can also indicate that a variable is influencing its own future value, such as is the case for CC in the running example. The number of DRDs depends on whether all components are connected, or not. Noteworthy is that multiple decisions can happen simultaneously, as an activity can set multiple variables at the same time. Control flow-based approaches do not incorporate this possibility.

Algorithm 3. Constructing the output DMN model

1: **procedure** BUILDDMNMODEL(D_{dm}, ID)
2:     **for** $d_1^a = (I_1, o_1, L) \in D_{dm}$ **do**
3:         **for** $d_2^a(I_1, o_1, L) \in D_{dm}$ **do**
4:             **if** $o_1 \in I_2 \wedge a_1 < a_2$ **then** $IR \leftarrow (a_1, a_2)$
5:             **else if** $o_2 \in I_1 \wedge a_1 > a_2$ **then** $IR \leftarrow (a_2, a_1)$
6:         add all $i \in I_1$ that were not added by other decisions
7:     **return** (D_{dm}, ID, IR) ▷ The DMN model as DRD

4.1.4   Step 4: Mine the Control Flow of the Decisions

The final step from the P-MInD approach exists in substituting all the occurrences of the activities, once classified, with the corresponding decision nodes from the DRD in the event log. I.e., According to which values are set in a certain instantiation of the activity in the event log, the appropriate decision variant of that activity setting that value replaces the generic activity label. If multiple variables are set at the same time, they are merged in one label. This way, the control flow over the decisions can be mined directly as well. It can also be used to verify the relations between the decisions, i.e., lines 4 and 6 in Algorithm 3. Any process mining technique that mines control flow, e.g., Inductive Miner [21] can be used towards this outcome. Hence, the event log forms the source for both the decision and the process model, which gets extended with decision information.

The P-MInD approach can be considered a framework for integrated decision and process mining as numerous placeholders are present in the steps. Both the inference of connections between the different inputs and outputs of decisions, as well as the control flow perspective can be adjusted according to the most appropriate algorithms. However, the 4-step approach provides a fundamental basis for obtaining an integrated model that contains a decision model that spans the whole control flow and in which long distance dependencies and loops in the control flow do not clutter the decision model.

4.2 Application to Running Example

Consider the running example in the case of it being recorded in an event log $\mathcal{L}$. The algorithm will first evaluate all events ($e \in \mathcal{L}$) and classify the activities. There are shifts ($S(v, a) > 0$) of variable values for activities RL, RC, DR, and SPA. In step 2, predictive models are built for all the variables for which the values shifted ($S(v, a) > s_t$). Models with a significant AUROC can be trained, i.e., D_{dm} gets $d^{DR} = (\{LS, RiC\}, RS, L_{DR \to RS})$ and $d^{SPA}(LS, CC, L_{SPA \to CC})$, in case no noise is present. The other activities with shifts are administrative activities and are considered as input nodes, i.e., ID gets $\{(RL, LS), (RC, RiC)\}$. In step 3, the models are constructed by connecting the input and output nodes of the different decision models where RL and RC serve as inputs for DR and LS serves as input for SPA. This way, the same DRD as in Fig. 3 is obtained, without the operational activities. Finally in step 4, all the decision activities in D_{dm} replace the activity instances of DR and SPA. Note that in this case, there are no multiple variants of the activities as they only decide on one variable.

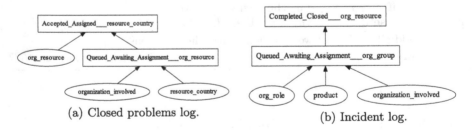

(a) Closed problems log.   (b) Incident log.

Fig. 4. DRDs mined from the 2013 BPI challenge logs with $s_t = 0.1$ and $minsup = 0.8$.

5   Implementation and Empirical Evaluation

In this section, an overview of the implementation and empirical evaluation is given. Additionally, a concise comparison with existing techniques is provided.

5.1   Implementation

The concepts of Sect. 4 are implemented in Java and can be used with any XES file [22]. A working prototype can be found at http://processmining.be/PMInD. The current models are built using the decision table plugin of the Weka data mining toolbench[1], however, any inference technique can be used to build the decision model. The output is displayed as a (set of) DRDs and can be set to display the decision logic layer in form of a decision table as well.

5.2   Evaluation

The approach was applied to the event logs made available for the 2013 BPI Challenge[2]. No extra pre-processing was performed, i.e., the unaltered files were used to create the output. The contents pertain to an incident management log at Volvo IT Belgium, in which cases are assigned a certain status according to their nature and urgency. This type of process is typically very decision-driven and provides a suitable example to illustrate how the decision model surrounding the process can be constructed. The log is used to evaluate the first three steps of the P-MInD framework and the output can be found in Fig. 4. A Petri net representing the control flow model for the closed problems log, mined with Inductive Miner (noise threshold 0.2) [21] and annotated with read and write operations (standard settings) [13] is depicted in Fig. 5.

The results for the incident and closed problems logs are shown. Small, two node-DRDs are excluded from the results (hence also the DRD for the open problems variant). In the DRDs, it can be seen how the variables are actually used to decide on other variables. Furthermore, it creates a picture of how the loops should be interpreted. Many decisions in the DRDs are re-initiated many times, and it might be that, e.g., the resource's country that deals with a certain case is determined, and later redetermined, as is the case for *Queued Awaiting Assignment* and *Accepted Assigned*. Finally, multiple decisions can occur at the same time. The downside of relating dependent variables, e.g., *resource_country* and *org_resource* are frequently set together and therefore should not be included in each other's predictive model, is present. However, the DRD still explains how the re-occurrence of the decision activity *Queued_awaiting_assignment* does influence the decisions regarding the variables' values over time. This is invisible to control flow-based approaches.

Other techniques for decision (point) analysis have a hard time to capture the intrinsics of the relations between the activities as their results are still

---

[1] http://weka.sourceforge.net/doc.dev/weka/classifiers/rules/DecisionTable.html.
[2] http://dx.doi.org/10.4121/uuid:a7ce5c55-03a7-4583-b855-98b86e1a2b07.

intertwined with the specific areas of a process model that contain XOR-gates. Indeed, the control flow needs many invisible activities and extra places (which constitute decision points) to tailor for the convoluted control flow, while the data, and hence the decisions, are driving the process. In Fig. 5, it is illustrated that the technique of [13] has a hard time deriving interesting guards from the control flow. In Fig. 4, the result of applying the approach from [4] shows that, as a result of the unclear interplay of control flow constructs, only a single DRD can be constructed because of the presence of loops and convoluted decision dependencies, as well as nominal data variables. They prevent the algorithm from finding relations between data decisions (attributes), and control flow decisions. In this respect, P-MInD is better capable of giving insights into how the process actually evolved. By applying step 4 and replacing activities by their decision variants, no clearer control flow could be retrieved than in Fig. 5, confirming that the process is decision-driven (Fig. 6).

5.3   Comparison with Existing Techniques and Limitations

Contrary to [2–4], P-MInD considers the activities as the main contributors of the decision model, i.e., the decisions are made by the activities, rather than focusing on how the control flow has decided on how and where the decisions are made. Unlike P-MInD, techniques based on decision point analysis are not able to discover long-distance decision dependencies (with the exception of [4]), as they only focus on the XOR-gates of a process model. Besides, P-MInD also supports the occurrence of multiple decisions at the same time, as activities are

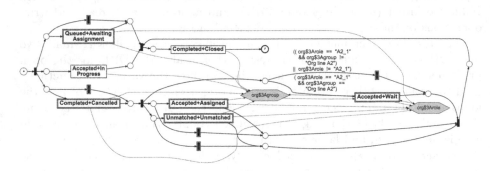

Fig. 5. Petri net mined with Inductive Miner to visualize the control flow and annotated with read and write operations.

Fig. 6. DRD mined with the approach proposed in [4].

able to set multiple variables simultaneously. Furthermore, the interplay between the *minsup* and s_t threshold provides a way to deal with noise, rather than only incorporating results of a perfectly fitting decision tree inference. The retrieval of reading and writing operations as in [13] is similar, however, the way in which they are related to the activities is different. In P-MInD, the variables are incorporated into the activities to form decisions. The focus is on the relation between the attributes through the activities, rather than towards determining the guards of the activities. While the approach of [4] is also capable of finding long distance dependencies and mining DRDs, it suffers from incorporating control flow, which can clutter up DRDs and is incapable of displaying loops. Furthermore, attributes are considered decisions, while P-MInD considers the activities as the drivers of decisions. Overall, P-MInD does not heavily rely on control flow information, although it is incorporated in steps 3 and 4 (by the $<$ relations). It adheres to the separation of concerns between decisions and processes. Hence, P-MInD can be categorised in the fourth decision mining quadrant of Fig. 1. Nevertheless, decision point analysis is compatible with P-MInD. By mining for the exact locations where certain decisions are made, the DRDs can be refined, or augmented with routing information.

The major limitation of the technique, however, stems from its independence of control flow. As illustrated before, it requires a more profound explanation of loops to avoid correlating dependent variables that are set at the same time, and does not use a strong way to incorporate sequence information in the DRD.

6   Conclusion and Future Work

This work revised the way in which a holistic decision model for process-driven environments can be retrieved. First of all, a classification of process activities was made to bridge the gap with decision model constructs. Next, an approach for retrieving DRDs based on the concept of operational, administrative, and decision activities was proposed. The approach was evaluated on the 2013 BPI Challenge log to illustrate the empirical usefulness of the framework. The results show that it is better capable of representing the decision layer of a process than existing techniques, as it does not solely rely on control flow, hence allowing for different insights into how data variables and decisions are related to activities, over long distance dependencies and loops as well.

In future endeavors, it will be investigated in what way the decision model can aid in refactoring the process model, according to the findings of [9]. This way, redesign can be suggested automatically. Furthermore, it will be tested which inference techniques are the most suitable to refine the retrieval of decision models, as P-MInD was only tested with decision table learning. Finally, while it is now assumed that the shifts hold over the whole model and can be conjoined for a global decision model, it will be investigated how an event log can be broken down according to the shifts, and the models that correspond to the particular decisions they are tied to.

References

1. OMG: Decision Model and Notation (2015)
2. Rozinat, A., van der Aalst, W.M.P.: Decision mining in ProM. In: Dustdar, S., Fiadeiro, J.L., Sheth, A.P. (eds.) BPM 2006. LNCS, vol. 4102, pp. 420–425. Springer, Heidelberg (2006). doi:10.1007/11841760_33
3. Batoulis, K., Meyer, A., Bazhenova, E., Decker, G., Weske, M.: Extracting decision logic from process models. In: Zdravkovic, J., Kirikova, M., Johannesson, P. (eds.) CAiSE 2015. LNCS, vol. 9097, pp. 349–366. Springer, Cham (2015). doi:10.1007/978-3-319-19069-3_22
4. Bazhenova, E., Buelow, S., Weske, M.: Discovering decision models from event logs. In: Abramowicz, W., Alt, R., Franczyk, B. (eds.) BIS 2016. LNBIP, vol. 255, pp. 237–251. Springer, Cham (2016). doi:10.1007/978-3-319-39426-8_19
5. de Leoni, M., Dumas, M., García-Bañuelos, L.: Discovering branching conditions from business process execution logs. In: Cortellessa, V., Varró, D. (eds.) FASE 2013. LNCS, vol. 7793, pp. 114–129. Springer, Heidelberg (2013). doi:10.1007/978-3-642-37057-1_9
6. Mannhardt, F., de Leoni, M., Reijers, H.A., van der Aalst, W.M.P.: Decision mining revisited - discovering overlapping rules. In: Nurcan, S., Soffer, P., Bajec, M., Eder, J. (eds.) CAiSE 2016. LNCS, vol. 9694, pp. 377–392. Springer, Cham (2016). doi:10.1007/978-3-319-39696-5_23
7. Vanthienen, J., Caron, F., De Smedt, J.: Business rules, decisions and processes: five reflections upon living apart together. In: Proceedings SIGBPS Workshop on Business Processes and Services (BPS 2013), pp. 76–81 (2013)
8. OMG: Business process model and notation (BPMN) 2.0 (2011)
9. Janssens, L., Bazhenova, E., Smedt, J.D., Vanthienen, J., Denecker, M.: Consistent integration of decision (DMN) and process (BPMN) models. In: CAiSE Forum. CEUR Workshop Proceedings. vol. 1612, pp. 121–128. CEUR-WS.org (2016)
10. Vanderfeesten, I., Reijers, H.A., van der Aalst, W.M.P.: Product based workflow support: dynamic workflow execution. In: Bellahsène, Z., Léonard, M. (eds.) CAiSE 2008. LNCS, vol. 5074, pp. 571–574. Springer, Heidelberg (2008). doi:10.1007/978-3-540-69534-9_42
11. Murata, T.: Petri nets: properties, analysis and applications. Proc. IEEE 77(4), 541–580 (1989)
12. Kim, A., Obregon, J., Jung, J.-Y.: Constructing decision trees from process logs for performer recommendation. In: Lohmann, N., Song, M., Wohed, P. (eds.) BPM 2013. LNBIP, vol. 171, pp. 224–236. Springer, Cham (2014). doi:10.1007/978-3-319-06257-0_18
13. de Leoni, M., van der Aalst, W.M.: Data-aware process mining: discovering decisions in processes using alignments. In: Proceedings of the 28th Annual ACM Symposium on Applied Computing, pp. 1454–1461. ACM (2013)
14. Petrusel, R., Vanderfeesten, I., Dolean, C.C., Mican, D.: Making decision process knowledge explicit using the decision data model. In: Abramowicz, W. (ed.) BIS 2011. LNBIP, vol. 87, pp. 172–184. Springer, Heidelberg (2011). doi:10.1007/978-3-642-21863-7_15
15. de Leoni, M., van der Aalst, W.M.P., Dees, M.: A general framework for correlating business process characteristics. In: Sadiq, S., Soffer, P., Völzer, H. (eds.) BPM 2014. LNCS, vol. 8659, pp. 250–266. Springer, Cham (2014). doi:10.1007/978-3-319-10172-9_16

16. Popova, V., Fahland, D., Dumas, M.: Artifact lifecycle discovery. Int. J. Cooper. Inf. Syst. **24**(01), 1550001 (2015)
17. Hull, R., et al.: Introducing the guard-stage-milestone approach for specifying business entity lifecycles. In: Bravetti, M., Bultan, T. (eds.) WS-FM 2010. LNCS, vol. 6551, pp. 1–24. Springer, Heidelberg (2011). doi:10.1007/978-3-642-19589-1_1
18. De Smedt, J., vanden Broucke, S.K.L.M., Obregon, J., Kim, A., Jung, J.-Y., Vanthienen, J.: Decision mining in a broader context: an overview of the current landscape and future directions. In: Dumas, M., Fantinato, M. (eds.) BPM 2016. LNBIP, vol. 281, pp. 197–207. Springer, Cham (2017). doi:10.1007/978-3-319-58457-7_15
19. van der Aalst, W., Weijters, T., Maruster, L.: Workflow mining: discovering process models from event logs. IEEE Trans. Knowl. Data Eng. **16**(9), 1128–1142 (2004)
20. Mannhardt, F., de Leoni, M., Reijers, H.A., van der Aalst, W.M.P.: Data-driven process discovery - revealing conditional infrequent behavior from event logs. In: Dubois, E., Pohl, K. (eds.) CAiSE 2017. LNCS, vol. 10253, pp. 545–560. Springer, Cham (2017). doi:10.1007/978-3-319-59536-8_34
21. Leemans, S.J.J., Fahland, D., van der Aalst, W.M.P.: Discovering block-structured process models from incomplete event logs. In: Ciardo, G., Kindler, E. (eds.) PETRI NETS 2014. LNCS, vol. 8489, pp. 91–110. Springer, Cham (2014). doi:10.1007/978-3-319-07734-5_6
22. Verbeek, H.M.W., Buijs, J.C.A.M., van Dongen, B.F., van der Aalst, W.M.P.: XES, XESame, and ProM 6. In: Soffer, P., Proper, E. (eds.) CAiSE Forum 2010. LNBIP, vol. 72, pp. 60–75. Springer, Heidelberg (2011). doi:10.1007/978-3-642-17722-4_5

Effect of Linked Rules on Business Process Model Understanding

Wei Wang[1][✉], Marta Indulska[2], Shazia Sadiq[1], and Barbara Weber[3]

[1] School of Information Technology and Electrical Engineering, The University of Queensland, Brisbane, Australia
w.wang9@uq.edu.au, shazia@itee.uq.edu.au
[2] University of Queensland Business School, The University of Queensland, Brisbane, Australia
m.indulska@business.uq.edu.au
[3] Department of Applied Mathematics and Computer Science, Technical University of Denmark, Kongens Lyngby, Denmark
bweb@dtu.dk

Abstract. Business process models are widely used in organizations by information systems analysts to represent complex business requirements and by business users to understand business operations and constraints. This understanding is extracted from graphical process models as well as business rules. Prior research advocated integrating business rules into business process models to improve the effectiveness of important organizational activities, such as developing shared understanding, effective communication, and process improvement. However, whether such integrated modeling can improve the understanding of business processes has not been empirically evaluated. In this paper, we report on an experiment that investigates the effect of linked rules, a specific rule integration approach, on business process model understanding. Our results indicate that linked rules are associated with better time efficiency in interpreting business operations, less mental effort, and partially associated with improved accuracy of understanding.

Keywords: Business process modeling · Business rule modeling · Cognitive research

1 Introduction

In the Business Process Management (BPM) life cycle, the success of business process (re)design, analysis, and simulation are all underpinned by the assumption that the business activities are well understood. This understanding is extracted from graphical process models, which mainly focus on the temporal or logical relationships between business activities, as well as business rules, which are constraints and mandates that control the behavior of the process and business activities. Lack of good understanding

---

This work is partially supported by ARC DP140103171 and China Scholarship Council.

© Springer International Publishing AG 2017
J. Carmona et al. (Eds.): BPM 2017, LNCS 10445, pp. 200–215, 2017.
DOI: 10.1007/978-3-319-65000-5_12

of a business process and business rules that constrain the process can give rise to many risks. Users may inadvertently breach required standards of operation or make ill-informed decisions. Different stakeholders, such as process designers, information systems developers, and process participants may have inconsistent, or even conflicting, understanding of the same process. Ultimately, such inconsistencies hinder the effectiveness of important organizational activities and introduce risks of noncompliant process execution.

While all graphical process models generally integrate some aspects of rules (e.g. through control flow of the process), business rules can be represented in an integrated manner or in a separated manner. When represented in an integrated manner, they are shown graphically in a process model, either as textual annotations [1], as graphical links to external rules [2], or diagrammatically using the native notation of the graphical model [3], e.g. through a combination of sequence flows, activities and gateways. When modeled in a separated manner, rules are captured in separate documents or rule engines, and the relationships between the business process models and the rules are not explicitly represented in the process models. Traditionally, due to limited support for representation of business rules in graphical process modeling techniques [4], organizations often store such representations in separate text documents, spreadsheets, or disconnected business rule repositories [5]. Over the past two decades, prior work has argued for the need to model business rules in an integrated manner with business processes [6, 7], and a variety of integration methods [1–3, 6, 8–11] and initial guidelines on rule integration [5] have been developed.

Arguments for such integration are typically based on an assumption of process improvement and shared understanding [5]. However, despite such arguments, and despite the different integration methods developed, if and to what extent such integration improves user understanding of the process models has not been investigated. In particular, while researchers have argued that integrated modeling can improve the understanding of business processes [5], this proposition has not been empirically evaluated. In this paper, we first present the theoretical foundation of the effects of rule integration on the cognitive activities of process model comprehension. With a focus on linked rules, a type of rule integration with process models, we then hypothesize the relationships between linked rules and process model understanding and report the results of our experiment to determine if linked rules can improve the understanding of process models.

2 Background and Related Work

A business process is a structured collection of activities that accomplishes a specific goal [12]. Such structures also involve business rules, which specify obligations, permissions, and restrictions that will limit the choice of approaches toward achieving a given goal [13].

Business process modeling and business rule modeling both focus on creating a representation of the organization's current and future practices. They are complementary approaches as they address distinct aspects of organizational practices. The overlap between business process models and business rules indicates a need to model the two related aspects together. Researchers argue that the integration of business rules into business process models can achieve better process model understanding [14–16], and improved governance, risk management and control [1, 17]. At the same time, however, researchers have identified a general lack of capability among process modeling languages to adequately represent business rules [4, 18, 19].

To solve this problem, a variety of integration methods and techniques have been developed since the publication of the first paper suggesting that business process and rule modeling approaches should be merged [20]. To name a few, McBrien et al. defined the structure of rules to couple business process models and rules [21]. Knolmayer et al. refined process modeling and linked the resulting models to workflow execution through layers of so-called Reaction Business Rules [22]. Kovacic et al. developed a meta-model to demonstrate how rules can link process, activity, events, data objects, and software components [13]. To summarize, three forms of integration of business process models and rules have been developed in literature viz. link integration, text integration, and diagrammatic integration. These approaches are summarized below and illustrated in Fig. 1.

Link integration. Link integration approaches incorporate information about the location of a related, externally documented, rule in a process model. Links can be static or automatic. In static link integration, the location information can be the section number and id, or the page number of the rule in a rulebook, thus allowing process users to locate the rule. Automatic link integration means the location information can be implemented as links, which will automatically navigate to the rule in the rule repository when the link is clicked. Notable contributions on link Integration are [2, 9].

Text integration. Text integration approaches represent the content of a rule textually in a business process model. For example, BPMN has a text annotation construct which allows users to put business rules into such an annotation construct in sentential format. Notable contributions on text integration are [1, 11].

Diagrammatic integration. While rules in link integration and text integration are represented in a sentential format, diagrammatic integration approaches represent rules in a diagrammatic format in a process model, using process modeling constructs such as sequence flows and gateways. A notable contribution on diagrammatic integration is [3].

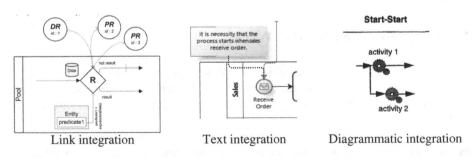

Link integration   Text integration   Diagrammatic integration

Fig. 1. Integration methods illustration

According to [23], the fundamental purpose of conceptual models is to improve users' understanding of the static and dynamic phenomena in a domain, and then to help developers and users to communicate and to serve as a basis for design. Process models are a typical type of conceptual model, and the factors affecting the understanding of process models have been well studied. Factors affecting the understanding of process models can be classified into two categories: process model factors and individual factors. Process model factors relate to the metrics of the process models, such as modularization [24], block structuredness [24], and complexity [25]. Individual factors, or personal factors, relate to process model users, such as an individual's domain knowledge [26], modeling knowledge [27], modeling experience [24], and education level [24]. Figl *et al.* [28] provided a comprehensive overview of the literature on process model comprehension.

The argument that rule integration can improve process model understanding is the foundation that has motivated the development of different integration methods and techniques. The evaluation of the argument is critical to progress this research field. However, despite a considerable number of integration methods have been introduced using existing process modeling constructs, and despite many factors that can effect process model understanding have been identified, the question of whether integrating business rules into process models can improve the understanding of process models has not been theoretically analyzed nor empirically evaluated.

3 Theoretical Background

The limitations of diagrammatic integration are widely known due to the expressibility limitations of process modeling languages [1]. Similarly the drawbacks of rule integration through text annotations are duplicate and potentially inconsistent rule representations [29]. Hence in this paper we focus on a specific form of rule integration, namely link integration – an approach that points the model to the relevant rule, rather than duplicating that rule in the process model in either text or graphical form.

Link integration approaches incorporate visual links that connect the relevant rules to a section of the model – i.e. the links are explicitly represented on the activities or gateways that the rules constrain. This approach thus makes the connections of rules and corresponding activities explicit, presumably reducing cognitive load required to mentally connect rules to the appropriate part of the process model [16]. When rules are modeled in a separated manner, on the other hand, they have to be semantically interpreted and manually matched by the model user to the relevant parts of the model. This is an error-prone process that requires the user to interpret the business rule against the background of the entire model to determine best fit. Accordingly, our first aim is to investigate the effect of link integration on process understanding accuracy, which means how well a process model is understood:

Hypothesis 1: Process models with linked rules are associated with better understanding accuracy compared with separated rules.

When rules are separated, all rules are organized as one set of rules, represented in some textual form (either plain text or in one of the business rule modeling languages).

Finding the relevant rules that constrain a specific activity or gateway requires a comprehensive search and semantic interpretation of the set (e.g. linearly down the entire list of rules), which takes more time to mentally connect rules and a process model.

Accordingly, our second aim is to investigate the effect of rule linking on process understanding efficiency, focusing on how much time it takes a participant to review the process model and related rules to demonstrate understanding accuracy.

Hypothesis 2: Process models with linked rules are associated with better understanding time efficiency compared with separated rules.

As extra cognitive activities such as search and semantic interpretation are needed with rule linking, our third aim is to investigate the mental effort:

Hypothesis 3: Process models with linked rules are associated with less mental effort needed for understanding.

Despite the benefits, link integration is not without limitations. First, people using linked rules may focus on the interactions of specific rules and process components, without a holistic understanding of the process model and rules as a prerequisite, thus may have inaccurate understanding. Second, it can cause the attention switching effect [30], which means that users need to split their attention among multiple sources of information and mentally integrate them. Given separated rules as a whole list, one can choose to learn and assimilate more rules before switching attention to a process model, thus to reduce attention switches and time needed. It is therefore not clear to which degree the additional cognitive cost in terms of attention switching counter-balances the improvement in understanding. Thus, a study is needed to investigate this effect of business process and rule integration. To this end, we propose an experimental approach to test our hypotheses.

4   Research Method

This study applies an experiment research method to explore differences between linked and separated business process models and rules. In this section, we introduce our experimental design and describe our instruments, experiment settings and participants.

4.1   Experiment Design

The experiment is a single factor experiment. In our experiment, the use of linked rules is the considered factor, with factor levels "present" and "absent". We used two groups, two factor levels, and two domains in our experiment. Each group was tested with two domains separately, and for each domain, the two groups had different factor levels.

We have three main considerations in our between-subject design. First, our experiment environment only allows us to have one participant to do the experiment at a time. Second, the understanding performance depends on an individual's cognitive competence and experience. Thus, group imbalance is a challenge for between-subject design. Third, we want to increase the generalization ability of the experiment in terms of domains, while controlling the learning effect.

Under these considerations, we designed our experiment as a balanced single factor experiment with repeated measurement, based on an experiment design from [24] which can increase the power of the experiment given the same number of participants [31]. The overall design is illustrated in Fig. 2. In this design, each participant will be tested for all factor levels and all domains, thus (1) more data will be collected than in a single run experiment, (2) two domains are tested to increase the generalizability of the results. The order of factor levels is reversed between groups, so the factor of order of treatment and learning effect are counterbalanced across groups. Please note that the forms of rule representation are inversed in the two runs. In the first run, Group 1 are given linked rules and Group 2 are given separated rules, while, in the second run, Group 1 are given separated rules and Group 2 are given linked rules.

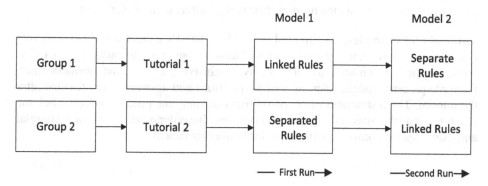

Fig. 2. Overall experiment approach

As illustrated in Fig. 3, when linked rules are present, link buttons (labeled with "R") will be shown on activities and gateways in a process model, when a link button is clicked, the rules that are connected to the activity/gateway via the link button will be displayed on the "Relevant Rules" area on the right of the screen. When linked rules are absent, no link buttons will be shown in a process model, and all rules will be displayed in the "Relevant Rules" area on the right side of the screen.

4.2 Measurements

To measure the accuracy of understanding we use the percentage of correct answers to comprehension questions. We use the time from the point that a process model is displayed on the screen, to the point that the last question for this process model is answered as the measurement of time efficiency. To measure mental effort we use both an objective measure and a perception measure. We used the eye-fixation duration for each model as the objective measure. Eye-fixation is the maintaining of the visual gaze on a single location. Vision is suppressed during the eye saccade, and new information is acquired only during the fixation. Eye-fixation duration was proved to surpass pupil size as a mental effort measure [32]. As measure of perception of required mental effort, we asked each participant to select the model he or she perceived more difficult.

4.3   Instruments

We briefly describe each part of the experimental instruments below.

Questionnaire.   We have a pre-experiment questionnaire and a post-experiment questionnaire. Questions for which the answers could be affected by participating in the experiment, such as the extent of familiarity with business process models and rules, and the extent of familiarity with the knowledge domains used in the experiment, were included in the pre-experiment questionnaire. To save a participant's mental effort before the experiment, objective questions which could not be affected by the participation in the experiment, such as a participant's major and which year he or she is in, were put into a post-experiment questionnaire, together with a question asking participants which model consumed most of their mental effort in the experiment.

Tutorial and examples.   The tutorial covered all BPMN elements and business rule concepts that participants would need to know to perform the tasks, e.g. activity, sequence, activity group, parallel gateway, exclusive gateway, and business rules. Example process models, rules, as well as questions and answers were provided after the tutorial. The instructions direct participants to study the process models, click the rule links, read the rules, and answer the questions. The order of treatments in the tutorial and examples are consistent with the order in the experiment.

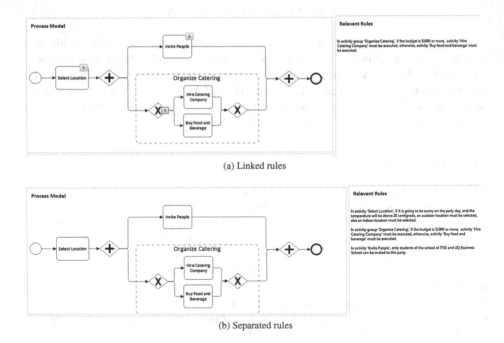

(a) Linked rules

(b) Separated rules

Fig. 3. Independent variable illustration

Treatment design. To limit the learning effect, only two process models were used, and only three questions were asked for each model. The information needed from a process model and rules to answer a question are independent from each other thus the information learned from a previous question has little contribution to the current question. We designed process model A based on previous experiments [33, 34], and designed process model B to keep the complexity of the two models as close as possible. The rules and questions of the two process models are designed with the same cognitive load level in mind. The rules covered common rule violations such as time constraints, route selections, and data logic. To assimilate what happens in practice, several rules can control a single activity, and a violation of any of the rules will lead to a breach. We kept a variety of metrics of the two sets of models, rules, and questions the same or as close as possible. A package of the entire experiment is available for download on Dropbox[1].

4.4 Settings

The pre-experiment and post-experiment questionnaires were implemented in Qualtrics[2]. The tutorial and experiment were implemented as an Eclipse RCP application[3]. The texts and diagrams were proved to be clearly visible from a distance of over 60 cm in the pilot test. As shown in Fig. 4, the screen was divided into three Areas, viz. Process Model Area, Relevant Rules Area and Questions Area. The complete process model and all the rules are displayed without the need of scrolling. No zooming is allowed in the application. All text and diagrams are in black and white so color blindness will not introduce bias to the experiment. We used Tobii Pro TX300, an eye tracker with a 23-inch screen of a resolution of 1920×1080 that captures gaze data at 300 Hz[4]. The experiment was set in a lab. The lab has no window and the rooftop lights are the only light source. The materials, eye-tracker, and lights had the same settings for all participants.

---

[1] The experiment can be downloaded from https://www.dropbox.com/s/g6jpb767m474vv2/experiment.rar?dl=0.

[2] Qualtrics is a web-based survey platform. See: www.qualtrics.com.

[3] Eclipse RCP is a platform for building applications. See: https://wiki.eclipse.org/Rich_Client_Platform.

[4] For more specifications, please see http://www.tobiipro.com/product-listing/tobii-pro-tx300.

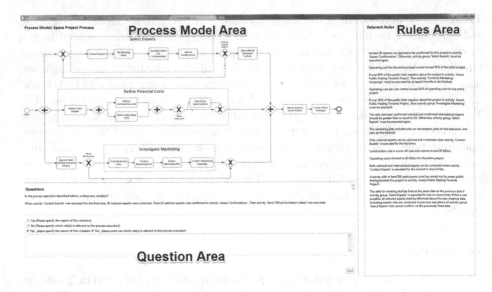

Fig. 4. Instrument Illustration

4.5 Participants

Students at an Australian university participated in this experiment voluntarily. Eight PhD students participated in the pilot tests. Fifty coursework students of an information systems course participated in the main experiment and were randomly assigned to two groups. Our sample size is considerable compared with other comparable experiments, which have sample sizes between 20–30 [32, 35]. All participants were required to have basic knowledge of flowcharts, UML or ER diagrams. We only used the most basic BPMN symbols and easily understandable daily English in the material, which did not require substantial experience from our participants. We did not put a time limit for each student, and all fifty students finished the experiment within an hour. Forty-eight students finished the experiment successfully, and the eye movements of two students in Group 1 failed to be properly recorded by the eye-tracker. Thus, we discarded the two samples in the analysis of eye-movement related data. As an incentive, each student was offered a $30 voucher for participation.

5   Results

For comparing categorical dependent variables between two groups such as answer correctness and the choice of mental effort, we use Chi-squared test, which can be used to compare categorical data [36]. For other numerical dependent variables, we first checked if a dependent variable is normally distributed using Shapiro-Wilk test at a significance level of 0.05 [36]. If data of both groups were normally distributed, we checked whether the data met the assumption of equal variance using dependent

Levene's test[5] at the significance level of 0.05, and then used the independent-sample t test. If data in any group were not normally distributed, we used the Mann-Whitney U test[6] across groups. We describe the results for each hypothesis in turn.

For Hypothesis 1, we ran Chi-square tests between the two groups, with the correctness of answers as the dependent variable, for the two models separately. Table 1 shows the Chi-square test results, which show that understanding accuracy was significantly correlated with the form of rule presentation in Model 2 ($p = 0.03$), but not in Model 1 ($p = 0.16$), which partially supports Hypothesis 1.

Table 1. Test of Hypotheses 1 – understanding accuracy

	Group	Correct	Incorrect	Row total	p
Correctness in Model 1	G1	55	20	75	0.16
	G2	47	28	75	
	Column total	102	48	150	
Correctness in Model 2	G1	47	28	75	0.03
	G2	59	16	75	
	Column total	106	44	150	

Conclusion 1: Linked rules are partially associated with an improved understanding accuracy.

For Hypothesis 2, the time spent of Group 2 in Model 2 was not normally distributed. We ran independent-sample Mann-Whitney tests between Group 1 and Group 2, with the time (from beginning to the end of answering the last question in each run) as the dependent variable. The test result of Hypothesis 2 is shown in Table 2. Table 2 shows that time used in each model is related to the form of rule presentation, supporting Hypothesis 2 at a significance level of 0.05.

Table 2. Test of Hypothesis 2: understanding efficiency

	Group	N	Mean	Std. Deviation	p (1-tailed)
Time used in Model 1	G1	23	368.76	110.23	0.015
	G2	25	481.18	218.10	
Time used in Model 2	G1	23	468.57	173.06	0.009
	G2	25	370.46	116.88	

Conclusion 2: Linked rules are associated with increases in understanding efficiency.

For Hypothesis 3, the eye-fixation durations in the two runs were not normally distributed. We therefore ran independent-sample Mann-Whitney tests for the two runs separately. The objective test of Hypothesis 3 is shown in Table 3. From Table 3 we can

---

[5] Levene's test is an inferential statistic used to assess the equality of variances for a variable calculated for two or more groups.

[6] The Mann-Whitney U test is used to compare differences between two independent groups when the dependent variable is not normally distributed.

see that the mental effort is associated with the type of rule presentation, supporting Hypothesis 3 at a significance level of 0.05.

Table 3. Test of Hypothesis 3: objective mental effort

	Group	N	Mean	Std. Deviation	p (1-tailed)
Fixation duration in Model 1	G1	23	322.98	100.30	0.024
	G2	25	411.43	188.22	
Fixation duration in Model 2	G1	23	409.68	159.94	0.007
	G2	25	318.53	102.31	

The results of the perception of mental effort are shown in Table 4. In Group 1, 0 participants selected Model 1 (linked rules), while 23 participants selected Model 2 (separated rules) as the model requiring *more* mental effort. Two participants selected *'equal'* as the answer. In Group 2, 11 participants selected Model 1 (separated rules), while 6 participants selected Model 2 (linked rules) as the model requiring *more* mental effort. Eight participants selected *'equal'* as the answer. From Table 4 we can intuitively see that participants indicate that models with separated rules require more mental effort, regardless of model content (model 1 or model 2).

Table 4. Perception of mental effort

	Group 1	Group 2
Model 1 requires more mental effort	0 (linked rules)	11 (separated rules)
Model 2 requires more mental effort	23 (separated rules)	6 (linked rules)
Equal	2	8

To statistically compare linked and separated rules, we coded the perception answers as follows: When a model with linked rules was selected as the model that required more mental effort, linked rules were assigned 2 points. When the model with separated rules was selected as the model that required more mental effort, separated rules were assigned 2 points. When a participant selected the two models as equal, both linked rules and separated rules were assigned 1 point. We used a *t test* for the difference in average mental effort perception between linked and separated rules. Table 5 shows that mental effort in linked rules is significantly smaller than in separated rules.

Table 5. Coded mental effort

	N	Coded mean	Std. Deviation	p (1-tailed)
Linked rules	50	0.44	0.70	0.000
Separated rules	50	1.56	0.70	

Conclusion 3: Linked rules are associated with reduced mental effort required for model understanding.

6   Discussion

Our results support Hypotheses 2 and 3, indicating that linked rules are associated with increases in understanding efficiency and reduced mental effort required for model understanding. While Hypothesis 1 has only partial support. For the results of Hypothesis 1, the p value for Model 1 was greater than 0.05, indicating a lack of statistical significance. To explore this result further, first we compared the two models, and the metrics comparison showed that the two sets of models, rules and questions are the same or close in all the metrics. Second, we investigated answer correctness and time spent of each model. The statistics showed that the two sets of models, rules, and questions had no significant difference (with $p = 0.647$ and $p = 0.822$ respectively). Thus, we concluded that there was no bias between Model 1 and Model 2. Finally, we broke down the correctness of answers to each question to explore the lack of statistical significance of the differences between linked and separated rules in Model 1. As shown in Fig. 5, the result of question 1 shows that the group with linked rules had lower understanding accuracy than the group with separated rules, which is against Hypothesis 1, while the correctness of all other 5 questions indicates the support of Hypothesis 1. We assume that one possible reason is that the participants had not learnt how to use linked rules well when they met the first question. Recall that we had to balance time with fatigue and tracking data accuracy and thus had a time constrain in the experiment, so we used a simple illustration of linked rules (See Fig. 3) in the training material, compared with the models and rules in the formal experiment what were much more complex and challenging. Thus, participants may not quickly find how to utilize rule links.

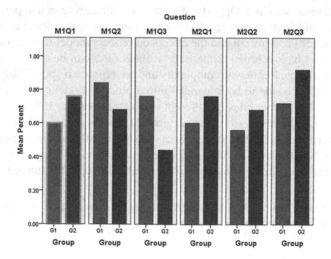

Fig. 5. Answer correctness breakdown to each question

Our study is not without limitations. In terms of internal validity, the different layout of screen areas could possibly affect the results. It is possible that the experiment results will be different if we change the location of each area. In terms of construct validity, we operationalized each construct in our study in limited ways. The questions were designed to test the understanding of the effect of business rules on business process models. Following [37], it would have been ideal if we had measured the perceived quality and efficiency of understanding, and asked questions only about a process model itself. Thus, our research results are limited to the treatments, measurements, and questions that we used. Finally, in terms of external validity, we cannot say that the process models, rules, and questions we used faithfully reflect those used in organizations in practice. Organizations may use more complex process models and lager number of rules and the tasks may be more challenging. The use of students as participants could also weaken the generalization ability of the results.

7   Conclusions and Outlook

In this paper, we have studied the relationship between rule integration and business process model understanding. Rules can be integrated into process models in a variety of ways, and in this paper, we report on our findings based on a specific form of rule integration, namely linked rules. We focused on 3 aspects of understanding: understanding accuracy, time efficiency, and mental effort. Our study results presented three conclusions: (1) The association between linked rules and understanding accuracy is partially supported. (2) Linked rules are significantly associated with improved time efficiency. (3) Linked rules are significantly associated with reduced mental effort. Our conclusions are drawn from an experiment design that utilized an eye-tracker. The design of the experiment provides a methodological contribution towards the study of process model understanding. Opportunities exist for future research to perform similar experiments on different rule integration methods such as annotation and diagrammatical integration [38] and investigate the effects on process model understanding.

Business rules have a broad scope, and business rules can be quite varied in many aspects such as change frequency, complexity and governance responsibility [38]. Thus, the best way for each rule to be integrated into a process model can be different. The characteristics of business rules or different rule categories can influence which integration method has the best performance in terms of process model understanding. Quite a few business rule classification frameworks such as [39, 40] exist in literature. Finding the connection between type of rule and the best corresponding integration approach to improve process model understanding will be a valuable topic for future research.

Acknowledgement. We would like to acknowledge the advice from Dr Thomas Taimre regarding the use of relevant statistical methods.

References

1. Cheng, R., Sadiq, S., Indulska, M.: Framework for business process and rule integration: a case of BPMN and SBVR. In: Abramowicz, W. (ed.) BIS 2011. LNBIP, vol. 87, pp. 13–24. Springer, Heidelberg (2011). doi:10.1007/978-3-642-21863-7_2
2. Sapkota, B., van Sinderen, M.: Exploiting rules and processes for increasing flexibility in service composition. In: 2010 14th IEEE International Enterprise Distributed Object Computing Conference Workshops (EDOCW), pp. 177–185. IEEE (2010)
3. Kappel, G., Rausch-Schott, S., Retschitzegger, W.: Coordination in workflow management systems — a rule-based approach. In: Conen, W., Neumann, G. (eds.) ASIAN 1996. LNCS, vol. 1364, pp. 99–119. Springer, Heidelberg (1998). doi:10.1007/BFb0027102
4. Recker, J., Rosemann, M., Green, P.F., Indulska, M.: Do ontological deficiencies in modeling grammars matter? MIS Q. **35**, 57–79 (2011)
5. Zur Muehlen, M., Indulska, M., Kittel, K.: Towards integrated modeling of business processes and business rules. In: Proceedings of the 19th Australasian Conference on Information Systems (ACIS)-Creating the Future: Transforming Research into Practice, Christchurch, New Zealand, pp. 690–697. Citeseer (2008)
6. Habich, D., Richly, S., Demuth, B., Gietl, F., Spilke, J., Lehner, W., Assmann, U.: Joining business rules and business processes. In: Proceedings of IT (2010)
7. Zur Muehlen, M., Indulska, M.: Modeling languages for business processes and business rules: a representational analysis. Inf. Syst. **35**, 379–390 (2010)
8. Kluza, K., Kaczor, K., Nalepa, G.J.: Enriching business processes with rules using the Oryx BPMN Editor. In: Rutkowski, L., Korytkowski, M., Scherer, R., Tadeusiewicz, R., Zadeh, L.A., Zurada, J.M. (eds.) ICAISC 2012. LNCS, vol. 7268, pp. 573–581. Springer, Heidelberg (2012). doi:10.1007/978-3-642-29350-4_68
9. Nalepa, G.J., Kluza, K., Kaczor, K.: Proposal of an inference engine architecture for business rules and processes. In: Rutkowski, L., Korytkowski, M., Scherer, R., Tadeusiewicz, R., Zadeh, L.A., Zurada, J.M. (eds.) ICAISC 2013. LNCS, vol. 7895, pp. 453–464. Springer, Heidelberg (2013). doi:10.1007/978-3-642-38610-7_42
10. Milanovic, M., Gasevic, D., Rocha, L.: Modeling flexible business processes with business rule patterns. In: 2011 15th IEEE International Enterprise Distributed Object Computing Conference (EDOC), pp. 65–74 (2011)
11. Governatori, G., Shek, S.: Rule based business process compliance. In: Proceedings of the RuleML2012@ ECAI Challenge, article 5 (2012)
12. Hammer, M., Champy, J.: Reengineering the corporation: a manifesto for business revolution. Bus. Horiz. **36**, 90–91 (1993)
13. Kovacic, A., Groznik, A.: The business rule-transformation approach. In: 26th International Conference on Information Technology Interfaces, vol. 1, pp. 113–117 (2004)
14. Rabova, I.: Methodology of the enterprise architecture creating and the role of the enterprise architecture in rural development. Agricultural Economics-Zemedelska Ekonomika **56**, 334–340 (2010)
15. Skersys, T., Tutkute, L., Butleris, R., Butkiene, R.: Extending BPMN business process model with SBVR business vocabulary and rules. Inf. Technol. Control **41**, 356–367 (2012)
16. Wang, W., Indulska, M., Sadiq, S.: Cognitive efforts in using integrated models of business processes and rules - semantic scholar. In: Proceedings of the 28th International Conference on Advanced Information Systems Engineering (CAiSE Workshop), Ljubljana, Slovenia. Springer (2016)

17. Ly, L.T., Rinderle-Ma, S., Göser, K., Dadam, P.: On enabling integrated process compliance with semantic constraints in process management systems. Inf. Syst. Front. **14**, 195–219 (2012)

18. Green, P.F., Rosemann, M.: Perceived ontological weaknesses of process modeling techniques: further evidence. In: Proceedings of the ECIS, pp. 312–321 (2002)

19. Herbst, H., Knolmayer, G., Myrach, T., Schlesinger, M.: The specification of business rules: a comparison of selected methodologies. In: Methods and Associated Tools for the Information Systems Life Cycle, pp. 29–46 (1994)

20. Krogstie, J., McBrien, P., Owens, R., Seltveit, A.H.: Information systems development using a combination of process and rule based approaches. In: Andersen, R., Bubenko, J.A., Sølvberg, A. (eds.) CAiSE 1991. LNCS, vol. 498, pp. 319–335. Springer, Heidelberg (1991). doi:10.1007/3-540-54059-8_92

21. McBrien, P., Seltveit, A.H.: Coupling process models and business rules. In: Sölvberg, A., Krogstie, J., Seltveit, A.H. (eds.) Information Systems Development for Decentralized Organizations. ITIFIP, pp. 201–217. Springer, Boston, MA (1995). doi:10.1007/978-0-387-34871-1_12

22. Knolmayer, G., Endl, R., Pfahrer, M.: Modeling processes and workflows by business rules. In: van der Aalst, W., Desel, J., Oberweis, A. (eds.) Business Process Management. LNCS, vol. 1806, pp. 16–29. Springer, Heidelberg (2000). doi:10.1007/3-540-45594-9_2

23. Burton-Jones, A., Meso, P.N.: Conceptualizing systems for understanding: an empirical test of decomposition principles in object-oriented analysis. Inf. Syst. Res. **17**, 38–60 (2006)

24. Reijers, H.A., Mendling, J., Dijkman, R.M.: Human and automatic modularizations of process models to enhance their comprehension. Inf. Syst. **36**, 881–897 (2011)

25. Mendling, J., Strembeck, M., Recker, J.: Factors of process model comprehension—findings from a series of experiments. Decis. Support Syst. **53**, 195–206 (2012)

26. Bera, P.: Does cognitive overload matter in understanding BPMN models? J. Comput. Inf. Syst. **52**, 59–69 (2012)

27. Recker, J.C., Dreiling, A.: Does it matter which process modelling language we teach or use? An experimental study on understanding process modelling languages without formal education. In: Toleman, M., Cater-Steel, A., Roberts, D. (eds.) Faculty of Science and Technology, pp. 356–366. University of Southern Queensland, Toowoomba, Australia (2007)

28. Figl, K.: Comprehension of procedural visual business process models–a literature review. Bus. Inf. Syst. Eng. **59**, 41–67 (2017)

29. Loucopoulos, P., Kadir, W.M.N.W.: BROOD: business rules-driven object oriented design. J. Database Manage. (JDM) **19**, 41–73 (2008)

30. Sweller, J., Chandler, P.: Why some material is difficult to learn. Cogn. Instr. **12**, 185–233 (1994)

31. Charness, G., Gneezy, U., Kuhn, M.A.: Experimental methods: between-subject and within-subject design. J. Econ. Behav. Organ. **81**, 1–8 (2012)

32. Meghanathan, R.N., van Leeuwen, C., Nikolaev, A.R.: Fixation duration surpasses pupil size as a measure of memory load in free viewing. Front. Hum. Neurosci. **8**, 1063 (2015)

33. Zugal, S., Pinggera, J., Weber, B., Mendling, J., Reijers, H.A.: Assessing the impact of hierarchy on model understandability – a cognitive perspective. In: Kienzle, J. (ed.) MODELS 2011. LNCS, vol. 7167, pp. 123–133. Springer, Heidelberg (2012). doi:10.1007/978-3-642-29645-1_14

34. Zugal, S.: Applying cognitive psychology for improving the creation, understanding and maintenance of business process models

35. Haji, F.A., Rojas, D., Childs, R., de Ribaupierre, S., Dubrowski, A.: Measuring cognitive load: performance, mental effort and simulation task complexity. Med. Educ. **49**, 815–827 (2015)

36. Box, G.E., Hunter, W.G., Hunter, J.S.: Statistics for experimenters: an introduction to design, data analysis, and model building. JSTOR (1978)

37. Campbell, D.T., Fiske, D.W.: Convergent and discriminant validation by the multitrait-multimethod matrix. Psychol. Bull. **56**, 81–105 (1959)
38. Wang, W., Indulska, M., Sadiq, S.: To integrate or not to integrate – the business rules question. In: Nurcan, S., Soffer, P., Bajec, M., Eder, J. (eds.) CAiSE 2016. LNCS, vol. 9694, pp. 51–66. Springer, Cham (2016). doi:10.1007/978-3-319-39696-5_4
39. Zoet, M., Versendaal, J., Ravesteyn, P., Welke, R.J.: Alignment of business process management and business rules. In: Proceedings of the 19th European Conference on Information Systems, Helsinki, Finland, p. 34 (2011)
40. Hashmi, M., Governatori, G., Wynn, M.T.: Normative requirements for business process compliance. In: Davis, J.G., Demirkan, H., Motahari-Nezhad, H.R. (eds.) ASSRI 2013. LNBIP, vol. 177, pp. 100–116. Springer, Cham (2014). doi:10.1007/978-3-319-07950-9_8

On the Performance Overhead of BPMN Modeling Practices

Ana Ivanchikj[✉], Vincenzo Ferme, and Cesare Pautasso

Software Institute, Faculty of Informatics, USI, Lugano, Switzerland
ana.ivanchikj@usi.ch

Abstract. Business process models can serve different purposes, from discussion and analysis among stakeholders, to simulation and execution. While work has been done on deriving modeling guidelines to improve understandability, it remains to be determined how different modeling practices impact the execution of the models. In this paper we observe how semantically equivalent, but syntactically different, models behave in order to assess the performance impact of different modeling practices. To do so, we propose a methodology for systematically deriving semantically equivalent models by applying a set of model transformation rules and for precisely measuring their execution performance. We apply the methodology on three scenarios to systematically explore the performance variability of 16 different versions of parallel, exclusive, and inclusive control flows. Our experiments with two open-source business process management systems measure the execution duration of each model's instances. The results reveal statistically different execution performance when applying different modeling practices without total ordering of performance ranks.

Keywords: BPMN 2.0 · Execution performance · Semantic equivalence

1 Introduction

As customer retention becomes strongly related to service execution time, velocity requirements have gone down from days to hours, minutes and seconds. This is especially true in fully automated Business Processes (BPs), where any additional millisecond of performance boost brings companies a competitive advantage and potential cost savings on the Cloud [2]. Assuming that model's execution semantics has already been optimized, could such boost be achieved by using what La Rosa et al. [25] call the "Alternative Representation Pattern", i.e., modeling the same execution semantics with different static structures? For instance, parallelism in BPMN 2.0 can be modeled explicitly by using a parallel gateway, or implicitly through multiple outgoing flows from an activity [23]. Although the modeling practices used in two such models are different in terms of the model's graph topology (size and used constructs), their execution semantics is the same, i.e., they both depict parallelism. As modelers freely pick which

© Springer International Publishing AG 2017
J. Carmona et al. (Eds.): BPM 2017, LNCS 10445, pp. 216–232, 2017.
DOI: 10.1007/978-3-319-65000-5_13

modeling practice to use, their choice should result in the same execution performance. Does this expectation hold in practice? Does the answer depend on the Business Process Management Systems (BPMSs)? These are open questions which have not received the due attention so far.

The first research question (RQ_1) we address in this work is: *Does the application of different modeling practices have significant impact on the duration of a BP instance execution?* Our null hypothesis (H_{RQ1}) is that *there is no statistically significant difference in the execution duration between instances of models which are semantically equivalent but structurally different*. We use trace equivalence [1] to define semantically equivalent models, i.e., models which produce the same traces (execution logs), regardless of the differences in their static structure (the control-flow constructs used). The data flow of the models remains unaltered. The second research question (RQ_2) we address is: *If H_{RQ1} is rejected, is there a total order between semantically equivalent but structurally different models, when ranked according to their performance?*

By answering RQ_1, BPMS vendors can decide whether there is a potential for performance improvement of their products based on alternative representations of deployed BPs. For instance, if an implicit parallel gateway executes significantly faster than an explicit one, the vendor can use the same implementation for both. The answer to RQ_2, on the other hand, indicates potential generalization opportunities of any identified optimization rules. For instance, it can show whether the execution of implicit gateways always ranks better than the execution of explicit gateways, regardless of the gateway type (e.g., parallel, inclusive, exclusive). Answering both questions is required before investing in further research towards automatic performance optimization by semantics preserving model transformations.

To this end, we propose a methodology for transforming an initial model into semantically equivalent models by using a predefined set of transformation rules. We also propose a statistical procedure to analyze the results of executing the equivalent models in order to answer the two research questions. To delimit the exploration space for this paper we have selected three scenarios which follow some of the modeling guidelines defined by Mendling et al. [21] and deal with the frequently used parallel and exclusive gateways [22], as well as with the inclusive gateway which has been found to hinder BP understandability [21]. We run the initial models as well as the derived semantically equivalent models on two open source BPMSs, Camunda and Activiti. The contribution of this paper consists of the methodology, the experimental results, and the analysis, which indicate that semantically equivalent models with different structure demonstrate statistically different execution behaviour, thus justifying further research into automated BP execution performance improvement.

The rest of this paper is organized as follows. In Sect. 2 we propose a methodology for defining the experiments needed for assessing the performance differences. We apply the methodology on three scenarios and discuss the results in Sect. 3, followed by a short survey of the related work in Sect. 4. We elaborate on the threats to the validity of our work in Sect. 5, while concluding the paper in Sect. 6.

2   Methodology

The methodology (Fig. 1) is divided into two parts: (1) define and perform experiments, which as such can be applied to obtain metrics needed for different research questions; and (2) statistically analyze the experiment results to test the H_{RQ1} hypothesis and rank the performance of the models to answer RQ_2.

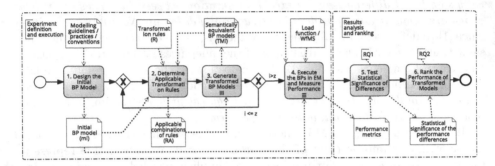

Fig. 1. Methodology for systematic exploration of the performance of structurally different but semantically equivalent models

1. **Initial Model.** The experiment definition process starts with the design of an initial model (m_i) (Fig. 1.1). Such initial model could follow existing best practices and conventions, or modeler's personal preferences.

2. **Applicable Rules.** In addition to the initial model, a set of transformation rules $R = \{r_1, r_2, r_3, ..., r_n\}$ are used as input to the second step of the methodology (Fig. 1.2). A transformation rule is an operation which adds/deletes elements (nodes, edges) to/from the BP model, provided that certain preconditions are met. In the context of this work, a transformation rule only affects the structure of the BP graph, leaving the execution semantics and the graph layout unchanged. The set of transformation rules can reflect different modeling practices and guidelines. Given R, all feasible combinations of the transformation rules should be iteratively applied starting from the initial model to generate semantically equivalent models until a fixed-point is reached. The output of the second step is $R_A(m_i) \subseteq R$, containing only the rules that are applicable to a given BP model (m_i in the first iteration), to produce a correctly deployable and executable, semantically-equivalent model.

3. **Model Transformation.** The model transformation function $f(r_j, m)$ is executed applying the transformation rule r_j to model m. Applying all rules found in $R_A(m_i)$ results in $TM_1 = \{t_j | (\forall r_j \in R_A(m_i)) [t_j = f(r_j, m_i)]\}$, the set of transformed models which are semantically, but not syntactically, equivalent to the initial model m_i and obtained by applying just once all the applicable transformation rules. The generation process (Fig. 1.2 and 1.3) is repeated for all elements of TM_1, generating a new set of equivalent models $TM_2 = \{t_{jk} | (\forall t_j \in TM_1)(\forall r_k \in R_A(t_j)) [t_{jk} = f(r_k, t_j)]\}$, by transforming all the models $t_j \in TM_1$

with the corresponding applicable rules $R_A(t_j)$. In other words, starting from the initial model m_i, only one rule is applied to generate the models in TM_1, two rules to generate TM_2, etc. This iteration stops when no new models are generated or after a given maximum number z of model generations. The output of the iterations is $EM = \{m_i\} \cup \bigcup_{i=1}^{z} TM_i$, the set of semantically equivalent models to be executed in the experiment.

4. **Model Execution.** Finally each of the models in EM is deployed and multiple instances are started always using the same load functions and test data. During the execution of the BP instances, performance data is collected and different metrics (M) are calculated (Fig. 1.4). Depending on the optimization goal, metrics may include time (e.g., duration of the BP instance/constructs, or throughput), or resource utilization (e.g., CPU/RAM).

5. **Statistical Significance.** Determining statistical significance (Fig. 1.5) requires statistical tests, the nature of which depends on different factors [19]. The first factor to consider is the number and the nature of the samples. In our case, one sample is comprised of the instances of a given executed model $em_i \in EM$. Thus, the number of samples to be analyzed depends on the cardinality of EM, and is greater than two. The samples are unpaired, i.e., independent, since the executed instances of different $em_i \in EM$ are not related. The difference between samples can go in any direction (increase or decrease), thus a two-tailed test is required. Next we need to consider the nature and the distribution of the collected data, i.e., in our case the metrics M. We are working with quantitative continuous performance data, which based on our experience in previous work [9,10,27], is not normally distributed, and thus requires the use of non-parametric tests where the original data is recorded in ranks. This type of tests are almost as powerful as parametric tests when the sample is large enough [19]. The **appropriate sample size** can be determined with the statistical power analysis [5], which uses as input the level of significance, the power and the expected effect size. The level of significance refers to the accepted level of probability that the observed result is a false positive, i.e., it is due to chance. It is usually conventionally set to 5% or 1% [4]. The power refers to the probability of false negative, i.e., accepting the H_0 when there is actually a difference between the results. The most commonly accepted level of power is 80% or above [4]. The expected effect size refers to the expected difference in the measured variable between the different groups. It can be determined based on pilot data, previous research if available, or an educated guess. As pilot data, we use the results of one test trial.

Considering all of the above factors, the appropriate test to run is the Kruskal-Wallis which is a non-parametric one-way ANOVA [6]. The H_0 of this test is that the distribution of the variable being tested is the same across the samples. The alternative hypothesis (H_1) is that the distribution of the same variable is different across the samples. Thus, rejecting the H_0 and accepting H_1, when $P - value < 0.05 = \alpha$, means that at least one sample stochastically dominates one other sample, i.e., in our work it means that there is **statistically significant difference** in the execution performance of at least two of the tested semantically equivalent models.

Given the non-deterministic nature of BPMSs, and software systems in general, using just one trial is not sufficient for obtaining reliable results. Having **multiple trials** means that each em_i is deployed multiple times in isolation, and each time multiple instances are executed. Running the Kruskal-Wallis test on multiple trials is sufficient to answer research question RQ_1.

6. **Model Ranking.** If we want to identify the pairs of models between which performance differences exist (RQ_2), an additional post-hoc test, such as Dunn's test [6], is necessary (Fig. 1.6). In this test the H_0 is that the probability of observing a randomly selected value from the first sample that is larger than a randomly selected value from the second sample equals one half, i.e., no sample dominates the other. Rejecting the H_0, implies that the first sample is dominated by the second sample. Dunn's test does not account for the number of samples, thus it needs to be adjusted by the Bonferroni correction [6]. In our methodology, we use Dunn's test to **rank the models** from the best performing one to the worst performing one in each trial, assigning the same rank to models where Dunn's H_0 cannot be rejected. To combine the results of the different trials, we calculate a base-case rank as a sum of the assigned ranks on the em_i in each performed trial using Dunn's test.

The sufficient number of trials is determined by the stability of the obtained ranking, which can be assessed using a sensitivity analysis [14], that investigates how the uncertainty in the output is related to the uncertainty in the input. In our case, the output is the order of the performance of the executed models as per the base-case rank. We test the **sensitivity** of that order on the input, i.e., the ranks from the individual trials. We use a deterministic one-at-a-time, also known as one-side, sensitivity analysis, where we vary the trials to be included in the calculation of the base-case rank. Namely, we remove one of the trials at a time in order to observe whether the aggregated order will change.

3   Use Case Scenarios

In this section we specify how we have applied the methodology to three scenarios, each comprised of sixteen semantically equivalent models generated using four transformation rules, and executed on two different BPMSs.

3.1   Transformation Rules

Based on literature review and analysis of the BPMN standard we have defined R to include the following transformation rules which preserve the execution semantics. These rules can be frequently applied on real-world models given that gateways are among BPMN's most frequently used constructs [22].

r_1 **Coalesce Joins** - *precondition*: existence of multiple nested join gateways
   of the same type; *rule*: collapse the multiple join gateways into a single one;

r_2 **Coalesce Splits** - *precondition*: existence of multiple nested split gateways
   of the same type; *rule*: collapse the multiple split gateways into a single split

gateway, adjusting the predicates when necessary to maintain the execution semantics of the model. Rules r_1, r_2 have been proposed as WFT-JC1 rule in [8];

r_3 **Use Inclusive Gateway** - *precondition*: existence of parallel, exclusive gateway or a combination of parallel and exclusive gateway; *rule*: replace the existing gateway(s) with an inclusive gateway adjusting the predicate logic accordingly. This rule is motivated by [21];

r_4 **Use Implicit Gateway** - *precondition*: existence of a parallel split, inclusive split or an exclusive join; *rule*: replace a parallel split with multiple outgoing flows from the preceding activity, an inclusive split with multiple conditional outgoing flows from the preceding activity and an exclusive join with multiple incoming flows to the succeeding activity, as per the BPMN standard [17, pp. 36–38], [23].

3.2   Executed Models

The three scenarios are aimed at testing the execution performance of semantically equivalent control flow structures expressing path parallelism (EX1_PCF), exclusive (EX2_ECF) and inclusive (EX3_ICF) path selection. The corresponding initial models (Fig. 2) are arbitrary and designed so that all four transformation rules are applicable. By adding the necessary inverse transformation rules and without loss of generality, any other of the semantically equivalent models could be used as an initial model.

If we do consider the process modeling guidelines [G] defined by Mendling et al. [21], we see that the initial models comply with: minimize the routing paths per element [G2], use one start and one end event [G3], match every split with a join of the same type [G4] and avoid OR gateways [G5]. [G4] and [G5] are also recommended by Koehler and Vanhatalo [18]. These initial models however do not use as few elements as possible [G1]. Still, some of the transformed models do follow [G1]. Use verb-object activity labels [G6] and decompose a model with more than 50 elements [G7] are out of the scope of this paper.

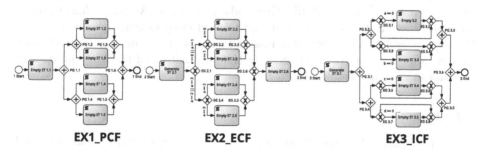

EX1_PCF   **EX2_ECF**   **EX3_ICF**

Fig. 2. Initial model m_i for each use case scenario

Starting from the three initial models, we have generated all the possible transformed models (Fig. 1.3) by applying combinations of up to four transformation rules, i.e., $z = 4$. The set of executed models $EM = \{m_i, t1, t2, ..., t1234\}$

contains 16 models, since multiple application of a single rule on a given model did not result with any structurally different models. By convention, the name of each model (e.g., $t123$) contains the index(es) of the transformation rule(s) applied to generate it (e.g., r_1, r_2, r_3). All the 48 transformed models are visualized at: http://benchflow.inf.usi.ch/bpm2017.

Given our goal of testing the performance impact of alternative control flow structures, all models in our experiments are fully automated, using mainly empty script tasks, and no manual or user tasks or service calls. This way we ensure that any identified performance bottleneck is due to the execution of the control flow, and not the tasks per se. Only the scripts that precede a decision gateway, inclusive or exclusive, contain code to generate random numbers, ensuring a uniform probability of executing any of the gateway's outgoing paths.

3.3   Load Function

The load start function is comprised of the load time, the ramp-up period, the number of users and the think time. To avoid BPMS's saturation, based on our previous experience with running experiments on Camunda [10], we simulate 500 users who gradually become active within 30 s (ramp-up period), sending BP instantiation requests each second (think time) for a period of 5 min (load time). With Activiti, we reduced the number of users to 50 and increased the load time to 15 min to ensure that a sufficiently high number of instances is started. This choice allows us to obtain performance data with few outliers after removing the warm up period instances.

We conducted a pilot study for each of the scenarios to determine the effect size using each model's mean and standard deviation. The effect size was necessary to calculate the minimal sample size with a level of significance of 5% ($\alpha = 0.05$) and power level of 80% ($power = 0.80$) using GPower[1]. We have verified that the number of executed BP instances (from 22'497 to 53'754 in Camunda and from 41'912 to 44'677 in Activiti), in each trial, was sufficient to make statistical inference of the results. We run the experiments (Fig. 1.4) on two widely-used open-source BPMSs, Camunda v.7.5.0 and Activiti v.5.21.0, using the BenchFlow framework set up in the testbed environment described in [10]. We were prevented from including more BPMSs at this point due to limitations in their Management APIs [11] making the automation of the large number of experiment runs unfeasible.

3.4   Results

For each scenario (3), trial (3), executed model (16) and BPMS (2) we collected the duration of each BP instance in milliseconds (ms) and run the corresponding statistical tests as described in Sect. 2. To ease the comparison of the different models' performance, in Table 1 (Camunda) and Table 2 (Activiti) we show the 95% confidence interval (CI) of the duration of the BP instances of all the executed models (ms) in each trial for each experiment.

---

[1] http://www.gpower.hhu.de.

The results from running the experiments on **Camunda** are presented in Table 1. The initial model in the **EX1_PCF** has an average duration between 2.51 and 2.59 ms with CI range of ±0.05. Overlapping CIs with the initial models are noticed for $t3$, $t4$, $t13$ and $t34$, while for the rest of the transformed models the CI goes down to 2.16 ± 0.03 for $t124$ in trial 1. The average duration of the initial model in **EX2_ECF** is between 1.53 and 1.55 ms with CI range of ±0.02. In this experiment only the model $t3$ has a significantly overlapping interval with the initial model, which means that their performance is very similar. The best performing model with CI of 1.32 ± 0.02 is in trial 3 for $t124$, as is the case in EX1_PCF. The average duration of the initial model in **EX3_ICF** is between 2.55 and 2.76 ms with CI range of ±0.03 to ±0.08. When the inclusive gateway is not used, i.e., for models not applying r_3, the magnitude of the variation in performance compared to the initial model is similar as in the other two experiments, with the best performing model remaining $t124$ with CI of 2.25 ± 0.03 ms in trial 1. However, CI values get much lower when r_3 is used, with CI of 1.78 ± 0.02 ms in trial 1 for the model that applies all rules ($t1234$).

Table 1. Camunda: 95% confidence intervals of the BP instance duration (ms)

	Parallel			Exclusive			Inclusive		
	Trial 1	Trial 2	Trial 3	Trial 1	Trial 2	Trial 3	Trial 1	Trial 2	Trial 3
m_i	2.59±0.05	2.55±0.05	2.51±0.05	1.53±0.02	1.54±0.02	1.55±0.02	2.55±0.03	2.76±0.08	2.73±0.06
$t1$	2.36±0.06	2.44±0.05	2.42±0.04	1.49±0.02	1.46±0.02	1.42±0.02	2.63±0.06	2.43±0.03	2.44±0.03
$t2$	2.39±0.05	2.45±0.05	2.39±0.04	1.46±0.02	1.49±0.02	1.46±0.02	2.67±0.08	2.53±0.04	2.64±0.07
$t3$	2.57±0.05	2.51±0.05	2.58±0.05	1.57±0.02	1.53±0.02	1.53±0.02	2.02±0.03	2.03±0.03	2.09±0.04
$t4$	2.49±0.06	2.45±0.04	2.42±0.04	1.46±0.02	1.49±0.02	1.47±0.02	2.57±0.04	2.66±0.05	2.68±0.08
$t12$	2.21±0.04	2.24±0.04	2.25±0.04	1.42±0.02	1.41±0.02	1.40±0.02	2.43±0.04	2.51±0.05	2.45±0.07
$t13$	2.49±0.06	2.44±0.03	2.45±0.04	1.47±0.02	1.48±0.02	1.42±0.02	1.90±0.03	1.98±0.03	2.03±0.05
$t14$	2.31±0.05	2.28±0.04	2.28±0.03	1.44±0.02	1.41±0.02	1.36±0.02	2.43±0.03	2.50±0.05	2.68±0.08
$t23$	2.37±0.04	2.48±0.05	2.41±0.05	1.48±0.02	1.47±0.02	1.46±0.02	1.89±0.03	1.97±0.03	2.05±0.04
$t24$	2.29±0.04	2.35±0.05	2.30±0.04	1.43±0.02	1.39±0.02	1.40±0.02	2.38±0.03	2.57±0.04	2.44±0.03
$t34$	2.48±0.06	2.49±0.04	2.58±0.06	1.47±0.02	1.43±0.02	1.46±0.02	1.99±0.03	2.05±0.03	2.03±0.03
$t123$	2.23±0.04	2.29±0.05	2.28±0.05	1.41±0.02	1.41±0.02	1.39±0.02	1.90±0.03	1.87±0.03	1.83±0.02
$t124$	2.16±0.03	2.21±0.04	2.23±0.04	1.38±0.02	1.33±0.02	1.32±0.02	2.25±0.03	2.34±0.03	2.48±0.06
$t134$	2.35±0.04	2.38±0.05	2.34±0.05	1.45±0.02	1.40±0.02	1.41±0.02	1.92±0.03	1.90±0.03	1.91±0.03
$t234$	2.33±0.03	2.39±0.04	2.34±0.04	1.40±0.02	1.43±0.02	1.42±0.02	1.92±0.03	1.90±0.03	1.98±0.04
$t1234$	2.20±0.03	2.21±0.04	2.28±0.04	1.36±0.02	1.35±0.02	1.37±0.02	1.78±0.02	1.81±0.02	1.87±0.03

The results from running the experiments on **Activiti** are presented in Table 2. The average values of the duration of the initial model in **EX1_PCF** are between 25.22 and 26.04 ms with CI range of ±0.21. None of the transformed models has an overlapping interval with the initial parallel flow model, and the CI of the duration goes down to 15.83 ± 0.12 ms for $t1234$ in trial 1. Exclusive control flow executes faster with average duration of the initial model in **EX2_ECF** between 2.01 and 2.14 ms with CI range between ±0.06 to ±0.07. In this experiment, although many of the transformed models have overlapping intervals with the initial model, still the best performing model $t124$ in trial 1 has a rather lower duration CI of 1.39 ± 0.02 ms. In **EX3_ICF** the average duration of the initial model returns closer to the one in EX1_PCF with values between 27.93 and 29.87 ms with CI range between ±0.24 and ±0.30. Overlapping intervals are only noticed for $t4$, and while the CI of the duration for models without inclusive gateway goes only down to 22.42 ± 0.21 ms in trial 2 for $t124$, for model $t1234$ in trial 2 it goes all the way down to 9.93 ± 0.14 ms.

Table 2. Activiti: 95% confidence intervals of the BP instance duration (ms)

	Parallel			Exclusive			Inclusive		
	Trial 1	Trial 2	Trial 3	Trial 1	Trial 2	Trial 3	Trial 1	Trial 2	Trial 3
m_i	26.04±0.21	25.33±0.21	25.22±0.21	2.12±0.07	2.01±0.06	2.14±0.07	29.87±0.30	27.93±0.24	29.57±0.27
$t1$	21.62±0.18	21.47±0.16	20.90±0.17	2.09±0.07	1.93±0.05	1.91±0.06	24.04±0.21	27.14±0.26	24.75±0.22
$t2$	21.34±0.17	20.38±0.14	20.55±0.15	2.02±0.06	1.98±0.06	1.95±0.05	25.43±0.24	27.29±0.26	26.75±0.26
$t3$	19.57±0.18	20.02±0.19	20.00±0.19	2.10±0.06	2.06±0.06	2.02±0.06	11.82±0.16	11.66±0.16	11.74±0.17
$t4$	25.32±0.21	24.55±0.20	24.66±0.20	2.06±0.06	1.92±0.05	1.99±0.06	28.92±0.26	28.24±0.26	29.64±0.27
$t12$	18.83±0.14	18.38±0.14	18.79±0.15	1.84±0.05	1.91±0.06	1.89±0.06	23.68±0.22	23.00±0.21	24.53±0.23
$t13$	18.26±0.15	17.39±0.14	17.42±0.15	1.92±0.06	1.93±0.06	2.02±0.06	11.38±0.18	11.19±0.18	11.11±0.15
$t14$	21.25±0.16	21.39±0.17	20.88±0.18	1.91±0.06	1.85±0.05	1.77±0.05	24.78±0.23	23.07±0.19	23.61±0.20
$t23$	16.52±0.14	16.37±0.14	15.62±0.12	2.11±0.06	1.93±0.05	2.12±0.07	10.56±0.14	10.90±0.16	11.06±0.16
$t24$	21.67±0.17	21.80±0.17	21.72±0.18	1.91±0.05	1.86±0.06	1.86±0.05	23.66±0.21	25.08±0.24	23.45±0.21
$t34$	19.06±0.17	18.06±0.15	19.13±0.17	2.11±0.07	1.89±0.05	1.93±0.05	11.14±0.15	11.02±0.15	11.06±0.15
$t123$	17.35±0.16	17.39±0.21	16.47±0.13	1.92±0.06	2.01±0.07	2.05±0.07	10.19±0.14	10.27±0.14	10.01±0.15
$t124$	19.09±0.14	17.86±0.12	18.64±0.14	1.39±0.02	1.45±0.03	1.59±0.04	22.46±0.21	22.42±0.21	23.31±0.24
$t134$	18.03±0.16	18.61±0.17	18.53±0.16	1.92±0.06	2.03±0.09	1.92±0.05	10.51±0.14	11.46±0.20	10.83±0.15
$t234$	18.04±0.16	17.36±0.16	17.08±0.15	1.89±0.05	1.98±0.07	2.01±0.07	10.53±0.15	10.51±0.15	10.71±0.16
$t1234$	15.83±0.12	15.75±0.13	15.86±0.13	1.72±0.04	1.67±0.04	1.62±0.04	10.03±0.13	9.93±0.14	9.99±0.13

On the raw data for the duration of each BP instance of all of the models belonging to the same experiment, we have run the Kruskal-Wallis Test (Fig. 1.5) for the significance of the differences between the models using IBM SPSS Statistics Version 24. Summary of the test's results is provided in Table 3. The *Total #* shows the total number of BP instances compared in the test, where larger *test statistic* values indicate larger differences between the compared models. The H_{RQ1} gets rejected if the values of the *Asymptotic Sig.* are smaller than 0.05.

Table 3. Kruskal-Wallis test summary results

		EX1_PCF			EX2_ECF			EX3_ICF		
		Trial 1	Trial 2	Trial 3	Trial 1	Trial 2	Trial 3	Trial 1	Trial 2	Trial 3
Camunda	Total #	739'845	721'265	775'940	765'393	817'415	850'040	487'213	494'020	544'059
	Test Statistic	6'036	5'686	3'989	2'852	3'290	3'695	27'512	29'646	29'530
	Asymptotic Sig.	0	0	0	0	0	0	0	0	0
Activiti	Total N	697'492	698'976	699'478	708'138	710'987	710'467	687'930	687'011	687'746
	Test Statistic	135'513	141'611	139'567	1'268	1'268	1'882	351'816	354'735	354'047
	Asymptotic Sig.	0	0	0	0	0	0	0	0	0

The post-hoc Dunn's test (Fig. 1.6) indicates the pairs of models with statistically significant differences in the duration of their instances. We used its results to rank each model, assigning the same rank to models where the performance differences were not significant (Table 4). The sensitivity analysis confirmed that running 3 trials in each experiment is sufficient for the desired rank stability.

3.5   Discussion

3.5.1   Statistically Significant Performance Differences (RQ_1)

The values of the asymptotic significance very close to 0 in the results of the statistical tests (Table 3) shows that there are statistically significant differences in the execution duration between instances of models which are semantically equivalent but structurally different. Thus, we can reject H_{RQ1}. The extent of performance differences between the models varies between experiments (EX1_PCF, EX2_ECF and EX3_ICF), as well as between BPMSs (Camunda and Activiti).

3.5.2   Total Order (RQ_2)

Given the cardinality of the executed models set, $|EM| = 16$, the theoretical maximal rank is 16 and would imply significantly different performance among all models. As can be seen in Table 4, execution of the models on Camunda results with 12 ranks for the parallel, 6 for the exclusive and 12 for the inclusive control flow experiment. Execution on Activiti results with 14 ranks for the parallel, 6 for the exclusive and 11 for the inclusive control flow experiment. Therefore, it is not possible to induce a total ordering (16 ranks) between all semantically equivalent models in a given experiment based on their execution duration. The actual number of ranks depends both on the BPMSs and on the experiment. In our use case the exclusive control flow models seem to have the most performance similarities, thus resulting with the smallest number of ranks (6 ranks in both BPMSs).

Table 4. Models ranked over their performance in three trials using Dunns' test

Camunda						Activiti					
EX1_PCF		EX2_ECF		EX3_ICF		EX1_PCF		EX2_ECF		EX3_ICF	
Model	Rank	Model	Rank	Model	Rank	Model	Rank	Model	Rank	Model	Rank
t124	*1*	*t124*	*1*	*t1234*	*1*	*t1234*	*1*	*t124*	*1*	*t1234*	*1*
t1234	2	*t1234*	*1*	*t123*	2	*t23*	2	*t1234*	2	*t123*	*1*
t12	3	*t12*	2	*t134*	3	*t123*	3	*t24*	2	*t234*	2
t123	3	*t123*	2	*t234*	3	*t234*	4	*t12*	3	*t23*	2
t14	4	*t24*	2	*t13*	4	*t13*	5	*t14*	3	*t13*	3
t24	5	*t234*	2	*t23*	4	*t134*	6	*t123*	4	*t134*	3
t234	6	*t14*	3	*t34*	5	*t34*	7	*t134*	4	*t34*	3
t134	6	*t134*	3	*t3*	6	*t3*	8	*t234*	4	*t3*	4
t2	7	*t1*	4	*t124*	7	*t12*	9	*t1*	5	*t124*	5
t1	7	*t13*	4	*t12*	8	*t124*	9	*t2*	5	*t12*	6
t23	8	*t2*	5	*t1*	9	*t14*	10	*t4*	5	*t14*	7
t4	8	*t23*	5	*t24*	9	*t1*	11	*t13*	5	*t24*	8
t13	9	*t34*	5	*t14*	9	*t2*	12	*t23*	5	*t1*	9
t34	10	*t4*	6	*t2*	10	*t24*	12	*t34*	5	*t2*	10
m_i	11	*t3*	7	*t4*	11	*t4*	13	m_i	6	m_i	11
t3	12	m_i	7	m_i	12	m_i	14	*t3*	6	*t4*	11

3.5.3   Experiments Performance Variability

To facilitate the visualization (Fig. 3) of the differences between the duration interval of the initial model and that of the transformed models, we have decided to use the acceptability index [26]. The **acceptability index** is calculated as $I(A, B) = \frac{m(B)-m(A)}{w(B)+w(A)}$ such that $A = [a_l, a_r]$ and $B = [b_l, b_r]$ are interval values,

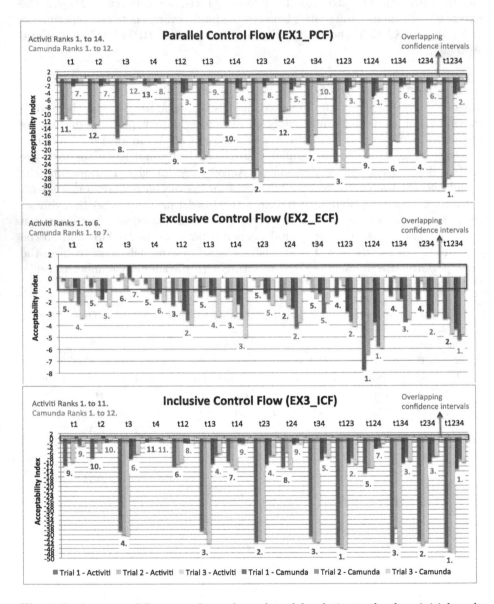

Fig. 3. Performance differences of transformed models relative to the three initial models with Activiti and Camunda (ranks and acceptability index of three trials)

where a_l, b_l and a_r, b_r stand for the left and right limits of the interval. $m(A)$ and $m(B)$, in our case, are the average duration of all the instances of the respective model, $w(A)$ and $w(B)$ are the half-width of the corresponding confidence interval. In this paper A always refers to the duration interval of the initial model m_i, and B refers to the duration interval of the transformed model $t1, ..., t1234$. Thus, negative values of the index show that on average the initial model's instances have longer execution than the instances of the respective transformed model, and vice-versa with positive index values. Index values between -1 and 1 indicate an overlap between the compared intervals A and B.

As visualized in Fig. 3, the differences between BPMSs are particularly evident in the parallel (EX1_PCF) and the inclusive (EX3_ICF) control flow experiments. In **EX1_PCF** all transformed models perform better than the initial model when executed on Activiti (c.f. Table 2). When EX1_PCF is run on Camunda, the differences are lower in magnitude (c.f. Table 1), but still statistically significant as shown by Dunns' test (c.f. Table 4). In Camunda, the only transformed model that performs worse than the initial model is $t3$ which uses the inclusive gateway with parallel gateway logic, i.e., with conditional statements which are always true. In Activiti on the other hand, $t3$ performs better than half of the models. As evident from Fig. 3, although applying the combination of all the transformation rules ($t1234$) seems to work well for both BPMSs, as mentioned earlier, in Camunda it is better not to use the inclusive gateway to implement parallelism since $t124$ ranks as the most performant one. Table 4 shows that, in Camunda all of the top performant models coalesce the split and join, i.e., contain the combination of r_1 and r_2, while in Activiti they coalesce the split while using the inclusive gateway to implement parallelism, i.e., they result from the combination of r_2 and r_3.

In **EX2_ECF**, both the difference between BPMSs and the difference between the transformation models (c.f. Fig. 3), is much smaller than in the other two experiments, with the confidence intervals, when applying just one transformation rule in Activiti, overlapping with initial model's confidence interval. The best performance is obtained by combining all rules, using either exclusive ($t124$) or inclusive gateway ($t1234$). These two models rank the same in Camunda, while in Activiti $t124$ ranks first, and $t1234$ second. The noticed trend in the first experiment for Camunda, to combine r_1 and r_2 in all top performing models, is also evident in this experiment.

The performance of the initial model in **EX3_ICF**, in both BPMSs, is similar to the one in EX1_PCF, with an even greater magnitude of differences between the initial and the transformed models in EX3_ICF (c.f. Fig. 3). The top two models ($t1234$ and $t123$) are the same in both BPMSs, with no statistically significant performance differences between them in Activiti as evident by their equal rank. The trend mentioned in EX1_PCF, i.e., to combine r_2 and r_3 in Activiti in the top performing models, reemerges in this experiment as well. In general, in EX1_PCF and EX3_ICF, the results indicate greater potential for performance improvement in Activiti than in Camunda.

3.5.4   Impact of Modeling Practices

Dumas et al. [7] show that there is no absolute truth about the impact of structuredness on understandability and that it depends on the number of gateways in
the model. Minimizing the overall model size [G1] and minimizing the number of
routing paths per element [G2] cannot be both fulfilled, as acknowledged by [21].
Another controversial guideline refers to the implicit gateway which according to
Koehler and Vanhatalo [18] simplifies BP model's visualization, while Recker's
empirial study [23] shows that explicit gateways improve the interpretational
fidelity, i.e., understandability.

Having this in mind, systematically applying the transformation rules (R)
described in this paper generates a variety of models which give priority to
different modeling practices. The graph layout guidelines and practices are not
taken into consideration.

The initial model in EX3_ICF follows the guidance of avoiding inclusive gateways [G5], by using a combination of parallel and exclusive gateways. However,
the results in both BPMSs, show that using r_3, i.e., an inclusive gateway, on
EX3_ICF models significantly improves the performance, as all the transformations applying r_3 outrank the ones that do not. As evident from Table 4, the
performance of the model combining all rules together $(t1234)$ is always ranked
as first or second in all experiments and in both BPMSs. When it is ranked second, $t124$ is ranked as first. While $t1234$ is indeed the smallest executed model
in size [G1], not all the best performing models are among the smallest ones.
For instance, in Activiti in EX1_PCF model $t23$ is second ranked, but it has
24 elements, as opposed to the smallest models in this experiment ($t1234$ and
$t124$) which have 18 elements. Furthermore, both $t1234$ and $t124$ do not comply
with other modeling guidelines, since their maximum number of routing paths
per element is 5, as opposed to 3 in the initial models [G2], they do not use
explicit gateways thus they are not structured [G4], and $t1234$ also uses inclusive gateways [G5]. Thus, in most of the cases when modeling for deployment,
if performance is important, priority should be given to minimizing the overall
model size [G1] with respect to [G2] or [G4].

However, as previously stated, modeling should be a discretional decision of
the modeler and we do not aim at changing modeling practices (or guidelines)
to improve the execution performance. Our goal is to elicit BPMS performance
improvements by enabling different BPMS vendors to test their products by
using the proposed methodology, with the same or different set of transformation
rules and models. If they notice that the acceptability indexes they calculate are
significantly lower than -1, dedicating time to implementing performance optimization can bring them competitive advantage. For instance, although further
work with greater number of initial models and with different BPMSs is required
to make generalization about the effect of the transformation rules included in
this paper, the initial results already provide some useful hints. Clearly, BPMS
vendors can boost the performance of their products by coalescing multiple splits
and joins of the same type. Activiti could also take inspiration of their implementation of the inclusive gateway, when implementing parallelism as r_3 provides a
non-negligible performance improvements.

4   Related Work

Previous studies have analyzed the impact of BP model structure on its under-standability [23,24] or error-proneness [20]. For example, Mendling et al. [21] provide seven modeling guidelines towards more comprehensive and syntacti-cally correct models, synthesized from empirical work linking model's structural characteristics with its understandability, error probability and label ambiguity.

We are not aware of any existing work studying the connection between the BPMN static control flow structure and its execution performance. On the other hand, there is extensive work on programming language compiler optimiza-tion based on transformations that reduce the number of instructions or max-imise parallelism. Bacon et al. [3] provide a comprehensive overview of compiler transformations, while Hoste et al. [15] discuss optimization space exploration strategies to provide for inevitable optimization trade-offs. Furthermore, in Data-base Management Systems (DBMSs), queries are optimized using transformation rules which preserve their execution semantics. Jarke and Koch [16] propose a framework for evaluation of query optimization, comprised of four steps: (1) find an internal query representation, (2) apply logical transformations, (3) define alternative sequences of elementary operations, and (4) find the cheapest alter-native among the ones proposed in step 3 and execute it. Taking inspiration from this existing work, we focus on transformation optimization strategies in BPMSs, and our initial goal is to assess whether different representations of the same BP significantly impact its execution performance. Gournaris [13] already points to DBs and data-centric flows as automated performance optimization opportunity in BPMN process models' execution. BPMN elements are mapped to annotated directed acyclic graphs used for optimal task ordering and task assignment based on statistical metadata, such as task duration, gathered from execution logs. While [13] targets optimal task execution, in this paper we focus on the control flow.

Work on BP models' equivalence [1] and modeling best practices [21] is related to what we do, since we use semantically equivalent models to study the effect that their structure has on their execution by a given BPMS. Eder et al. [8] propose a set of basic operations (e.g., moving or confluence of gate-ways) to transform a given BP model represented as a structured graph to a semantically equivalent model. Gert et al. [12] propose a language independent algorithm for determining semantical equivalence of fragments of structured or unstructured BP models, motivated by the industry need of BP model change management. They use a non-exhaustive set of rules for rewriting BPs into a nor-mal form, later used for fragment comparison. While we use the operations and rules mentioned in existing work [8,12], we also take into consideration BPMN-specific transformations, such as replacing explicit gateways with implicit ones. Our approach also differs in the goal of the use case which requires such trans-formations.

5   Threats to Validity

Construct Validity - We conduct our experiments on a single version of two BPMSs in a standalone deployment and only in their default configuration, since it is the configuration usually utilized by potential users when evaluating system's performance. Although each model is executed in isolation from the other models, all the instances of the same model are executed together.

Internal Validity - The experiments we perform are inherently subject to variability in the obtained metrics value, due to the many factors impacting the runtime of a software system. We mitigate this variability by defining load functions that do not overload the BPMSs [27], performing multiple trials for each of the models, and verifying the variance among trials in order to provide reliable measures validated by significance testing.

External Validity - The results we obtained present limited generalizability since: they depend on the behaviour of different BPMSs, or the same BPMS under a different load function; the size and the number of the initial models are rather small; and all models are realized by script tasks. We plan further experiments to improve and delimit the generalizability of our results w.r.t. different BPMS, load functions, initial model sizes and used BPMN 2.0 elements.

6   Conclusion and Future Work

In this work we study and compare the execution performance of semantically equivalent BP models with different control flow structures. To do so, we propose a methodology for deriving such models based on an initial model and a set of semantics-preserving transformation rules. The models are executed on different BPMSs measuring the corresponding process instance duration, which is statistically analyzed to identify and characterize significant performance differences. By applying the methodology on three scenarios (parallel, exclusive and inclusive control flows), we identify significantly different performance among the models in both BPMSs (RQ_1). However, in all experiments, it was not possible to establish a total order among all 16 semantically equivalent models (RQ_2).

The observed performance variability is more evident in Activiti (acceptability index up to -38.53) than in Camunda (up to -13.07). We discover that following certain modeling guidelines, e.g., avoiding the use of inclusive gateways when implementing inclusive control flow execution semantics, has a negative performance impact on the model's execution duration. These are only initial but promising results, measured with load functions designed to avoid system saturation: 500 users for Camunda and 50 for Activiti. Further experiments are necessary to investigate the impact of the load function on the observed performance differences. However, these results are already sufficient to demonstrate the existence of statistically significant differences in the execution of semantically equivalent models designed following different modeling practices.

Our research efforts will further explore the execution performance improvement opportunities by using larger initial models, larger sets of transformation rules and more BPMSs. We are currently comparing each of the transformation rules individually to draw conclusions on which of them are good candidates for optimization rules. These initial results pave the way towards automatic BP model performance optimization by means of semantics-preserving transformation rules that can be applied when a BP model is deployed on a specific BPMS.

Acknowledgements. This work is partially funded by the "BenchFlow" project (DACH Grant Nr. 200021E-145062/1).

References

1. Aalst, W.M.P., Medeiros, A.K.A., Weijters, A.J.M.M.: Process equivalence: comparing two process models based on observed behavior. In: Dustdar, S., Fiadeiro, J.L., Sheth, A.P. (eds.) BPM 2006. LNCS, vol. 4102, pp. 129–144. Springer, Heidelberg (2006). doi:10.1007/11841760_10
2. Abbott, M.L., Fisher, M.T.: The Art of Scalability. Pearson, Upper Saddle River (2009)
3. Bacon, D.F., Graham, S.L., Sharp, O.J.: Compiler transformations for high-performance computing. ACM Comput. Surv. (CSUR) **26**(4), 345–420 (1994)
4. Cohen, J.: A power primer. Psychol. Bull. **112**(1), 55 (1992)
5. Dattalo, P.: Determining Sample Size: Balancing Power, Precision, and Practicality. Oxford University Press, New York (2008)
6. Dinno, A.: Nonparametric pairwise multiple comparisons in independent groups using dunns test. Stata J. **15**, 292–300 (2015)
7. Dumas, M., Rosa, M., Mendling, J., Mäesalu, R., Reijers, H.A., Semenenko, N.: Understanding business process models: the costs and benefits of structuredness. In: Ralyté, J., Franch, X., Brinkkemper, S., Wrycza, S. (eds.) CAiSE 2012. LNCS, vol. 7328, pp. 31–46. Springer, Heidelberg (2012). doi:10.1007/978-3-642-31095-9_3
8. Eder, J., Gruber, W., Pichler, H.: Transforming workflow graphs. In: Konstantas, D., Bourrières, J.P., Léonard, M., Boudjlida, N. (eds.) Interoperability of Enterprise Software and Applications, pp. 203–214. Springer, London (2006)
9. Ferme, V., Ivanchikj, A., Pautasso, C.: A framework for benchmarking BPMN 2.0 workflow management systems. In: Motahari-Nezhad, H.R., Recker, J., Weidlich, M. (eds.) BPM 2015. LNCS, vol. 9253, pp. 251–259. Springer, Cham (2015). doi:10.1007/978-3-319-23063-4_18
10. Ferme, V., Ivanchikj, A., Pautasso, C.: Estimating the cost for executing business processes in the cloud. In: La Rosa, M., Loos, P., Pastor, O. (eds.) BPM 2016. LNBIP, vol. 260, pp. 72–88. Springer, Cham (2016). doi:10.1007/978-3-319-45468-9_5
11. Ferme, V., et al.: Workflow management systems benchmarking: unfulfilled expectations and lessons learned. In: Proceedings of ICSE 2017, May 2017
12. Gerth, C., et al.: Detection of semantically equivalent fragments for business process model change management. In: Proceedings of SCC, pp. 57–64. IEEE (2010)
13. Gounaris, A.: Towards automated performance optimization of BPMN business processes. In: Ivanović, M., et al. (eds.) ADBIS 2016. CCIS, vol. 637, pp. 19–28. Springer, Cham (2016). doi:10.1007/978-3-319-44066-8_2
14. Hamby, D.: A review of techniques for parameter sensitivity analysis of environmental models. Environ. Monit. Assess. **32**(2), 135–154 (1994)

15. Hoste, K., Eeckhout, L.: Cole: compiler optimization level exploration. In: Proceedings of CGO, pp. 165–174. ACM (2008)
16. Jarke, M., Koch, J.: Query optimization in database systems. ACM Comput. Surv. (CsUR) **16**(2), 111–152 (1984)
17. Jordan, D., Evdemon, J.: Business Process Model And Notation (BPMN) Version 2.0. OMG. http://www.omg.org/spec/BPMN/2.0/
18. Koehler, J., Vanhatalo, J.: Process anti-patterns: how to avoid the common traps of business process modeling. IBM WebSph. Dev. Tech. J. **10**(2), 4 (2007)
19. Marusteri, M., Bacarea, V.: Comparing groups for statistical differences: how to choose the right statistical test? Biochemia Medica **20**(1), 15–32 (2010)
20. Mendling, J.: Metrics for Process Models: Empirical Foundations of Verification, Error Prediction, and Guidelines for Correctness. LNBIP, vol. 6. Springer, Heidelberg (2008). doi:10.1007/978-3-540-89224-3
21. Mendling, J., Reijers, H.A., van der Aalst, W.M.: Seven process modeling guidelines (7PMG). Inf. Softw. Technol. **52**(2), 127–136 (2010)
22. Muehlen, M., Recker, J.: How much language is enough? Theoretical and practical use of the business process modeling notation. In: Bellahsène, Z., Léonard, M. (eds.) CAiSE 2008. LNCS, vol. 5074, pp. 465–479. Springer, Heidelberg (2008). doi:10.1007/978-3-540-69534-9_35
23. Recker, J.: Empirical investigation of the usefulness of gateway constructs in process models. Eur. J. Inf. Syst. **22**(6), 673–689 (2013)
24. Reijers, H.A., Mendling, J.: A study into the factors that influence the understandability of business process models. IEEE Trans. Syst. Man Cybern. Part A Syst. Hum. **41**(3), 449–462 (2011)
25. Rosa, M.L., et al.: Managing process model complexity via concrete syntax modifications. IEEE Trans. Ind. Inf. **7**(2), 255–265 (2011)
26. Sengupta, A., Pal, T.K.: On comparing interval numbers. Eur. J. Oper. Res. **127**(1), 28–43 (2000)
27. Skouradaki, M., Ferme, V., Pautasso, C., Leymann, F., Hoorn, A.: Micro-Benchmarking BPMN 2.0 workflow management systems with workflow patterns. In: Nurcan, S., Soffer, P., Bajec, M., Eder, J. (eds.) CAiSE 2016. LNCS, vol. 9694, pp. 67–82. Springer, Cham (2016). doi:10.1007/978-3-319-39696-5_5

Process Knowledge

Weak, Strong and Dynamic Controllability of Access-Controlled Workflows Under Conditional Uncertainty

Matteo Zavatteri[1]($\boxtimes$), Carlo Combi[1], Roberto Posenato[1], and Luca Viganò[2]

[1] Dipartimento di Informatica, Università di Verona, Verona, Italy
matteo.zavatteri@univr.it
[2] Department of Informatics, King's College London, London, UK

Abstract. A workflow (WF) is a formal description of a business process in which single atomic work units (tasks), organized in a partial order, are assigned to processing entities (agents) in order to achieve some business goal(s). A workflow management system must coordinate the execution of tasks and WF instances. Usually, the assignment of tasks to agents is accomplished by external constraints not represented in a WF. An access-controlled workflow (ACWF) extends a classical WF by explicitly representing agent availability for each task and authorization constraint. Authorization constraints model which users are authorized for which tasks depending on "who did what". Recent research has addressed temporal controllability of WFs under conditional and temporal uncertainty. However, controllability analysis for ACWFs under conditional uncertainty has never been addressed before. In this paper, we define weak, strong and dynamic controllability of ACWFs under conditional uncertainty, we present algorithmic approaches to address each of these types of controllability, and we synthesize execution strategies that specify which user has been (or will be) assigned to which task.

Keywords: Access-controlled workflow · Uncertainty · Dynamic controllability · AI-based security

1 Introduction

Context and motivation. A *workflow schema* (or simply *workflow*, *WF*) is a formal description of a business process in which single atomic work units (*tasks*), organized in a partial order, are assigned to processing entities (*agents*) in order to achieve some business *goal(s)*. A *workflow management system (WfMS)* must coordinate the execution of tasks and WF instances. Usually, the assignment of agents to tasks considers external constraints not represented in a WF [11].

An *access-controlled workflow (ACWF)* extends a classical WF by adding users and authorization constraints. Users are authorized for tasks whereas authorization constraints say which users remain authorized for which tasks depending on who did what.

© Springer International Publishing AG 2017
J. Carmona et al. (Eds.): BPM 2017, LNCS 10445, pp. 235–251, 2017.
DOI: 10.1007/978-3-319-65000-5_14

The conceptual modeling of WFs underlying business processes has been receiving increasing attention over the last years and many technical aspects have been discussed, including WF flexibility, structured vs. unstructured modeling, change management, authorization models, and temporal features and constraints (see, e.g., [4,15,17]). Recently, attention has been devoted to the issue of expressing *temporal features* of WFs, such as task-duration constraints, temporal constraints between non-consecutive tasks, deadlines and so on. Moreover, properties of such temporal WF models have been defined and analyzed. The most interesting property is *dynamic controllability*, which ensures that a WF can be executed satisfying all the given temporal constraints without the WfMS restricting and/or controlling task durations but only assuming that each such duration is within a designed range (*temporal uncertainty*) [4]. The authors of [4] also tackled dynamic controllability under another uncertainty, *conditional uncertainty*, represented by the fact that some subsets of tasks have to be executed if and only if some conditions (abstracted as Boolean propositions) are true. Similarly to what happens for uncontrollable task durations, the truth-value assignments to such propositions are *out of control*. For instance, when a patient enters the ER, the severity of his condition is not known a priory but it is established by a physician, while the WF is being executed. Since such a condition discriminates what tasks have, or have not, to be executed, the system must be able to get to the end of the WF satisfying all relevant temporal constraints regardless of which tasks have to be executed and which task durations have to be satisfied. However, to the best of our knowledge, controllability analysis of (non-temporal) ACWFs (e.g., those presented in [17]) remains unexplored.

Contributions. Towards this aim, our contributions are four-fold: (1) We define *ACWFs under conditional uncertainty* as a structured extension of a fragment of BPMN. (2) We define weak, strong and dynamic controllability of an ACWF under conditional uncertainty. (3) We provide an encoding from WF-paths into binary constraint networks and exploit directional consistency algorithms to check the consistency of a single (unconditional) path. (4) We present algorithmic approaches for the three kinds of controllability and we synthesize execution strategies that specify which user has been (or will be) assigned to which task.

Organization. Section 2 introduces a motivating example that we will use throughout the paper. Section 3 gives essential background on structured WFs and constraint networks. Section 4 introduces ACWFs under conditional uncertainty, and Sect. 5 discusses weak, strong and dynamic controllability of such ACWFs. Section 6 gives an encoding from WF-paths into constraint networks. Section 7 discusses the algorithmic approaches to the three kinds of controllability. Section 8 discusses related work. Section 9 draws conclusions and discusses future work.

2   A Motivating Example

As a running, motivating example coming from the health-care domain, we consider (an excerpt of) a simplified triage WF schema taken from [2]. We slightly

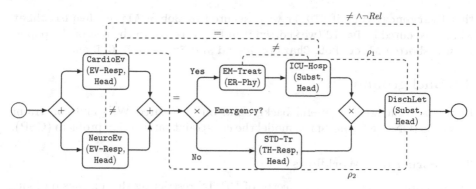

Fig. 1. An excerpt of a simplified triage WF schema. Head = {B}, EV-Resp = {A,C}, ER-Phy = {D,E}, Subst = {C,D}, TH-Resp = {A,E}.

modified it and added some access control to get a few features of interest that we want to discuss. Figure 1 shows an ACWF under conditional uncertainty having 6 tasks, 5 roles (EV-Resp, ER-Phy, Subst, TH-Resp and Head) and 5 users (Alice, Bob, Charlie, David and Eve). EV-Resp contains Alice and Charlie, ER-Phy contains David and Eve, Subst contains Charlie and David, TH-Resp contains Alice and Eve, Head contains only Bob. Head is senior to all roles besides ER-Phy. Thus, Head is authorized for all tasks except for EM-Treat.

The ACWF starts with a cardiological evaluation (CardioEv) and a neurological evaluation (NeuroEv) where an evaluation-responsible (EV-Resp) establishes whether the patient is in need of immediate medical attention. The order of these two tasks does not matter. However, this example requires that CardioEv and NeuroEv are not executed simultaneously. Hereinafter, we will assume, without loss of generality, that in the considered executions CardioEv happens before NeuroEv. The physicians executing these two tasks must be different.

After this initial parallel block terminates (and consequently the level of emergency has been assessed), the flow of execution splits in two (mutually-exclusive) paths. If the patient needs immediate medical attention (Emergency? = ⊤), then an ER-physician (ER-Phy) executes an emergency treatment (EM-Treat), and afterwards a substitute (Subst) takes care of the ICU-Hospitalization process (ICU-Hosp). This last physician must be the same who did CardioEv and must be different from the one who has just concluded EM-Treat. Instead, if Emergency? = ⊥, then a therapy-responsible (TH-Resp) carries out a standard therapy. This physician must be the same who did CardioEv.

Regardless of which WF-path (see Sect. 5) has been taken, the process concludes with a discharge letter (DischLet) released by a Subst who must be different from, and not a relative of, the physician who did CardioEv (we shorten it in Fig. 1 as ≠ ∧¬*Rel*). Alice and Bob, who are married, are the only relatives in this example. Furthermore, assume that Emergency? = ⊤; if ICU-Hosp was executed by Bob (respectively, Charlie), then DischLet will be executed by Charlie (respectively, Bob); we shorten it in Fig. 1 as ρ_1. Conversely, assume that `Emergency?` $= \perp$; if `STD-Tr` was executed by `Bob` or `Alice`, then `DischLet` will be executed by `David` (we shorten it in Fig. 1 as ρ_2). In the rest of the paper, we will shorten `Alice`, `Bob`, `Charlie`, `David` and `Eve` as A, B, C, D and E.

3   Background

In this section, we give useful background on structured WFs and constraint networks (CNs), a formalism to model the constraint satisfaction problem (CSP).

3.1   Structured Workflows

In this paper, we consider an excerpt of BPMN restricting the analysis on loop-free WFs and following the structured approach of the conceptual model Nest-Flow [3], where the specification of a WF is given by a *WF schema*, a directed graph (also called *WF graph*) where nodes correspond to *activities* and arcs represent the control flow defining dependencies between the order of execution of such activities. There exist two different types of activity: *tasks* (rounded boxes) and *connectors* (diamonds). Tasks represent elementary work units that cannot be decomposed further and that will be executed by external agents. Connectors (or *gateways* in BPMN) represent internal activities executed by the WfMS to achieve a correct and coordinated execution of tasks.

Since we focus on access control, connectors are restricted to being of two types: *total* (+) and *conditional split* (×). A connector is conditional when it splits a single flow of execution in exactly two mutually-exclusive branches or it joins two mutually-exclusive branches into a single one. A connector is total when it splits the flow of the execution into $n > 1$ parallel branches or it joins $n > 1$ incoming parallel branches into a single outgoing flow.

Figure 2 depicts the basic components of a business process. Each component can be thought of as a symbol in a context-free grammar. In particular, PROCESS (Fig. 2a) can be thought of as the starting symbols of such a grammar embedding the non-terminal symbol block $\langle B \rangle$. A non-terminal block can be a Sequence (Fig. 2b), a Parallel (Fig. 2e) or a Choice (Fig. 2f), whereas a terminal block can be a Task (Fig. 2c) or Skip (Fig. 2d).

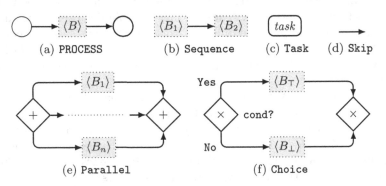

Fig. 2. A fragment of a structured BPMN.

3.2 Constraint Networks and the Constraint Satisfaction Problem

Definition 1 (CN, CSP and consistency [7]). *A Constraint Network (CN)* $\mathcal{R}$ *is a triple* $\langle X, D, C \rangle$, *where* $X = \{x_1, \ldots, x_n\}$ *is a finite set of variables,* $D = \{D_1, \ldots, D_n\}$ *is a set of associated domains* $D_i = \{v_1, \ldots, v_j\}$ *(one for each variable), and* $C = \{C_1, \ldots, C_k\}$ *is a finite set of constraints. Each* C_i *is a relation* R_i *defined over a* scope *of variables* $S_i \subseteq X$ *(i.e., if* $S_i = \{x_{i_1}, \ldots, x_{i_r}\}$, *then* $R_i \subseteq D_{i_1} \times \cdots \times D_{i_r}$). *A* Constraint Satisfaction Problem (CSP) *is the problem of assigning a value* $v_i \in D_i$ *to each variable* $x_i \in X$ *such that all constraints in* C *are satisfied. If this is possible, then the CN is* consistent.

CSP is NP-hard [7]. A CN is *binary* if all constraints have scope cardinality ≤ 2 (in general, k-*ary* if all constraints have scope cardinality $\leq k$) [7,13]. Let R_{ij} be a shortcut to represent a binary relation having scope $S = \{x_i, x_j\}$. A binary CN is *minimal* if any tuple $(v_i, v_j) \in R_{ij} \in C$ belongs to at least one global solution for the underlying CSP [13]. Thus, a minimal CN models an n-ary relation whose scope is X and whose tuples represent the set of all solutions. Besides for a few restricted classes of CNs, the general process of computing a minimal network is NP-hard [13]. Furthermore, even considering a binary minimal network, the problem of generating an arbitrary solution is NP-hard if there is no total order on the variables [10].

Therefore, a first crude technique is that of searching for a solution by exhaustively enumerating (and testing) all possible solutions and stopping as soon as one satisfies all constraints in C. This initial idea entailed the employment of more sophisticated techniques such as backtracking combined to heuristics such as *node, arc* and *path consistency* (pruning techniques) [12].

A variable x_i is *node-consistent* if for each $v \in D_i$ we have that $v \in R_{x_i}$. A CN is node-consistent if each variable is node-consistent. A variable x_i is *arc-consistent* with respect to a second variable x_j if for each $v \in D_i$ there exists $u \in D_j$ such that $(v, u) \in R_{ij}$. A CSP instance is arc-consistent if every variable is arc-consistent with respect to any other variable. A pair of variables (x_i, x_j) is *path-consistent* with respect to a third variable x_k if for any assignment $x_i = v, x_j = u$, where $v \in D_i$ and $u \in D_j$, there exists $k \in D_k$ such that $(v, k) \in R_{ik}$ and $(k, u) \in R_{kj}$. A CSP instance is path-consistent if any pair of variables are path-consistent with any other third variable.

k-*consistency* guarantees that any (locally consistent) assignment to any subset of $(k-1)$-variables can be extended to a kth (still unassigned) variable such that all constraints between these k-variables are satisfied. *Strong* k-*consistency* is k-consistency for each j such that $1 \leq j \leq k$ [9]. As a result, 1, 2 and 3-consistency are node, arc and path consistency, respectively.

Directional consistency has been introduced to speed up the process of synthesizing a solution for a constraint network limiting backtracking [8]. In a nutshell, given a total order on the variables of a CN, the network is *directional-consistent* if it is consistent with respect to the given order that dictates the assignment order of variables. In [8], an *adaptive-consistency* (ADC) *algorithm* was provided as a directional consistency algorithm adapting the level of

---

Algorithm 1. ADC($\mathcal{R}, d$)

Input: A constraint network $\mathcal{R} = \langle X, D, C \rangle$ and an ordering $d = \{x_1, \ldots, x_n\}$
Output: A backtrack-free network (along d) if $\mathcal{R}$ is consistent, inconsistent otherwise.
1 Partition the constraints as follows:
2 **for** $i \leftarrow n$ **downto** 1 **do**
3 $\quad \lfloor$ Put in $Bucket(x_i)$ all unplaced constraints mentioning x_i

4 **for** $p \leftarrow n$ **downto** 1 **do**
5 $\quad$ Let $j \leftarrow |Bucket(x_p)|$ and S_i be the scope of $R_{S_i} \in Bucket(x_p)$
6 $\quad A \leftarrow \bigcup_{i=1}^{j} S_i \setminus \{x_p\}$
7 $\quad R_A \leftarrow \pi_A (\bowtie_{i=1}^{j} R_{S_i})$
8 $\quad$ **if** R_A *is not empty* **then**
9 $\quad \quad \lfloor$ $Bucket(x') \leftarrow Bucket(x') \cup \{R_A\}$, where $x' \in A$ is the "latest" variable w.r.t. d.
10 $\quad$ **else**
11 $\quad \quad \lfloor$ **return** inconsistent

12 **return** $\mathcal{R}' = \langle X, D, \bigcup_{i=1}^{n} Bucket_i \rangle$

---

k-consistency needed to guarantee a backtrack-free search once the algorithm terminates, if the network is not inconsistent (see Algorithm 1). The input of ADC is a CN $\mathcal{R} = \langle X, D, C \rangle$ along with an order d for the set X. At each step the algorithm *adapts* the level of consistency to guarantee that if the network passes the test, a backtrack-free solution can be generated. If the network is inconsistent, the algorithm detects it before the solution generation process starts. ADC initializes a $Bucket(x_i)$ for each variable x_i and first processes all the variables top-down (i.e., from last to first in the order d) by filling each bucket with all (still unplaced) constraints $R_{S_i} \in C$ whose scope S_i contains x_i. Then, it processes again the variables top-down and, for each variable x_i, it computes a new scope A consisting of the union of all scopes of the relations in $Bucket(x_i)$ neglecting x_i itself. After that, it computes a new relation R_A by joining all $R_{S_i} \in Bucket(x_i)$ and projecting with respect to A ($\bowtie$ and π are the classical relational algebra operators). In this way, it enforces the appropriate level of consistency. If the resulting relation is empty, then $\mathcal{R}$ is inconsistent; otherwise, the algorithm adds R_A to the bucket of the latest variable in A (with respect to the order d), and goes on by processing the next variable. Finally, it returns a network $\mathcal{R}'$ (possibly tighter than the initial $\mathcal{R}$) whose set of constraints is equal to the union of all (tighten) relations in the buckets. Note that ADC takes as input a k-ary CN $\mathcal{R}$ and returns a k'-ary CN $\mathcal{R}'$, where $k' \geq k$ (an example of a binary CN turned into a ternary one can be found in [7, Chap. 4]).

Time and space complexity of ADC are $\mathcal{O}(n(2z)^{w^*+1})$ and $\mathcal{O}(nz^{w^*})$, respectively, where $n = |X|$, $z = \max_{i=1,\ldots,n} |D_i|$ and w^* is the induced width of the graph along the order of processing [7, Chap. 4]. Informally, w^* represents the maximum number of variables that can be affected by the value assumed by another variable.

4   ACWFs Under Conditional Uncertainty

In this section, we extend the fragment of (the structured) BPMN we discussed in Sect. 3.1 by injecting a role-based access-control model (RBAC, [16]) and formalizing authorization constraints at user level into the process specification. We call this language *access-controlled BPMN (ACBPMN)*. Before we begin, we point out our assumptions on `Choice` blocks.

Given a set P of propositional letters, a *label* ℓ is any (possibly empty) conjunction of literals, where a literal is either a propositional letter $p \in P$ or its negation $\neg p$. The *empty label* is denoted by $\boxdot$. The *label universe of* P, denoted by P^*, is the set of all possible labels representing all possible (finite) conjunctions of literals we can obtain from P. Two labels $\ell_1, \ell_2 \in P^*$ are *consistent* if and only if their conjunction $\ell_1 \wedge \ell_2$ is satisfiable.

We assume that each conditional split connector of a `Choice` block is associated to a unique proposition $p \in P$ whose truth value assignment is *not decided* by the WfMS but by some run-time condition. We call such property *conditional uncertainty*. That is, the WfMS is only able to *observe* such a truth value assignment once the conditional split connector has been executed. We also assume that all tasks of the WF are implicitly labeled by labels according to the nesting levels of the `Choice` blocks. This is implicit since the WF is structured. For example, in Fig. 2f, the label of each task in $B_\top$ contains c, whereas that of each task in $B_\bot$ contains $\neg c$, where c is a Boolean proposition modeling `cond?`. In case of nested conditional blocks (suppose that $B_\top$ is another `Choice` block having sub-blocks $F_\top$ and $F_\bot$ and split connector associated to f), the labels of tasks in $F_\top$ are $c \wedge f$, whereas those of $F_\bot$ are $c \wedge \neg f$, and so on.

We formalize the label of each task as a function $L\colon T \to P^*$, where T is the set of tasks. Getting back to our example, we have that $P = \{e\}$, where e abstracts `Emergency`, the condition associated to the (unique) conditional connector. The labels of the tasks are: $L(\texttt{CardioEv}) = L(\texttt{NeuroEv}) = \boxdot$, $L(\texttt{EM-Treat}) = L(\texttt{ICU-Hosp}) = e$, $L(\texttt{STD-Tr}) = \neg e$ and $L(\texttt{DischLet}) = \boxdot$, meaning that all tasks labeled by $\boxdot$ are always executed, and the remaining ones are conditionally executed depending on the truth value of their labels.

Definition 2 (Role-Based Access-Control Models for Workflows). *The RBAC part of a business process consists of a finite set of roles $R = \{r_1, \ldots, r_l\}$, a finite set of users $U = \{u_1, \ldots, u_m\}$ and a finite set of tasks $T = \{t_1, \ldots, t_n\}$. Roles are associated to both users and tasks, acting as an interface between them. Formally, $UA \subseteq U \times R$ is the many-to-many user-to-role assignment relation, whereas $TA \subseteq T \times R$ is the many-to-many task-to-role assignment relation. Thus, $(u, r) \in UA$ means that u belongs to role r, whereas (t, r) means that task t can be executed by any user (i.e., agent) belonging to role r. We write $users(r) = \{u \mid (u, r) \in UA\}$ for the set of users belonging to a role r, and $roles(t) = \{r \mid (t, r) \in TA\}$ for the set of roles authorized for a task t. We abuse notation and write $users(t) = \{u \mid (u, r) \in UA \wedge r \in roles(t)\}$ for the set of users authorized for a task.*

Figure 3 shows the two main extensions to the language given in Fig. 2. We label a task t by a finite set of roles $\{r_1, \ldots, r_e\} \subseteq R$ (Fig. 3a) meaning that $(t, r_1), \ldots, (t, r_e) \in TA$. Assigning roles to tasks models "who does what".

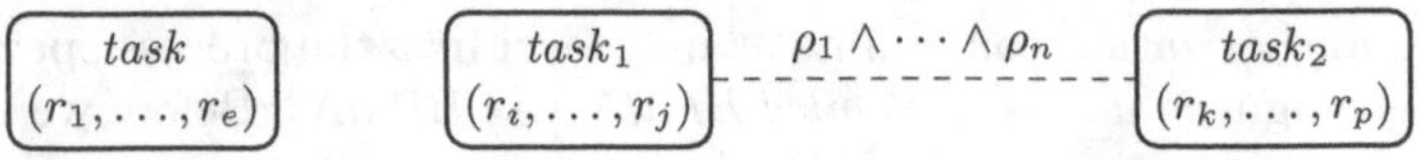

(a) Task   (b) AuthorizationConstraint

Fig. 3. Injecting roles (a) and authorization constraints (b).

However, classical RBAC models are unable to specify security policies at user level such as *separation of duties (SoD)* and *binding of duties (BoD)*.[1] To address such an issue, we express *authorization constraints* between pairs of tasks t_1, t_2, having consistent labels[2] $L(t_1)$ and $L(t_2)$, as a conjunction of binary relations $\rho_1 \wedge \cdots \wedge \rho_n$ over users ($\rho_i \subseteq U \times U$) such that if $u_1 \in users(t_1)$ and $u_2 \in users(t_2)$ and the pair (u_1, u_2) also belongs to all ρ_i, then any execution assigning t_1 to u_1 and t_2 to u_2 *satisfies* the authorization constraint. In Fig. 1 $L(\texttt{CardioEv}) = \boxdot$ and $L(\texttt{ICU-Hosp}) = e$ are consistent; thus, in the execution where the conditional split connector assigns $\top$ to e, both of these two tasks must be executed, and the authorization constraint labeled by $=$ (connecting them) satisfied.

In ACBPMN, we draw *authorization constraints* as undirected dashed edges (Fig. 3b) connecting pairs of tasks and label them by a conjunction of binary relations as we have just discussed. Figure 1 is an ACWF under conditional uncertainty expressing which roles are authorized for which tasks, and which are the authorization constraints. Again, no authorization constraint is specified between $\texttt{EM-Treat}$ and $\texttt{STD-Tr}$ as $L(\texttt{EM-Treat}) = e$ and $L(\texttt{STD-Tr}) = \neg e$ are inconsistent with each other.

5   Controllability of ACWFs Under Conditional Uncertainty

We give the semantics for *weak*, *strong* and *dynamic controllability* of ACWFs under conditional uncertainty. Our goal is that of synthesizing execution strategies specifying which user to assign to which task so that all relevant authorization constraints will eventually be satisfied no matter which (uncontrollable) truth values the conditional connectors assign to their associated propositions. Tasks and constraints are *relevant* if the WfMS must consider them during execution.

A *scenario* (or *interpretation function*) $s: P \to \{\top, \bot\}$ is a complete assignment of truth values to the propositions associated to all conditional split connectors in the Choice blocks. $\mathcal{I}$ models the set of all possible scenarios.

In our example, there are two possible (mutually-exclusive) scenarios $s_1(e) = \top$ and $s_2(e) = \bot$, modeling the patient's urgency level. Once a scenario is selected, the WF-path corresponding to the considered scenario is the projection of the initial WF with respect to the scenario.

---

[1] SoD is a security policy saying that a subset of tasks must be carried out by different users, whereas BoD says that a subset of tasks must be carried out by the same user.

[2] Two tasks with consistent labels must be considered in at least one WF execution.

Definition 3 (WF-path). *A* workflow path (WF-path) *is the projection of an ACWF with respect to a given scenario s; i.e., a new (unconditional) ACWF obtained as a copy of the considered ACWF, where all tasks and authorization constraints having label not consistent with s are removed.*

The ACWF in Fig. 1 consists of the following two WF-paths

(1) `CardioEv → NeuroEv → EM-Treat → ICU-Hosp → DischLet`
(2) `CardioEv → NeuroEv → STD-Tr → DischLet`

where (1) is the result of the projection according to scenario s_1, and (2) to s_2. For both WF-paths, the order of `CardioEv` and `NeuroEv` does not matter.

Definition 4 (Schedule). *A schedule, for a subset of tasks $T' \subseteq T$, is a mapping $\psi \colon T' \to U$ assigning users to those tasks. The set of all possible schedules is represented by Ψ. A schedule is* consistent *if the assignments it makes eventually satisfy all relevant authorization constraints.*

An *execution strategy* is a mapping $\sigma \colon \mathcal{I} \to \Psi$ from scenarios to schedules such that the domain of the resulting ψ consists of all tasks belonging to the WF-path arising from the projection of the ACWF with respect to s. If $\psi = \sigma(s)$ is also consistent, then we say that σ is *viable*. We denote the user u assigned to the task t by the strategy σ in the scenario s as $[\sigma(s)]_t = u$.

The first kind of controllability is *weak controllability*, which ensures that we can execute the WF satisfying all user and authorization constraints whenever we have full information on the uncontrollable part *before* starting the execution. That is, whenever we know what truth values the `Choice` blocks will assign to the associated propositions.

Definition 5 (Weak Controllability). *An ACWF under conditional uncertainty is* weakly controllable *if there exists a viable execution strategy.*

Dealing with such a controllability is quite complex as it *always* requires one to predict how all uncontrollable parts will behave before starting the execution (i.e., to predict which the scenario will be). This lead us to considering the opposite case, the one in which we want to synthesize a strategy returning a static schedule working for all combinations of truth value assignments (i.e., for all scenarios) *before* starting the execution. Thus, the second kind of controllability is *strong controllability*, which ensures that we can preassign users to tasks *before* starting, being guaranteed that such an assignment will always satisfy all constraints whatever the truth value assignments.

Definition 6 (Strong Controllability). *An ACWF under conditional uncertainty is* strongly controllable *if there exists a viable execution strategy σ such that for any pair of scenarios $s_1, s_2 \in \mathcal{I}$ and any shared task t belonging to the WF-paths arising from the projections of s_1, s_2, we have that $[\sigma(s_1)]_t = [\sigma(s_2)]_t$.*

Strong controllability is, however, "too strong". If an ACWF is not strongly controllable, it could be still executable deciding which user to commit to which task by *reacting* to the uncontrollable parts as soon as they become known. To achieve this purpose, we introduce *dynamic controllability*. An ACWF is dynamically controllable if it admits a viable execution strategy able to modify its associated schedule (if needed) whenever a truth value for a still unknown proposition becomes known due to the execution of the associated conditional split connector. Since the truth values of propositions are *revealed incrementally*, in what follows we introduce the formal definition of *history* that we then use to define dynamic controllability.

Definition 7 (History). *The history $\mathcal{H}(t, s)$ of a task t in the scenario s consists of the set of truth value assignments to the propositions made by the conditional split connectors executed before task t in s.*

Take Fig. 1 as an example and consider the scenario $s(e) = \top$. It holds that $\mathcal{H}(\texttt{CardioEv}, s) = \emptyset$ before executing the conditional split connector and $\mathcal{H}(\texttt{EM-Treat}, s) = \{e\}$ after executing the conditional split connector.

Definition 8 (Dynamic Controllability). *An ACWF is* dynamically controllable *if there exists a viable execution strategy σ such that for any pair of scenarios $s_1, s_2 \in \mathcal{I}$ and any task t belonging to the WF-path arising from the projection s_1, it holds that if $\mathcal{H}(t, s_1) = \mathcal{H}(t, s_2)$, then t belongs to the WF-path arising from the projection of s_2, and $[\sigma(s_1)]_t = [\sigma(s_2)]_t$.*

That is, an ACWF is dynamically controllable if there exists an execution strategy assigning users to incoming tasks consistently (with the previous assignments), knowing only the values of propositions associated to the conditional split connectors already executed.

6   Encoding WF-Paths into Constraint Networks

As we have discussed previously, a WF-path is the projection of a particular scenario. Uncontrollable parts no longer exist in this projection.

We focus on tasks only, implicitly considering that the WfMS takes care of executing the connectors and that no constraints between connectors exist.

To encode a WF-path into a CN suitable for our purposes, we first turn the current partial order between tasks into a *total* one $d = (t_1, \ldots, t_n)$, where $t_1, \ldots, t_n \in T$. In this way, we can guarantee an efficient (dynamic) user assignment during execution as we exploit directional consistency. Recall that a dynamic assignment of values (i.e., users) to variables (i.e., tasks) no matter the order in which the variables are chosen (i.e., tasks executed by users) has already been proved to be NP-hard in [10]. Instead, following an order when assigning the variables has been proved to be linear [7,8] after checking the consistency of the initial CN, which remains NP-hard.

Indeed, the initial checking does not worry about computing a solution, but it only focuses on proving that at least a solution exists by tightening the network (this is the hard part).

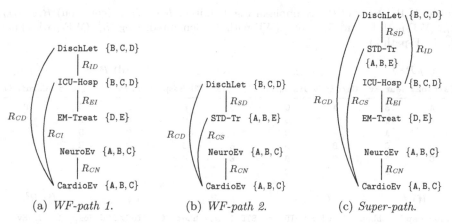

(a) *WF-path 1.*   (b) *WF-path 2.*   (c) *Super-path.*

Fig. 4. (Binary) Constraint networks for WF-path 1 (a), WF-path 2 (b), and the super-path modeling the whole WF turned unconditional (c). The order of tasks has to be interpreted bottom-up (↑).

In this paper, we accept that the initial check of consistency on these equivalent CNs is NP-hard in favor of guaranteeing an efficient (dynamic) execution. Of course, the more users, tasks and Choice blocks a WF specifies, the more this approach becomes intractable in its preliminary phase.

We start from a single WF-path and we encode it in an equivalent CN $\mathcal{R} = \langle X, D, C \rangle$. We fill X with the set of tasks belonging to the WF-path under analysis. Considering WF-path 1, we have that $X = \{$CardioEv, NeuroEv, EM-Treat, ICU-Hosp, DischLet$\}$. We now restrict the current partial order so that it becomes total. In this WF-path, the unordered tasks are CardioEv and NeuroEv, and, as we previously said in Sect. 2, we assume that CardioEv executes first. Thus, the order is $d = ($CardioEv, NeuroEv, EM-Treat, ICU-Hosp, DischLet$)$, where CardioEv and DischLet are the first and the last task, respectively. As for classic ordered CNs, we show the graphical representation of the ordered tasks of WF-path 1 in Fig. 4a. We interpret the figure bottom-up (↑). For each variable x_i modeling the task t_i, the corresponding domain of the variable consists of all users authorized for t_i, i.e., $D_{t_i} = users(t_i)$. For WF-path 1, $D_{\text{CardioEv}} = D_{\text{NeuroEv}} = \{A, B, C\}$, $D_{\text{EM-Treat}} = \{$D, E$\}$, $D_{\text{ICU-Hosp}} = D_{\text{DischLet}} = \{B, C, D\}$. We show these domains on the right of the variables in Fig. 4a.

We initialize the set C of constraints as follows. For each pair of tasks t_1, t_2 connected by an authorization constraint (labeled by $\rho_1 \wedge \cdots \wedge \rho_n$) in the original WF-path, we add to C the relation $R_{t_1 t_2} = \rho_1 \cap \cdots \cap \rho_n$ as the set of all tuples (u_i, u_j) satisfying the resulting relation where $u_i \in D_{t_1}$ and $u_j \in D_{t_2}$. Any tuple (u_i, u_j) of $R_{t_1 t_2}$ means that if task t_1 is assigned to user u_i and task t_2 is assigned to user u_j, then the original authorization constraint between t_1 and t_2 is satisfied. For instance, we encode $\neq \wedge \neg Rel$ between CardioEv and NeuroEv as the binary relation R_{CN} (Table 1a). To ease reading, we shorten the subscript of each relation with the first letters of the two tasks it constrains.

Table 1. Relations of the example in Fig. 1. Initial: R_{CN}, R_{CD} (common) R_{CI}, R_{EI}, R_{ID} (WF-path 1), and R_{CS}, R_{SD} (WF-path 2). Generated: R_{CE}, R_C^1 (WF-path 1) and R_C^2 (WF-path 2).

(a) R_{CN}		(b) R_{CD}		(c) R_{EI}		(d) R_{CE}		(e) R_C^1
CardioEv	NeuroEv	CardioEv	DischLet	EM-Treat	ICU-Hosp	CardioEv	EM-Treat	CardioEv
A	B	A	C	D	B	B	D	B
A	C	A	D	D	C	B	E	C
B	A	B	C	E	B	C	D	
B	C	B	D	E	C	C	E	
C	A	C	B	E	D			
C	B	C	D					

(f) R_{ID}		(g) R_{CS}		(h) R_{SD}		(i) R_{CI}		(j) R_C^2
ICU-Hosp	DischLet	CardioEv	STD-Tr	STD-Tr	DischLet	CardioEv	ICU-Hosp	CardioEv
B	C	A	A	A	D	B	B	A
C	B	B	B	B	D	C	C	B

We do the same with the authorization constraints between CardioEv and ICU-Hosp (R_{CI}, Table 1i), CardioEv and DischLet (R_{CD}, Table 1b), EM-Treat and ICU-Hosp (R_{EI}, Table 1c), and ICU-Hosp and DischLet (R_{ID}, Table 1f). We proceed similarly for WF-path 2 (we show its CN in Fig. 4b).

7   Weak, Strong and Dynamic Controllability Checking

We now address the algorithmic part for weak, strong and dynamic controllability defined in Sect. 5.

7.1   Weak Controllability Checking

Weak Controllability Checking (WC-checking) simply checks that each CN encoding a WF-path is consistent. Once we have encoded each WF-path in a CN, we can simply employ any algorithm for directional consistency we like and run it on each one of these networks as input. If these CNs are *all* consistent, then the WF is weakly controllable.

In our example, we use ADC (discussed in Sect. 3.2). Figure 5a shows the run of ADC for the CN of WF-path 1 (Fig. 4a), whereas Fig. 5b shows the run for the CN of WF-path 2 (Fig. 4b). More precisely, for WF-path 1 we start from the equivalent CN $\mathcal{R} = \langle X, D, C \rangle$ along with the order d we discussed in the second half of Sect. 6. ADC processes the variables (i.e., tasks) top down ($\downarrow$) by filling the corresponding buckets with the relations in C (Table 1) as we discussed at the end of Sect. 3.2. Initial relations appear on the left of $\|$, whereas new ones generated by the algorithm appear on the right. ADC starts by processing $Bucket(\texttt{DischLet})$ by computing the new scope $A = \{C, D, I\} \setminus \{D\} = \{C, I\}$. It then computes (i.e., infers the implicit possibly missing) relational constraints $R_{CI} = \pi_{C,I}(R_{CD} \bowtie R_{ID})$. Since we will assign ICU-Hosp after CardioEv (as in

$Bucket(\texttt{DischLet}) : R_{CD}, R_{ID} \|$

$Bucket(\texttt{ICU-Hosp}) : R_{CI}, R_{EI} \|$

$Bucket(\texttt{EM-Treat}) : \| R_{CE}$

$Bucket(\texttt{NeuroEv}) : R_{CN} \|$

$Bucket(\texttt{CardioEv}) : \| R_C^1$

(a) *WF-path1.*

$Bucket(\texttt{DischLet}) : R_{CD}, R_{SD} \|$

$Bucket(\texttt{STD-Tr}) : R_{CS} \|$

$Bucket(\texttt{NeuroEv}) : R_{CN} \|$

$Bucket(\texttt{CardioEv}) : \| R_C^2$

(b) *WF-path2.*

Fig. 5. ADC run on WF-path1 and WF-path2.

WF-path 1 ICU-Hosp is the latest variable in the set {ICU-Hosp, CardioEv}), we add R_{CI} to $Bucket(\texttt{ICU-Hosp})$. Actually, since $Bucket(\texttt{ICU-Hosp})$ already contains R_{CI}, adding the new one is equivalent to tightening the existing one to the intersection between itself and this new one. But since this intersection results in the same R_{CI}, ADC does nothing. Also, since the generated $R_{CI} \neq \emptyset$, ADC goes ahead by processing $Bucket(\texttt{ICU-Hosp})$. This time, the scope of the generated constraint is $A = \{C, I, E\} \setminus \{I\} = \{C, E\}$. Therefore, ADC computes $R_{CE} = \pi_{C,E}(R_{CI} \bowtie R_{EI})$ (Table 1d) and adds it to $Bucket(\texttt{EM-Treat})$ since the bucket is empty and EM-Treat comes after CardioEv. Since $R_{EI} \neq \emptyset$, ADC now takes into account EM-Treat, by computing the new scope $A = \{C, E\} \setminus \{E\} = \{C\}$. It then computes $R_C^1 = \pi_C(R_{CE})$ (Table 1e) and adds to $Bucket(\texttt{CardioEv})$. This (node-consistency) constraint rules out A for CardioEv. Again, $R_C^1 \neq \emptyset$ so ADC processes $Bucket(\texttt{NeuroEv})$ resulting in no tightening for $Bucket(\texttt{CardioEv})$ (it just generates (again an) $R_C^1 = \pi_C(R_{CN})$, which does not imply any new tightening for R_C^1 in $Bucket(\texttt{CardioEv})$). Since ADC did not come across any empty relation, the CN in Fig. 4a is consistent.

We proceed similarly for WF-path 2. This time ADC eventually generates R_C^2 (Table 1j) ruling out C for CardioEv.

Our example is weakly controllable. A viable strategy is σ is $\sigma(e) = \psi_e$ and $\sigma(\neg e) = \psi_{\neg e}$, where $\psi_e(\texttt{CardioEv}) = $ C, $\psi_e(\texttt{NeuroEv}) = $ A, $\psi_e(\texttt{EM-Treat}) = $ E, $\psi_e(\texttt{ICU-Hosp}) = $ C, $\psi_e(\texttt{DischLet}) = $ B (WF-path 1), whereas $\psi_{\neg e}(\texttt{CardioEv}) = $ A, $\psi_{\neg e}(\texttt{NeuroEv}) = $ C, $\psi_{\neg e}(\texttt{STD-Tr}) = $ A, $\psi_{\neg e}(\texttt{DischLet}) = $ D (WF-path 2).

The time complexity of WC-checking is $\mathcal{O}(2^m n(2z)^{w^*+1})$. Indeed, it corresponds to the complexity of ADC multiplied for the number of possible different WF-paths, which in the worst case are 2^m when considering a sequence of m Choice blocks. However, the complexity of the execution is linear in the number of the tasks (the strategy σ has already been synthesized).

7.2   Strong Controllability Checking

Strong Controllability Checking (SC-checking) does not need to unfold all WF-paths and test them independently. SC-checking works as follows. We first turn every conditional split connector × into a total one + (i.e., all Choice blocks become Parallel ones). Then, we encode this (now unique) *super-path* into a CN exactly as we discussed in Sect. 6 for a single WF-path and run again any algorithm for directional consistency in order to synthesize a strategy $\sigma(s)$ working *no matter s*. We show the corresponding CN in Fig. 4c.

Our example is *not* strongly controllable. Indeed, any viable strategy σ implies $\psi(\texttt{CardioEv}) = \texttt{B}$, the only *conservative* choice for CardioEv not having any information on which WF-path we will have to take. This implies in turn that both $\psi(\texttt{ICU-Hosp}) = \texttt{B}$ and $\psi(\texttt{STD-Tr}) = \texttt{B}$ (recall that the authorization constraints between CardioEv and ICU-Hosp (WF-path 1) and between CardioEv and STD-Tr (WF-path 2) are both labeled by =). Now, since there are two authorization constraints connecting ICU-Hosp, STD-Tr to DischLet labeled by ρ_1 and ρ_2, we can see that there is no valid user for DischLet as $\psi(\texttt{ICU-Hosp}) = \texttt{B}$ implies $\psi(\texttt{DischLet}) = \texttt{C}$, whereas $\psi(\texttt{STD-Tr}) = \texttt{B}$ implies $\psi(\texttt{DischLet}) = \texttt{D}$. The time complexity of SC-checking coincides with that of ADC after turning the WF unconditional (linear in the number of tasks and constraints).

7.3 Dynamic Controllability Checking

Dynamic Controllability Checking (DC-checking) refines the WC-checking by reasoning on the labels of tasks shared by different WF-paths. In our example, WF-path 1 and WF-path 2 share CardioEv, NeuroEv, and DischLet tasks. These tasks must always be executed since $L(\texttt{CardioEv}) = L(\texttt{NeuroEv}) = L(\texttt{DischLet}) = \boxdot$ (in the initial ACWF under conditional uncertainty).

Approaches such as keeping the intersection of the users authorized for those tasks with respect to different WF-paths are in general wrong. In our example, the authorized users for DischLet are $\{\texttt{C}, \texttt{B}\}$ (WF-path 1), and $\{\texttt{D}\}$ only (WF-path 2). Thus, $\{\texttt{C}, \texttt{B}\} \cap \{\texttt{D}\} = \emptyset$ (indeed, the ACWF is not strongly controllable).

The intuition is that given a WF-path along with its total order among tasks, for each pair of tasks t_1 and t_2 such that t_1 is before t_2, we have that $L(t_2)$ must also contain $L(t_1)$. Getting back to our example, we have that $L(\texttt{DischLet}) = e$ because $L(\texttt{ICU-Hosp}) = e$ if we focus on WF-path 1, and $L(\texttt{DischLet}) = \neg e$ because $L(\texttt{STD-Tr}) = \neg e$ if we focus on WF-path 2. Therefore, our idea is that of *conditionally intersecting* the set of authorized users for tasks shared by WF-paths if and only if the labels of these tasks are consistent.

As a result, a "fixed-point" DC-checking algorithmic approach works in rounds until no tightening is possible. First, we encode each WF-path into a CN using the encoding provided in Sect. 6 as we did for weak controllability. Then, every round is as follows. (1) We run ADC on every CN. If one of these CNs is inconsistent, then the ACWF is not dynamically controllable (this is the *certificate* of "no"). (2) For each task t shared by more than one WF-path, we put in *Bucket(t)* of each WF-path (containing t) all the relations appearing in the same bucket of all other WF-paths (containing t) provided that the labels for t in these different WF-paths are consistent.

In our example, we have that, after running ADC on WF-path 1 and WF-path 2 as input we have generated R_{CE}, and $R_C^1 = \{\texttt{B}, \texttt{C}\}$ (WF-path 1), and $R_C^2 = \{\texttt{A}, \texttt{B}\}$ (WF-path 2). Therefore, in the CN encoding WF-path 1 we add R_C^2 to *Bucket(CardioEv)* and in the CN encoding WF-path 2 we add R_C^1 to

Bucket(CardioEv). That is, for both CNs we tighten the unary constraint getting $R_C^{1,2} = R_C^1 \cap R_C^2 = \{B\}$ (ruling out both A *and* C since no dynamic strategy assigning one of these users to CardioEv exists). We do the same with R_{CN} for both NeuroEv's buckets resulting in no tightening. We now run again ADC on the two WF-paths since in the previous round we made at least one modification (we tightened *Bucket*(CardioEv) in all CNs). In round 2 no tightening occurs, so the DC-checking terminates. Since no inconsistent WF-path has been detected, the ACWF under conditional uncertainty is dynamically controllable.

Dynamic controllability ensures that we can synthesize a strategy $\sigma(s)$ both offline and *online* (i.e., during execution) no matter which s is being generated.

We synthesize a strategy offline as we did for WC-controllability. The difference is that the choices of users for shared tasks have already been restricted.

We (efficiently) synthesize a strategy online by generating a solution as follows. We start by considering *all* those WF-paths containing the initial task t and we assign a user u such that u satisfies all the relations in *Bucket*(t) (for the first task *Bucket*(t) only contains unary relations enforcing node-consistency). If the ACWF starts with a Choice block, we choose the WF-paths to start from according to the truth value observed (in case of nested Choice blocks we proceed recursively). Then, we execute the next tasks moving from the considered set of WF-paths to others (more specific) according to the scenario being generated and picking a user satisfying all relations in their corresponding buckets (that is, a users satisfying all the constraints between that task and all tasks which have already been executed).

Our example is dynamically controllable. A viable strategy $\sigma(s)$ must initially consider both WF-paths, so the only conservative choice is $\sigma(\square) = \psi_\square$, where $\psi_\square$(CardioEv) = B. Then, if the conditional connector assigns $\top$ to e the strategy knows it has to go through WF-path 1. Hence, $\sigma(e) = \psi_e$ where ψ_e(EM-Treat) = D , ψ_e(ICU-Hosp) = B and ψ_e(DischLet) = C. Instead, if $e = \bot$, then $\sigma(\neg e) = \psi_{\neg e}$ where $\psi_{\neg e}$(STD-Tr) = B and $\psi_{\neg e}$(DischLet) = D. Since every execution takes into consideration either ICU-Hosp or STD-Tr (as the tasks belong to WF-paths which are mutually-exclusive) no authorization constraint is violated. The time complexity of DC-checking is $\mathcal{O}(2^{(m+1)}n(2z)^{w^*+1})$, where 2^m is the number of possible different WF-paths in the worst case and $\mathcal{O}(n(2z)^{w^*+1})$ is the complexity of ADC algorithm (see Sect. 3).

8 Related Work

The problem of verifying WF features related to the assignment of agents to tasks is known in literature as WF satisfiability and resiliency [17]. More specifically, the *workflow satisfiability problem (WSP)* is the problem of finding an assignment of users to tasks such that the execution of the WF gets to the end satisfying all authorization constraints. The *workflow resiliency problem* is WSP under the uncertainty that a maximum number of users may become (temporally) absent before or during execution. WSP does not address conditional uncertainty. In this work, we exploited controllability analysis to deal with a *dynamic* WSP, where we decide, during execution, which users to assign to which tasks depending on how the uncontrollable conditional part is behaving.

Other related work lies in the area of temporal networks, where some extensions injecting users and authorization constraints into the specification have been provided. In [6], *simple temporal networks with uncertainty (STNUs, [14])* are extended with security constraints in order to model temporal role-based ACWFs in which authorization constraints and temporal constraints mutually influence one another. Controllability checking has not been addressed for such an extension. *Access-Controlled Temporal Networks (ACTNs)* [5] address users and (conditional) temporal authorization constraints. DC-checking is done via timed game automata. WC-checking and SC-checking have not been addressed for ACTNs. Moreover, the analysis phase in [5] (i.e., the checking for DC before starting) stops as soon as *one* (offline) dynamic execution strategy is synthesized. In this work, DC-checking does not synthesize a particular dynamic strategy, but it handles all possible strategies satisfying the initial constraints.

In [1], Cabanillas et al. address the resource allocation for business processes. They consider an RBAC environment and they do not impose any particular order on activities. They also address loops. However, the authors clearly state that their work is unable to address *History-Based Allocation* of resources.

9   Conclusions and Future Work

We defined ACWFs under conditional uncertainty by injecting access control into BPMN. We then gave the semantics for weak, strong and dynamic controllability of ACWFs under conditional uncertainty and discussed the algorithmic approaches to both address these decision problems and synthesize execution strategies. We also discussed the complexity of WC, SC, and DC-checking. For classic (unconditional) ACWFs if the authorization constraints are monotone (e.g., $=$) the checking is done in n^3 where n is the number of tasks [13]. The same holds for non-monotone relations if each task has no more that 2 users authorized. In general, the problem is NP-hard. Adding conditions can only make it worse. However, ADC considerably speeds up the analysis (for the general CSP) allowing one to compute the necessary level of consistency to guarantee that a solution can be generated without backtracking. Once this check is done, every solution (i.e., assignment of users to tasks) can be generated in polynomial time. This means that the runtime phase is tractable.

As future work, we plan to implement and verify our approach on real-world ACWFs, where, as in the clinical domain, sophisticated security policies need to be specified, managed and enforced while executing the WF. We also plan to benchmark the proposed algorithms once we have carried out a more thorough analysis on how to compute the most conservative order for tasks. This is because for some WF instances the answer "dynamic controllable" or "uncontrollable" depends on how the tasks have been ordered.

References

1. Cabanillas, C., Resinas, M., del Río-Ortega, A., Cortés, A.R.: Specification and automated design-time analysis of the business process human resource perspective. Inf. Syst. **52**, 55–82 (2015). doi:10.1016/j.is.2015.03.002

2. Cimatti, A., Hunsberger, L., Micheli, A., Posenato, R., Roveri, M.: Dynamic controllability via timed game automata. Acta Informatica **53**(6–8), 681–722 (2016). doi:10.1007/s00236-016-0257-2

3. Combi, C., Gambini, M., Migliorini, S.: The NestFlow interpretation of workflow control-flow patterns. In: Eder, J., Bielikova, M., Tjoa, A.M. (eds.) ADBIS 2011. LNCS, vol. 6909, pp. 316–332. Springer, Heidelberg (2011). doi:10.1007/978-3-642-23737-9_23

4. Combi, C., Gambini, M., Migliorini, S., Posenato, R.: Representing business processes through a temporal data-centric workflow modeling language: an application to the management of clinical pathways. IEEE Trans. Syst. Man Cybern. Syst. **44**(9), 1182–1203 (2014). doi:10.1109/TSMC.2014.2300055

5. Combi, C., Posenato, R., Viganò, L., Zavatteri, M.: Access controlled temporal networks. In: Proceedings of the 9th International Conference on Agents and Artificial Intelligence (ICAART), pp. 118–131 (2017). doi:10.5220/0006185701180131

6. Combi, C., Viganò, L., Zavatteri, M.: Security constraints in temporal role-based access-controlled workflows. In: CODASPY 2016, pp. 207–218 (2016). doi:10.1145/2857705.2857716

7. Dechter, R.: Constraint Processing. Kaufmann, San Francisco (2003)

8. Dechter, R., Pearl, J.: Network-based heuristics for constraint-satisfaction problems. Artif. Intell. **34**(1), 1–38 (1987). doi:10.1016/0004-3702(87)90002-6

9. Freuder, E.C.: A sufficient condition for backtrack-free search. J. ACM **29**, 24–32 (1982)

10. Gottlob, G.: On minimal constraint networks. Artif. Intell. **191–192**, 42–60 (2012). doi:10.1016/j.artint.2012.07.006

11. Hollingsworth, D.: The workflow reference model (1995). http://www.wfmc.org/standards/model.htm

12. Mackworth, A.K.: Consistency in networks of relations. Artif. Intell. **8**(1), 99–118 (1977). doi:10.1016/0004-3702(77)90007-8

13. Montanari, U.: Networks of constraints: fundamental properties and applications to picture processing. Inf. Sci. **7**, 95–132 (1974). doi:10.1016/0020-0255(74)90008-5

14. Morris, P.H., Muscettola, N., Vidal, T.: Dynamic control of plans with temporal uncertainty. In: IJCAI 2001, pp. 494–502 (2001)

15. Reijers, H., Mendling, J.: Modularity in process models: review and effects. In: Dumas, M., Reichert, M., Shan, M.-C. (eds.) BPM 2008. LNCS, vol. 5240, pp. 20–35. Springer, Heidelberg (2008). doi:10.1007/978-3-540-85758-7_5

16. Sandhu, R.S., Coyne, E.J., Feinstein, H.L., Youman, C.E.: Role-based access control models. IEEE Comput. **29**(2), 38–47 (1996). doi:10.1109/2.485845

17. Wang, Q., Li, N.: Satisfiability and resiliency in workflow authorization systems. ACM Trans. Inf. Syst. Secur. **13**(4) (2010). doi:10.1145/1880022.1880034

An Eye into the Future: Leveraging A-priori Knowledge in Predictive Business Process Monitoring

Chiara Di Francescomarino[1(✉)], Chiara Ghidini[1], Fabrizio Maria Maggi[2], Giulio Petrucci[1,3], and Anton Yeshchenko[2]

[1] FBK-IRST, Via Sommarive 18, 38050 Trento, Italy
{dfmchiara,ghidini,petrucci}@fbk.eu
[2] University of Tartu, Ulikooli 18, 50090 Tartu, Estonia
{f.m.maggi,anton.yeshchenko}@ut.ee
[3] University of Trento, Via Sommarive 14, 38050 Trento, Italy

Abstract. Predictive business process monitoring aims at leveraging past process execution data to predict how ongoing (uncompleted) process executions will unfold up to their completion. Nevertheless, cases exist in which, together with past execution data, some additional knowledge (a-priori knowledge) about how a process execution will develop in the future is available. This knowledge about the future can be leveraged for improving the quality of the predictions of events that are currently unknown. In this paper, we present two techniques - based on Recurrent Neural Networks with Long Short-Term Memory (LSTM) cells - able to leverage knowledge about the structure of the process execution traces as well as a-priori knowledge about how they will unfold in the future for predicting the sequence of future activities of ongoing process executions. The results obtained by applying these techniques on six real-life logs show an improvement in terms of accuracy over a plain LSTM-based baseline.

Keywords: Predictive Process Monitoring · Recurrent Neural Networks · Linear Temporal Logic · A-priori Knowledge

1 Introduction

Predictive business process monitoring [19] is a research topic aiming at developing techniques that use event logs extracted from information systems in order to predict how ongoing (uncompleted) process executions (a.k.a. cases) will unfold up to their completion. A recent stream of work [12,13,23,28] has been focused on the provision of techniques able to predict the future path (continuation) of an ongoing case, a type of predictions that can be used to provide valuable input for planning and resource allocation. These predictions are generally based on: (i) the *sequence of activities* already executed in the case; (ii) the *timestamp* indicating when each activity in the case was executed; and (iii) the *values of data attributes* after each execution of an activity in the case.

© Springer International Publishing AG 2017
J. Carmona et al. (Eds.): BPM 2017, LNCS 10445, pp. 252–268, 2017.
DOI: 10.1007/978-3-319-65000-5_15

What motivates this paper is the surmise that past event logs, or more in general knowledge about the past, is not the only important source of knowledge that can be leveraged to make predictions. In many real life situations, cases exist in which, together with past execution data, some case-specific additional knowledge (a-priori knowledge) about the future is available and can be leveraged for improving the predictive power of a predictive process monitoring technique. Indeed, this additional a-priori knowledge is what characterizes the future context of execution of the process that will affect the development of the currently running cases. Think for instance to the temporary unavailability of a surgery room which may delay or even rule out the possibility of executing certain activities in a patient treatment process. While it is impractical to retrain the predictive algorithms to take into consideration this additional knowledge every time it becomes available, it is also reasonable to assume that considering it in some way would improve the accuracy of the predictions on an ongoing case.

In light of this motivation, in Sect. 5, we provide two techniques based on Recurrent Neural Networks with Long Short-Term Memory (LSTM) cells [16] able to leverage a-priori knowledge about process executions for predicting the sequence of future activities of an ongoing case. The proposed algorithms are opportunely tailored in a way that the a-priori knowledge is not taken into account for training the predictor. In this way, the a-priori knowledge can be changed on-the-fly at prediction time without the need to retrain the predictive algorithms. In particular, we introduce:

- a NOCYCLE technique which is able to leverage knowledge about the structure of the process execution traces, and in particular about the presence of repetitions of sequences (i.e., cycles), to improve a plain LSTM-based baseline so that it does not fall into a local minimum, a phenomenon already hinted in [28] but not yet solved;
- an A-PRIORI technique which takes into account a-priori knowledge together with the knowledge that comes from historical data.

In Sect. 6, we present a wide experimentation carried out using six real-life logs and aimed at investigating whether the proposed algorithms increase the accuracy of the predictions. The outcome of our experiments is that the application of these techniques provides an improvement up to 50% in terms of prediction accuracy over the baseline. In addition to the core part (Sects. 5 and 6), the paper contains an introduction to some background notions (Sect. 2), a detailed illustration of the research problem (Sect. 4), related work (Sect. 3) and concluding remarks (Sect. 7).

2   Background

In this section, we report the background concepts useful for understanding the remainder of the paper.

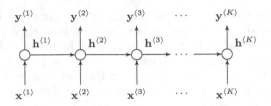

Fig. 1. Recurrent Neural Network

2.1 Event Logs and Traces

An event log is a set of traces, each representing the execution of a process (case instance). Each trace consists of a sequence of activities, each referring to the execution of an activity in a finite activity set A.

Definition 1 (Trace, Event Log). *A trace* $\sigma = \langle a_1, a_2, ...a_n \rangle \in A^*$ *over A is a sequence of activities. An event log $L \in \mathcal{B}(A)$ is a multi-set of traces over the activity set A.*

A *prefix of length* k of a trace $\sigma = \langle a_1, a_2, ...a_n \rangle \in A^*$, is a trace $p_k(\sigma) = \langle a_1, a_2, ...a_k \rangle \in A^*$ where $k \leq n$; the *suffix of the prefix of length* k is defined as the remaining part of σ, that is, $s_k(\sigma) = \langle a_k + 1, a_k + 2, ...a_n \rangle \in A^*$. For example, the prefix of length 3 of $\langle a, c, r, f, s, p \rangle$ is $\langle a, c, r \rangle$, while the suffix of this prefix is $\langle f, s, p \rangle$.

A *cycle* in a trace $\sigma \in A^*$ is a sequence of activities repeated at least twice in σ (with adjacent repetitions). For example, trace $\langle a, b, a, b, a, b, c, d, e, f, g, e, f, g, c, d \rangle$ contains two cycles: $\langle a, b \rangle$ (3 repetitions) and $\langle e, f, g \rangle$ (2 repetitions).

2.2 RNNs and LSTM

Artificial Neural Networks (or just Neural Networks, NNs) are a well known class of discriminative models. In classification tasks, they are used to model the probability of a given input to belong to a certain class, given some features of the input. We can describe them in mathematical terms as follows:

$$p(\mathbf{y}|\mathbf{x}) = f_{NN}(\mathbf{x}; \theta). \tag{1}$$

In (1), $\mathbf{x}$ is the feature vector that represents the input, $\mathbf{y}$ is a random variable representing the output class labels, f_{NN} is the function modeled by the neural network, and θ is the set of parameters of such a function to be learnt during the training phase.

Recurrent Neural Networks (RNNs, see Fig. 1) are a subclass of Neural Networks. We illustrate them with the help of an example in which the classification task concerns in assigning the correct part of speech – noun, verb, adjective, etc. – to words. If we take the word *"file"* in isolation, it can be both a noun and a verb. Nonetheless, this ambiguity disappears when we consider it in an actual sentence. Therefore, in the sentence *"I have to file a complain"* it acts as a verb, while in the sentence *"I need you to share that file with me"* it acts as a noun.

This simple example shows that for some tasks the classification at a certain time-step t depends not only on the current input (i.e., *"file"*) but also on the input (i.e., the part of the sequence) seen so far. The tasks that share this characteristic are said to be *recurrent*. Natural Language tasks are a typical example of recurrent phenomena.

In mathematical terms, let us write $\mathbf{x}^{\langle 1 \rangle}, ..., \mathbf{x}^{\langle K \rangle}$ to indicate an input sequence of K time-steps, represented by the superscript between angle brackets. In this way, at each time-step t, the conditional probability of a given input to belong to a certain class is described by

$$p(\mathbf{y}^{\langle y \rangle}|\mathbf{x}^{\langle t \rangle}, ..., \mathbf{x}^{\langle 1 \rangle}) = f_{RNN}(\mathbf{x}^{\langle 1 \rangle}, ..., \mathbf{x}^{\langle t \rangle}; \theta). \tag{2}$$

RNNs have been proven to be extremely appropriate for modeling sequential data (see [15]). As shown in Fig. 1, they typically leverage recurrent functions in their hidden layers, which are, in turn, composed of hidden states. Let $\mathbf{h}^{\langle t \rangle}$, with

$$\mathbf{h}^{\langle t \rangle} = h(\mathbf{x}^{\langle t \rangle}, \mathbf{h}^{\langle t-1 \rangle}; \theta_h); \tag{3}$$

be the activation of the hidden state at the t-th time-step. h is a so-called *cell function*, parameterized over a set of parameters θ_h to be learnt during the training, and accepting as inputs the current input $\mathbf{x}^{\langle t \rangle}$ and its value at the previous time-step $\mathbf{h}^{\langle t-1 \rangle}$. The activation of the hidden state is then mapped (using a linear map) into a continuous vector of the same size as the number of output classes. All the elements in such a vector are greater than zero and their sum is equal to one. Therefore, this vector can be seen as a probability distribution over the output space. All these constraints can be easily achieved specifying the generic Eq. (2) by means of a softmax function:

$$p(\mathbf{y}^{\langle y \rangle}|\mathbf{x}^{\langle t \rangle}, ..., \mathbf{x}^{\langle 1 \rangle}) = softmax(\mathbf{W}\mathbf{h}^{\langle t \rangle} + \mathbf{b}); \tag{4}$$

where the weight matrix $\mathbf{W}$ and the bias vector $\mathbf{b}$ are parameters to be learnt during the training phase.

Among the different cell functions h (see Eq. (3)) explored in literature, Long Short-Term Memory (LSTM) [16] shows a significant ability to maintain the memory of its input across long time spans. This property makes them extremely suitable to be used in RNNs that have to deal with input sequences with complex long-term dependencies such as the ones we consider in this paper.

2.3   RNNs with LSTM for Predictive Process Monitoring

In order to provide predictions on the suffix of a given prefix (of a running case), state-of-the-art approaches for predictive process monitoring use RNNs with LSTM cells. The most recent and performing approach in this field [28] relies on an encoding of activity sequences that combines features related to the activities in the sequence (the so called *one-hot encoding*) and features related to the time characterizing these activities. Given the set $A = \{a_{1_A}, ... a_{m_A}\}$ of all possible activities, an ordering function $idx : A \rightarrow \{1, ..., |A|\} \subseteq \mathbb{N}$ is defined on it, such that $a_{i_A} <> a_{j_A}$ if and only if $i_A <> j_A$, i.e., two activities have the same A-index if and only if they are the same activity. For instance, if $A = \{a, b, c\}$, we have $idx : A \rightarrow \{1, 2, 3\}$ and $idx(a) = 1$, $idx(b) = 2$ and $idx(c) = 3$. Each activity $a_i \in \sigma$ is encoded as a vector (A_i) of length $|A| + 3$ such that the first $|A|$ features are all set to 0, except the one occurring at the index of the current activity $idx(a_i)$, which is set to 1. The last three features of the vector pertain to time: the first one relates to the time increase with respect to the previous activity, the second reports the time since midnight (to distinguish between working and night time), and the last one refers to the time since the beginning of the week.

A trace is encoded by composing the vectors obtained from all activities in the trace into a matrix. During the training phase, the encoded traces are used for building the LSTM model. During the testing phase, a (one-hot encoded) prefix of a running case is used to query the learned model, which returns the predicted suffix by running an inference algorithm. Algorithm 1 reports the inference algorithm introduced in [28] and based on RNN with LSTM cells for predicting the suffix of a given prefix $p_k(\sigma)$ of length k. The algorithm takes as input the prefix $p_k(\sigma)$, the LSTM model $lstm$ and a maximum number of iterations max and returns as output the complete trace (the prefix and the predicted suffix). First, the prefix $p_k(\sigma)$ is encoded by using the one-hot encoding (line 5). The resulting matrix is then used for feeding the LSTM model and getting the probability distribution over different possible symbols that can occur in the next position of the trace (line 6). The symbol with the highest probability is hence selected from the ranked probabilities (line 7). Then, a new trace is obtained by concatenating the current prefix with the new predicted symbol (line 8). In order to predict the second activity, the one-hot encoding of the new prefix is computed and used to recursively feed the network. The procedure is iterated until the predicted symbol is the end symbol or a maximum number of iterations max is reached (line 10).

Algorithm 1. Inference algorithm for predicting the suffix of $p_k(\sigma)$

1: **function** PREDICTSUFFIX($p_k(\sigma)$, $lstm$, max)
2:     $h = 0$
3:     $trace = p_k(\sigma)$
4:     **do**
5:         $trace_{encoded} = $ ENCODE($trace$)
6:         $next_symbol_probs = $ PREDICTNEXTSYMBOLS($lstm$, $trace_{encoded}$)
7:         $next_symbol = $ GETSYMBOL($next\_symbol\_prob$, $trace_{encoded}$)
8:         $trace = trace \cdot next\_symbol$
9:         $h = h + 1$
10:     **while** ($next\_symbol <> end\_symbol$) and ($h < max$)
11:     **return** $trace$
12: **end function**

2.4 Linear Temporal Logic

In our approach, the a-priori knowledge that describes how a running case will develop in the future is formulated in terms of Linear Temporal Logic (LTL) rules [22]. LTL is a modal logic with modalities devoted to describe time aspects. Classically, LTL is defined for infinite traces. However, to describe the characteristics of a business process, we use a variant of LTL defined for finite traces (since business processes are supposed to complete eventually). We assume that activities occurring during the process execution fall into the set of atomic propositions. LTL rules are constructed from these atoms by applying the temporal operators $\bigcirc$ (next), $\Diamond$ (future), $\square$ (globally), and $\sqcup$ (until) in addition to the usual boolean connectives. Given a formula φ, $\bigcirc\varphi$ means that the next time instant exists and φ is true in the next time instant (strong next). $\Diamond\varphi$ indicates that φ is true sometimes in the future. $\square\varphi$ means that φ is true always in the future. $\varphi \sqcup \psi$ indicates that φ has to hold at least until ψ holds and ψ must hold in the current or in a future time instant.

3 Related Work

The literature related to predictive business process monitoring can be roughly classified according to the type of predictions that is provided. A first group of works focuses on the time perspective. In [2], the authors present a set of approaches in which annotated transition systems, containing time information extracted from event logs, are used to: (i) check time conformance; (ii) predict the remaining processing time of incomplete cases; (iii) recommend appropriate activities to end users working on these cases. In [14], an approach for predicting business process performances is presented. The approach is based on context-related execution scenarios discovered and modeled through state-aware performance predictors. In [24], the authors use stochastic Petri nets to predict the remaining execution time of a process execution. In [20], the authors present a technique for predicting the delay between the expected and the actual arrival time of cases pertaining to a transport and logistics process. In [25], queue theory is used to predict possible delays in process executions.

Another set of works in the literature focuses on approaches that generate predictions and recommendations to reduce risks. For example, in [6], the authors present a technique to support process participants in making risk-informed decisions with the aim of reducing the process risks. Risks are predicted by traversing decision trees generated from logs of past process executions. In [21], the authors make predictions about time-related process risks by identifying and leveraging statistical indicators observable in event logs that highlight the possibility of transgressing deadlines. In [27], an approach for Root Cause Analysis through classification algorithms is presented.

A third group of prediction approaches predicts the outcome (e.g., the satisfaction of a business objective) of a case. In [19] a framework is introduced, which is able to predict the fulfillment (or the violation) of a boolean predicate in a running case, by looking at: (i) the sequence of activities already performed in the case; and (ii) the data payload of the last activity of the running case. The framework, which provides accurate results at the expense of a high runtime overhead, has been enhanced in [9] by introducing a clustering preprocessing step in which cases sharing a similar activity history are clustered together. A classifier for each cluster is trained with the data payload of the traces in the cluster. In [17], the authors compare different feature encoding approaches where traces are treated as complex symbolic sequences, that is, sequences of activities each carrying a data payload consisting of attribute-value pairs. In [29], unstructured information contained in text messages exchanged during process executions has been leveraged for improving the prediction accuracy.

The problem investigated in this paper falls into a fourth and last set of works, i.e., into the set of very recent efforts aiming at predicting the sequence of future activities given the activities observed so far. In [23], Polato et al. propose several techniques for predicting the remaining time and the sequence of future activities in an ongoing case using simple regression, regression with contextual information, and data-aware transition systems. Other approaches [12,13,28] make use of RNNs with LSTM cells. In particular, Evermann et al. [12,13] propose an RNN with two hidden layers trained with back propagation, while Niek et al. [28] leverage LSTM and an encoding based on activities and timestamps (illustrated in detail in Sect. 2.3) to provide predictions on the next activities and their timestamps. Differently from all these works, this paper investigates how to take advantage of possibly existing a-priori knowledge for making predictions on the sequence of future activities.

4   The Problem

Predictive business process monitoring methods use past process executions, stored in event logs, in order to build predictive systems that work at runtime to make predictions about the future. Among the different interesting and appealing types of predictions about the future of an ongoing case, such as the remaining time or the fulfilment of a predicate, we can find the prediction of the sequence of future activities. This type of predictions can be useful in the scenario where some planning and resource allocation are needed for the running case. For instance, the hospital management can be highly interested in predicting the future activities of patients to be able to best organize machines and resources of a hospital.

Nonetheless, predicting sequences of activities is a quite complex and challenging task, as the longer the sequence is, the more difficult is to predict the most far-away activities. While predicting the sequence of future activities entirely from past execution data may be difficult, in real world scenarios, we often observe that some a-priori knowledge about the future of the running process executions exists and could hence be leveraged to support the predictive methods and improve their accuracy. For instance, in the hospital example, new medical guidelines may provide new knowledge on the fact that two treatments are not useful if used together in order to cure a certain disease, or that a certain screening is required in order to perform a specific surgery, or also that if a patient is allergic to a specific treatment she will never go to take it.

This a-priori knowledge can be expressed in terms of LTL rules. For instance, in the hospital example, LTL can be used for defining the following rules:

1. `treatmentA` and `treatmentB` cannot be both used within the same course of cure of a patient:

$$\neg(\Diamond\texttt{treatmentA} \wedge \Diamond\texttt{treatmentB}) \tag{5}$$

2. `screeningC` is a pre-requisite to perform `surgeryD`:

$$(\neg\texttt{surgeryD} \sqcup \texttt{screeningC}) \vee \Box(\neg\texttt{surgeryD}) \tag{6}$$

3. `treatmentB` cannot be performed on this course of cure (e.g., because the patient is allergic to it):

$$\neg\Diamond\texttt{treatmentB} \tag{7}$$

In this paper, we aim at understanding whether and how a-priori knowledge can be leveraged in order to improve the accuracy of the prediction of the (sequence of the) next activity(ies) of an ongoing case in a reasonable amount of time. For instance, in the example of the hospital, being aware of the fact that `treatmentA` and `treatmentB` can never be executed together could help in ruling out a prediction of `treatmentB` whenever we have already observed `treatmentA` and vice versa.

Formally, given a prefix $p_k(\sigma) = \langle a_1, ..., a_k \rangle$ of length k of a trace $\sigma = \langle a_1, ..., a_n \rangle$ and some knowledge $\mathcal{K}(\sigma)$ on σ, the problem we want to face is to identify the function f such that $f(p_k(\sigma), \mathcal{K}(\sigma)) = s_k(\sigma)$.

5   The Solution

Predicting the suffix of a given prefix is a problem that is tackled by state-of-the-art approaches that make use of LSTM-based RNNs [12,13,28]. We hence start from these approaches and build on top of them to take into account a-priori knowledge.

Before presenting our approach, we need to observe that a basic solution that can be used to leverage a-priori knowledge for making predictions is the one provided by the inclusion of the a-priori knowledge in the data used for training the prediction model. However, this solution would raise a main practical problem: since the a-priori knowledge can in principle change from case to case, this would require to retrain the model for each prediction, thus hampering the scalability of the predictive system. A smarter approach is hence required for taking into account a-priori knowledge when predicting the future path of an ongoing case.

In the next sections, we first introduce an enhancement, called NOCYCLE, of state-of-the-art approaches for overcoming the issues encountered with traces characterized by a high number of cycles (Sect. 5.1). We then describe A-PRIORI, an algorithm that allows us to take into account a-priori knowledge expressed in terms of LTL rules (Sect. 5.2). In both cases, we use the RNN architecture with LSTM cells and training system proposed in [28], while we extend and enhance the prediction phase. The A-PRIORI algorithm for accounting for a-priori knowledge and the enhancement for dealing with cycles are then combined into the A-PRIORI* technique.

5.1   Learning from Trace Structures

By experimenting the LSTM approach on different event logs, we found that event logs with traces containing a high number of repetitions of cycles perform worse than others, as also observed in [28]. This is mainly due to the fact that frequent repetitions of a cycle cause an increase in the probability distribution of the back-loop, i.e., the connection between the last and the first element of the cycle. To overcome this problem, we propose to equip Algorithm 1 with an additional function in charge of weakening such a back-loop probability. This function is composed of two parts: in the first part, the current trace is analyzed in order to discover possible cycles; in the second part, the cycle discovery is used for preventing the prediction of further repetitions of the cycle. More in detail:

1. For each prefix $p_k(\sigma) = \langle a_1 a_2 \ldots a_k \rangle$ of size k, the algorithm checks if there are j ($j >= 2$) consecutive occurrences of a cycle $c = \langle a_{c_1} \ldots a_{c_s} \rangle$, such that the last activity of the prefix corresponds to the last activity of the cycle $idx(a_k) = idx(a_{c_s})$;
2. j is then used to correct the distribution over different possible activities that can occur in the next position by decreasing the probability of the first activity of the cycle a_{c_1} to occur again. To decrease this probability, the algorithm uses a coefficient, function of the number of cycle repetitions j, as a weight to adjust the probability distribution. Examples of formulas that can be used for this purpose are j^2 or e^j.

Algorithm 2 reports the pseudo-code of the NOCYCLE technique. Similarly to Algorithm 1 presented in Sect. 2.3, it takes as input a prefix $p_k(\sigma)$, the trained LSTM model $lstm$, and the maximum number max of iterations allowed. Then, it returns as output the complete trace (the prefix and the predicted suffix). In particular, the algorithm adds to the state-of-the-art Algorithm 1 the WEAK-ENPROB procedure described above to find cycles in the trace and decrease the probability of the first activity of the cycle to occur again at the end of a repetition. The resulting vector of weakened probabilities is hence used for getting the next symbol as in the basic procedure.

5.2   Learning from A-priori Knowledge

The overall idea for leveraging a-priori knowledge for predictive monitoring is simple: (i) we use the LSTM approach to get the possible predictions for an ongoing trace; (ii) we rank them according to the likelihood of the prediction; and (iii) we select the first prediction that is compliant with the LTL rules describing the a-priori knowledge. However, although RNN inference algorithms are not computationally expensive per se, building all the possible predicted suffixes could be costly and inefficient.

Therefore, the alternative investigated in this paper leverages, on top of state-of-the-art LSTM techniques, the approach classically used in statistical sequence-to-sequence predictions in translation tasks [30], i.e., the *beamSearch* algorithm. The beamSearch is a heuristic algorithm based on graphs that explores the search space by expanding only the most promising branches. Then, in the testing

---

Algorithm 2. NOCYCLE extension for predicting the suffix of $p_k(\sigma)$

1: **function** PREDICTSUFFIXNOCYCLE($p_k(\sigma)$, *lstm*, *max*)
2:     $h = 0$
3:     *trace* $= p_k(\sigma)$
4:     **do**
5:         *trace*$_{encoded}$ = ENCODE(*trace*)
6:         *next_symbol_prob* = PREDICTNEXTSYMBOLS(*lstm*, *trace*$_{encoded}$)
7:         *weak_next_symbol_prob* = WEAKENPROB (*trace*, *next_symbol_prob*)
8:         *next_symbol* = GETSYMBOL(*weak_next_symbol_prob*, *trace*$_{encoded}$)
9:         *trace* = *trace* · *next_symbol*
10:         $h = h + 1$
11:     **while** (*next_symbol* <> *end_symbol*) and ($h < max$)
12:     **return** trace
13: **end function**

---

phase, to predict a certain suffix, we use a new inference algorithm (A-PRIORI), which explores the probability space using beamSearch to cut the branches of the LSTM model which bring to predictions that are not compliant with the a-priori knowledge.

Algorithm 3 reports the pseudo-code describing the A-PRIORI algorithm. It takes as input the prefix $p_k(\sigma)$, the available a-priori knowledge $\mathcal{K}(\sigma)$, and the trained LSTM model *lstm*, together with three parameters: (i) *bSize*, which is the maximum number of next symbols predicted by the LSTM model and used to construct the possible predicted suffixes at each iteration; (ii) *maxSize*, which is the maximum number of branches that can be explored by A-PRIORI at the same time; and (iii) *max*, which is the maximum number of allowed iterations.

Algorithm 3. A-PRIORI algorithm for predicting the suffix of $p_k(\sigma)$

1: **function** A-PRIORI ($p_k(\sigma)$, $\mathcal{K}(\sigma)$, *lstm*, *bSize*, *maxSize*, *max*)
2:     $h = 0$
3:     *prefixes* $= \{p_k(\sigma)\}$
4:     **while** ($h \leq max$) and (not ISEMPTY(*prefixes*)) **do**
5:         *candidates_next* = PREDICTPREFNEXTSYMBOLS(*lstm*, *prefixes*, *bSize*)
6:         *top_candidates* = TOPRANK(*candidates_next*, *maxSize*)
7:         EMPTY(*prefixes*)
8:         **for all** *candidate* in *top_candidates* **do**
9:             **if** LAST_SYMBOL(*candidate*) <> *end_symbol* **then**
10:                 PUSH(*candidate*, *prefixes*)
11:             **else**
12:                 **if** CHECK(*candidate*, $\mathcal{K}$) **then**
13:                     **return** *candidate*
14:                 **end if**
15:             **end if**
16:         **end for**
17:         $h = h + 1$
18:     **end while**
19: **end function**

Intuitively, the algorithm iterates over a priority queue of prefixes, which is initialized with the input prefix $p_k(\sigma)$ (line 3) and is used for regulating the number of branches to be explored. For each prefix in $prefixes$, $bSize$ possible next activities are predicted using the model $lstm$ and, for each prefix, $bSize$ new traces are obtained by concatenating the prefix with the corresponding $bSize$ predicted next activities (line 5). In this way, the algorithm generates $|prefixes| *$ $bSize$ traces. In order to limit the search space, the algorithm ranks the predicted traces based on their estimated probability[1] and takes only the top $maxSize$ ones (line 6). For each of these traces (line 8), if the last symbol predicted is not the end symbol, the trace is added to $prefixes$ (line 10). Otherwise, if the trace is complete, the algorithm checks if it is compliant to the LTL rules in $\mathcal{K}(\sigma)$ (line 12). In this case, the trace is returned (line 13). The algorithm is then iterated until the queue of prefixes is empty or the maximum number of iterations max is reached (line 4).

5.3 Implementation

Algorithms 2 and 3 (and their combination) have been implemented in Python 2.6. In particular, the Keras [5] and TensorFlow [3] libraries have been used for neural networks. The LTL checker for checking the compliance of traces with respect to LTL rules is instead based on automata and written in Java. The Py4J library has been used as a gateway to access Java code from Python. The full source code is available on github at https://github.com/yesanton/ ProcessSequencePrediction.

6   Evaluation

In this section, we provide an evaluation of our predictive business process monitoring techniques based on a-priori knowledge. In detail, we check: (i) whether the NOCYCLE algorithm leveraging knowledge about the structure of the process execution traces (and in particular about the presence of cycles) actually improves the accuracy of the predictions; and (ii) whether the combination of NOCYCLE with A-PRIORI, the A-PRIORI* algorithm, is able to leverage a-priori knowledge to improve the performance of the LSTM model.

6.1   Event Logs

For the evaluation of the techniques, we used six real-life event logs. Four of them were provided for the BPI Challenge (BPIC) 2011 [1], 2012 [10], 2013 [26], and 2017 [11], respectively. We also used two additional event logs, one pertaining to an environmental permit application process ("WABO"), used in the context of the CoSeLoG project [4] (*EnvLog* for short in this paper), and another containing cases from a ticketing management process of the help desk of an Italian

---

[1] Note that, in order to prevent overflow in the computation, the estimated probability for sequences of activities is computed as the sum of the logarithm of the probabilities of the next activities rather than as the product of the probabilities of the next activities.

Table 1. The event logs

Log	#Tr.	#Act.	avg-TL	avg-CR	Spars.
EnvLog	937	381	41.562	0.14	0.3191
HelpDesk	3804	9	3.6	0.22	0.0024
BPIC11	911	424	54.168	5.05	0.4654
BPIC12	9658	6	7.5	1.35	0.0006
BPIC13	7554	13	8.675	1.45	0.0017
BPIC17	31508	26	17.826	0.46	0.0008

software company (*Helpdesk*[2] for short). Note that all the logs have been filtered. In particular, *BPIC12*, *BPIC13*, *EnvLog* and *HelpDesk* are the ones used in [28], in order to ease the comparison of our techniques with the state-of-the-art. Similarly, *BPIC11* and *BPIC17* have been filtered by removing outlier traces with respect to the average trace length.

The characteristics of these logs are summarized in Table 1. For each log, we report the total number of traces, the number of activity labels (i.e., the size of the activity set of the log), the average trace length (avg-TL), the average number of repetitions of all cycles in the log (avg-CR), and the ratio between the number of activity labels and the number of traces, indicating the sparsity of the activity labels over the log.

6.2   Experimental Procedure

In order to evaluate the techniques presented in this paper, we adopted the following procedure. For each event log:

1. We divided the event log in two parts: a **training set** composed of 67% of traces of the whole event log used for building the LSTM models and a **testing set** composed of the remaining 33% used for testing the predictions of suffixes.
2. We derived the a-priori knowledge on the traces of the testing set as follows. We randomly selected 10% of traces of the testing set. We used the DeclareMiner ProM plug-in [18] to discover LTL rules satisfied in all these traces. Then, we defined 2 conjunctive rules describing a *strong a-priori knowledge* and a *weak a-priori knowledge*, which respectively strongly and weakly constrain the traces. In particular, we discovered rules of type $\Diamond A$ (which imposes the occurrence of A) for defining the weak a-priori knowledge and rules of type $\Box(A \rightarrow \Diamond B) \wedge \Diamond A$ (which imposes the occurrence of both A and B and that every occurrence of A is followed by an occurrence of B) for defining the strong a-priori knowledge. For the weak a-priori knowledge, we randomly selected from the discovered rules one, two or three[3] rules of type $\Diamond A$ and we composed them into a single conjunctive formula. Similarly, for the strong a-priori knowledge, we randomly selected one, two or three rules from the discovered rules of type $\Box(A \rightarrow \Diamond B) \wedge \Diamond A$ and we composed them

---

[2] https://data.mendeley.com/datasets/39bp3vv62t/1.

[3] The number of rules selected has been determined empirically to allow them to be satisfied in around 50% of the traces of the testing set.

into a single conjunctive formula. We followed this systematic procedure for defining the a-priori knowledge, to limit the bias of the selected rules while guaranteeing that they are satisfied in a reasonable number of traces in the testing set. The schematic form of the rules used in the evaluation is reported in Table 2, where - for the sake of readability - we replace the original activity names with single characters. Starting from strong and weak a-priori knowledge, we built a *strong a-priori testing set* and a *weak a-priori testing set*, respectively composed of the subsets of traces of the testing set satisfying strong and weak a-priori knowledge.

3. From each trace in the testing sets, we extracted 4 prefixes of lengths corresponding to the 4 integers in the interval $[mid - 2, mid + 2]$, where mid is half of the median of the trace lengths. Then, we compared NOCYCLE and A-PRIORI* against a baseline provided by the technique presented in [28], when predicting the suffixes of these prefixes.[4] For each technique, we computed: (i) the length of the predicted suffixes; and (ii) their similarity with the prediction ground truth measured using the Damerau-Levenshtein similarity [7].

Table 2. The a-priori knowledge

Log	A-priori Strong	A-priori Weak
EnvLog	$\square(a \rightarrow \Diamond b) \wedge \Diamond a \wedge \square(c \rightarrow \Diamond d) \wedge \Diamond c$	$\Diamond a \wedge \Diamond c$
HelpDesk	$\square(e \rightarrow \Diamond f) \wedge \Diamond e$	$\Diamond e$
BPIC11	$\square(g \rightarrow \Diamond h) \wedge \Diamond g \wedge \square(i \rightarrow \Diamond l) \wedge \Diamond i \wedge \square(m \rightarrow \Diamond n) \wedge \Diamond m$	$\Diamond i \wedge \Diamond h \wedge \Diamond o$
BPIC12	$\square(p \rightarrow \Diamond q) \wedge \Diamond p$	$\Diamond p$
BPIC13	$\square(r \rightarrow \Diamond s) \wedge \Diamond r \wedge \square(t \rightarrow \Diamond r) \wedge \Diamond t$	$\Diamond s \wedge \Diamond r$
BPIC17	$\square(u \rightarrow \Diamond v) \wedge \Diamond u$	$\Diamond u$

The experiments have been performed both on a GPU Tesla K40c and on a conventional laptop CPU on Code i5. As for the LSTM training settings we used the ones identified by Tax et al. [28] as the most performing ones for facing the problem of predicting sequences of future activities.[5] The time required for training the LSTM models is about 2 min per epoch using the GPU and 15 min using the CPU. The inference time for NOCYCLE is about 0.1–2 seconds per trace (depending on the log), whereas the inference time for A-PRIORI* is 4 times higher on average.

6.3   Results and Discussion

Tables 3 and 4 report, for each event log, the performances of the two techniques we propose on the strong a-priori and weak a-priori testing sets. The results for

---

[4] We set $bSize$ to 3 and, for the coefficient in charge of weakening the probabilities of activities in a cycle, we used the exponential formula (e^j, where j is the number of cycle repetitions).

[5] We used an architecture characterized by two LSTM layers. The algorithm used is the Adam learning algorithm with categorical cross entropy loss and the dropout coefficient has been set to 0.2.

both testing sets are compared with the baseline presented in [28]. For each log, we provide the average Damerau-Levenshtein similarity between the predicted sequence (in square brackets, its average length) and the ground truth (in column 5 its average length). The best average Damerau-Levenshtein similarity for each log is emphasized in gray. Column 6 reports the number of traces tested while column 7 specifies the range of the prefix lengths used for the specific event log.

Table 3. Prediction results on the strong a-priori testing set

Log	Baseline	NOCYCLE	A-PRIORI*	Groundtruth	Tested	Prefix
EnvLog	0.250 [17.40]	0.250 [17.4]	0.070 [95.00]	29.40	80	19 − 22
HelpDesk	0.551 [1.44]	0.551 [1.44]	0.816 [2.53]	3.00	576	2 − 5
BPIC11	0.204 [199.00]	0.281 [199.00]	0.276 [196.3]	117.11	144	13 − 16
BPIC12	0.071 [47.07]	0.387 [6.86]	0.408 [9.02]	10.95	1 548	2 − 5
BPIC13	0.116 [100.80]	0.502 [14.71]	0.516 [17.75]	7.15	3 209	2 − 5
BPIC17	0.448 [11.78]	0.448 [11.78]	0.439 [21.90]	16.01	10 153	6 − 9

Table 4. Prediction results on the weak a-priori testing set

Log	Baseline	NOCYCLE	A-PRIORI*	Groundtruth	Tested	Prefix
EnvLog	0.246 [18.22]	0.246 [18.22]	0.068 [95.00]	31.31	108	19 − 22
HelpDesk	0.551 [1.44]	0.551 [1.44]	0.816 [2.53]	3.00	576	2 − 5
BPIC11	0.220 [199.00]	0.292 [199.00]	0.287 [197.01]	112.66	450	13 − 16
BPIC12	0.100 [48.08]	0.263 [6.81]	0.273 [7.84]	8.33	3 179	2 − 5
BPIC13	0.130 [95.19]	0.459 [14.92]	0.476 [7.45]	5.85	4 364	2 − 5
BPIC17	0.448 [11.78]	0.448 [11.78]	0.424 [24.92]	16.01	10 153	6 − 9

The tables show that the proposed algorithms outperform the baseline in most of the logs. The presence of cycles in the logs has a strong impact on the performance of the NOCYCLE algorithm. In particular, if the logs have an average number of cycle repetitions smaller than 0.5, as in the case of EnvLog, HelpDesk and BPIC17, then NOCYCLE does not show any improvement over the baseline. Therefore, we can conclude that NOCYCLE correctly deals with the presence of cycles in the logs to improve the predictions.

A-PRIORI* performs worse on logs EnvLog and BPIC11. The reason for this can be explained by the fact that, in these two logs, activity labels are sparse with an unusually high number of labels with respect to the number of traces. Indeed, Table 1 shows that the ratio between the number of activity labels and the number of traces (column 6) for these logs is higher with respect to the other logs. We can also notice that the availability of highly constraining rules in the a-priori knowledge improves the performance of A-PRIORI*. Therefore, we can conclude that A-PRIORI* is able to correctly leverage a-priori knowledge in a way that it performs better when the activity set of the log is not particularly large (and the log does not contain sparse behaviors) and when the a-priori knowledge constrains more the process behavior.

7   Conclusions

In this paper, we have presented two techniques based on RNNs with LSTM cells able to leverage knowledge about the structure of the process execution traces as well as a-priori knowledge about their future development for predicting the sequence of future activities of an ongoing case. In particular, we show that, by opportunely tailoring LSTM-based algorithms, it is possible to take into account a-priori knowledge at prediction time without the need to retrain the predictive algorithms in case new knowledge becomes available. The results of our experiments show that NOCYCLE correctly deals with the presence of cycles in the logs and A-PRIORI* is able to correctly leverage a-priori knowledge in a way that it performs better with logs characterized by a low degree of sparsity of activity labels and when the a-priori knowledge constrains the behavior of the process more.

Future work will include: (i) dealing with more complex forms of a-priori knowledge. In particular, we aim at leveraging a-priori knowledge on activities and on their data payload, as well as dynamic knowledge that can evolve in the future of an ongoing case; (ii) extending the proposed algorithms to leverage a-priori knowledge also for other types of predictions; (iii) extending the experimental evaluation especially focusing on the investigation of metrics for evaluating the influence on the predictions of the different degrees of freedom/strength of the a-priori knowledge; and (iv) inserting the presented techniques in predictive business process monitoring frameworks such as the ones discussed in [8,9].

Acknowledgments. This research has been partially carried out within the Euregio IPN12 KAOS, which is funded by the "European Region Tyrol-South Tyrol-Trentino"(EGTC) under the first call for basic research projects.

References

1. 3TU Data Center: BPI Challenge 2011 Event Log (2011). doi:10.4121/uuid:d9769f3d-0ab0-4fb8-803b-0d1120ffcf54
2. van der Aalst, W.M.P., Schonenberg, M.H., Song, M.: Time prediction based on process mining. Inf. Syst. **36**(2), 450–475 (2011)
3. Abadi, M., Agarwal, A., Barham, P., Brevdo, E., Chen, Z., Citro, C., Corrado, G.S., Davis, A., Dean, J., Devin, M., Ghemawat, S., Goodfellow, I., Harp, A., Irving, G., Isard, M., Jia, Y., Jozefowicz, R., Kaiser, L., Kudlur, M., Levenberg, J., Mané, D., Monga, R., Moore, S., Murray, D., Olah, C., Schuster, M., Shlens, J., Steiner, B., Sutskever, I., Talwar, K., Tucker, P., Vanhoucke, V., Vasudevan, V., Viégas, F., Vinyals, O., Warden, P., Wattenberg, M., Wicke, M., Yu, Y., Zheng, X.: TensorFlow: Large-scale machine learning on heterogeneous systems (2015). software available from tensorflow.org. http://tensorflow.org/
4. Buijs, J.: Environmental permit application process ("wabo"), coselog project - municipality 4 (2014). doi:10.4121/uuid:e8c3a53d-5301-4afb-9bcd-38e74171ca32
5. Chollet, F.: Keras (2015). https://github.com/fchollet/keras
6. Conforti, R., de Leoni, M., La Rosa, M., van der Aalst, W.M.P.: Supporting risk-informed decisions during business process execution. In: Salinesi, C., Norrie, M.C., Pastor, Ó. (eds.) CAiSE 2013. LNCS, vol. 7908, pp. 116–132. Springer, Heidelberg (2013). doi:10.1007/978-3-642-38709-8_8
7. Damerau, F.J.: A technique for computer detection and correction of spelling errors. Commun. ACM **7**(3), 171–176 (1964)
8. Di Francescomarino, C., Dumas, M., Federici, M., Ghidini, C., Maggi, F.M., Rizzi, W.: Predictive business process monitoring framework with hyperparameter optimization. In: Nurcan, S., Soffer, P., Bajec, M., Eder, J. (eds.) CAiSE 2016. LNCS, vol. 9694, pp. 361–376. Springer, Cham (2016). doi:10.1007/978-3-319-39696-5_22

9. Di Francescomarino, C., Dumas, M., Maggi, F.M., Teinemaa, I.: Clustering-based predictive process monitoring. IEEE Trans. Serv. Comput. **PP**(99), 1–18 (2016)
10. van Dongen, B.: Bpi challenge 2012 (2012). doi:10.4121/uuid: 3926db30-f712-4394-aebc-75976070e91f
11. van Dongen, B.: Bpi challenge 2017 (2017). doi:10.4121/uuid: 5f3067df-f10b-45da-b98b-86ae4c7a310b
12. Evermann, J., Rehse, J.-R., Fettke, P.: A deep learning approach for predicting process behaviour at runtime. In: Dumas, M., Fantinato, M. (eds.) BPM 2016. LNBIP, vol. 281, pp. 327–338. Springer, Cham (2017). doi:10.1007/978-3-319-58457-7_24
13. Evermann, J., Rehse, J.R., Fettke, P.: Predicting process behaviour using deep learning. Decision Support Systems (2017)
14. Folino, F., Guarascio, M., Pontieri, L.: Discovering context-aware models for predicting business process performances. In: Meersman, R., Panetto, H., Dillon, T., Rinderle-Ma, S., Dadam, P., Zhou, X., Pearson, S., Ferscha, A., Bergamaschi, S., Cruz, I.F. (eds.) OTM 2012. LNCS, vol. 7565, pp. 287–304. Springer, Heidelberg (2012). doi:10.1007/978-3-642-33606-5_18
15. Goodfellow, I., Bengio, Y., Courville, A.: Sequence Modeling: Recurrent and Recursive Nets. In: Deep Learning, pp. 373–420. MIT Press, Cambridge (2016)
16. Hochreiter, S., Schmidhuber, J.: Long short-term memory. Neural Comput. **9**(8), 1735–1780 (1997)
17. Leontjeva, A., Conforti, R., Di Francescomarino, C., Dumas, M., Maggi, F.M.: Complex symbolic sequence encodings for predictive monitoring of business processes. In: Motahari-Nezhad, H.R., Recker, J., Weidlich, M. (eds.) BPM 2015. LNCS, vol. 9253, pp. 297–313. Springer, Cham (2015). doi:10.1007/978-3-319-23063-4_21
18. Maggi, F.M., Bose, R.P.J.C., van der Aalst, W.M.P.: Efficient discovery of understandable declarative process models from event logs. In: Ralyté, J., Franch, X., Brinkkemper, S., Wrycza, S. (eds.) CAiSE 2012. LNCS, vol. 7328, pp. 270–285. Springer, Heidelberg (2012). doi:10.1007/978-3-642-31095-9_18
19. Maggi, F.M., Di Francescomarino, C., Dumas, M., Ghidini, C.: Predictive monitoring of business processes. In: Jarke, M., Mylopoulos, J., Quix, C., Rolland, C., Manolopoulos, Y., Mouratidis, H., Horkoff, J. (eds.) CAiSE 2014. LNCS, vol. 8484, pp. 457–472. Springer, Cham (2014). doi:10.1007/978-3-319-07881-6_31
20. Metzger, A., Franklin, R., Engel, Y.: Predictive monitoring of heterogeneous service-oriented business networks: the transport and logistics case. In: Proceedings of the 2012 Annual SRII Global Conference, SRII 2012, pp. 313–322. IEEE Computer Society, Washington, DC (2012)
21. Pika, A., van der Aalst, W.M.P., Fidge, C.J., ter Hofstede, A.H.M., Wynn, M.T.: Predicting deadline transgressions using event logs. In: Rosa, M., Soffer, P. (eds.) BPM 2012. LNBIP, vol. 132, pp. 211–216. Springer, Heidelberg (2013). doi:10. 1007/978-3-642-36285-9_22
22. Pnueli, A.: The temporal logic of programs. In: 18th Annual Symposium on Foundations of Computer Science, Providence, Rhode Island, USA, 31 October - 1 November 1977, pp. 46–57. IEEE Computer Society (1977)
23. Polato, M., Sperduti, A., Burattin, A., de Leoni, M.: Time and activity sequence prediction of business process instances. CoRR abs/1602.07566 (2016)
24. Rogge-Solti, A., Weske, M.: Prediction of remaining service execution time using stochastic petri nets with arbitrary firing delays. In: Basu, S., Pautasso, C., Zhang, L., Fu, X. (eds.) ICSOC 2013. LNCS, vol. 8274, pp. 389–403. Springer, Heidelberg (2013). doi:10.1007/978-3-642-45005-1_27
25. Senderovich, A., Weidlich, M., Gal, A., Mandelbaum, A.: Queue mining for delay prediction in multi-class service processes. Inf. Syst. **53**, 278–295 (2015)

26. Steeman,   W.:   Bpi   challenge   2013   (2013).   doi:10.4121/uuid: a7ce5c55-03a7-4583-b855-98b86e1a2b07
27. Suriadi, S., Ouyang, C., van der Aalst, W.M.P., ter Hofstede, A.H.M.: Root cause analysis with enriched process logs. In: La Rosa, M., Soffer, P. (eds.) BPM 2012. LNBIP, vol. 132, pp. 174–186. Springer, Heidelberg (2013). doi:10.1007/ 978-3-642-36285-9_18
28. Tax, N., Verenich, I., La Rosa, M., Dumas, M.: Predictive business process monitoring with LSTM neural networks. In: Dubois, E., Pohl, K. (eds.) CAiSE 2017. LNCS, vol. 10253, pp. 477–492. Springer, Cham (2017). doi:10.1007/978-3-319-59536-8_30
29. Teinemaa, I., Dumas, M., Maggi, F.M., Di Francescomarino, C.: Predictive business process monitoring with structured and unstructured data. In: La Rosa, M., Loos, P., Pastor, O. (eds.) BPM 2016. LNCS, vol. 9850, pp. 401–417. Springer, Cham (2016). doi:10.1007/978-3-319-45348-4_23
30. Tillmann, C., Ney, H.: Word reordering and a dynamic programming beam search algorithm for statistical machine translation. Comput. Linguist. **29**(1), 97–133 (2003)

Analysis of Knowledge-Intensive Processes Focused on the Communication Perspective

Pedro Henrique Piccoli Richetti, João Carlos de A,R. Gonçalves$^{(\boxtimes)}$,
Fernanda Araujo Baião, and Flávia Maria Santoro

Department of Applied Informatics, Federal University of the State of Rio de Janeiro,
Pasteur Av. 458, Rio de Janeiro 22290-240, Brazil
{pedro.richetti,joao.goncalves,
fernanda.baiao,flavia.santoro}@uniriotec.br

Abstract. Knowledge-intensive Processes (KiPs) are unstructured processes that demand an understanding beyond control flow and data. Being knowledge-centric and varying at each instance, KiPs demand new perspectives for proper process analysis. Most KiPs have strong collaboration characteristics, where interactions among participants are crucial to achieve process goals. Process participants perform activities and collaborate with each other, driven by their Beliefs, Desires and Intentions; therefore, the analysis of these elements is vital to the correct understanding, modeling and execution of a KiP. This research proposes a method based on Speech Act Theory and Process Mining to discover the flow of speech acts related to Beliefs, Desires and Intentions from event logs, and shows how this relation fosters process performance analysis. The approach was evaluated through a case study in a real life scenario, and results showed that relevant insights in forms of speech acts flow patterns were discovered and related to performance issues of the KiP.

Keywords: Knowledge-intensive Process · Speech act · Process performance measures

1 Introduction

The research on Knowledge-intensive Processes (KiPs) has gained focus as an emerging area within Business Process Management, since many modern business functions have been acknowledged as knowledge-intensive, mainly due to collaborative interactions among process participants and flexibility to perform the work, making the process less predictable than routine structured work [8]. Examples of KiPs include healthcare processes, disaster management, information and communication technology troubleshooting and air traffic control.

F.A. Baião–partially funded by the CNPq brazilian research council, project 309069/2013-0.

F.M. Santoro–partially funded by the CNPq brazilian research council, project 307377/2011-3.

© Springer International Publishing AG 2017
J. Carmona et al. (Eds.): BPM 2017, LNCS 10445, pp. 269–285, 2017.
DOI: 10.1007/978-3-319-65000-5_16

Human knowledge and involvement are key to KiPs execution [10]. However, it is not trivial to properly understand and represent these two aspects in the context of a business process [18]. The involvement of process participants become evident when they collaborate with each other via interactions, in which knowledge is exchanged to achieve the process goal. Process participants have their own beliefs, desires and intentions that motivate them to act to perform their work. When process participants interact, all three elements are present in their communications, and in most of the time people communicate using natural language. In this work, we take this as an opportunity to analyze human interactions as conversations, supported by the Speech Act Theory [2]. According to this theory, an illocutionary act holds the pragmatics of an utterance and is characterized by a distinct illocutionary point [19].

We argue that these illocutionary points can be correlated to beliefs, desires and intentions, which opens a path to analyze speech acts that may represent part of human knowledge and involvement in KiPs, as previously defined in the Knowledge-intensive Process Ontology (KiPO) [18]. KiPO comprises well-founded definitions which enable us to precisely define the notions of agents, the interactions and knowledge-flow among them, and how the mental moments that are inherent to them (Beliefs, Desires, Intentions and Feelings) influence (or even drive) their decisions and the control-flow of the activities executed in each KiP instance.

The problem to be investigated is the difficulty imposed to analyze how human knowledge and involvement influence a KiP execution when this information is present only in unstructured natural language resources. Existing approaches on discovery of speech acts deal with email classification to track intentions of message's senders [5], discovery of business process choreography diagrams from message logs [13], discovery of speech act categories in dialogue-based, multi-party educational games [17], recognition of suggestions and complaints in software development online discussions [16], applications using social media such as Twitter [26,28] and to understand knowledge-sharing process in online Question and Answer communities [27]. None of these works advance in the analysis of the impact of the discovered speech acts on process performance.

The ultimate goal of this work is to verify to what extent beliefs, desires, intentions and feelings of process participants influence process execution. In this direction we propose an approach, based on the automatic discovery of speech acts from message logs and the usage of process mining, to analyze process performance from the illocutionary points perspective. We conducted an empirical study with a real-life KiP from an Information Technology Outsourcing Company that performs an incident troubleshooting process, in order to evaluate the proposal. A pre-requisite to employ this approach is the need to have an information system supporting the subject process and capable of registering conversations during process execution.

The paper is organized as follows. Section 2 presents background knowledge, including an Ontology that describes the elements of KiP, the Speech Act Theory and Process Mining. Section 3 describes our methodology and definitions proposed. Section 4 presents the experimental scenario and results. Section 5 concludes the paper and discusses future work.

2   Background and Related Work

2.1   Speech Act Theory and Automatic Discovery of Speech Acts

Firstly proposed by Austin [2], the Speech Act Theory looks beyond the literal meaning of utterances within a conversation and considers how context and intention contribute to their meaning. One of the main focus of the theory is the analysis of the intended communicative act of an utterance (i.e., what the utterance was meant to achieve). According to this theory, speech acts may be analyzed on three levels: a locutionary act (the utterance itself); an illocutionary act (the social action of utterance, its intended significance: whether it contains a request, an order, or a promise, etc.); and a perlocutionary act (the actual effects of the speech act, for example the act of fulfilling the uttered request).

Searle [19] refined the theory, by defining an illocutionary act as an act that one performs in producing an utterance, such as the act of asserting a proposition, asking someone a question, or directing someone to do something.

Searle and Vanderveken [20] further defined the illocutionary act as the minimal unit of human conversation, such as statements, questions or commands; thus, whenever a speaker utters a sentence in an appropriate context, with certain intentions, he/she performs one or more illocutionary acts. An illocutionary act is formally defined as having an illocutionary force F and a propositional content P (in the form of "F(P)"), which respectively denotes the speaker's intention on making the utterance and the meaning of a clause or sentence that is constant of the illocutionary act. According to the authors, an illocutionary act may be decomposed into three different speech acts: an utterance act (simply uttering an expression), a propositional act (the act of expressing a propositional content) and, if the illocutionary act is successful, a possible perlocutionary act.

A taxonomy of Speech Acts was proposed by Searle and Vanderveken [20], composed of five main classes:

- Assertives: commit a speaker to believing the expressed proposition;
- Directives: cause the hearer to take a particular action;
- Commissives: commit a speaker to doing some future action;
- Expressives: express the speaker's attitudes and emotions towards the proposition;
- Declaratives: change the social sphere in accordance with the proposition of the declaration.

Bach and Harnish [3] expand this initial classification, taking Austin's idea of a division of speech acts between constatives and performatives. Constatives are speech acts that can be reduced to true/false statements and Performatives are speech acts that do not conform to true/false statements, being more oriented towards the performance of an action and the "felicity conditions" of the action

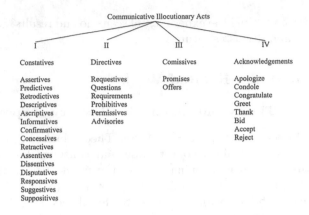

Fig. 1. Bach and Harnish's taxonomy of speech acts [3].

itself (i.e. the conditions for its successful performance). For our study, we adopt Bach and Harnish classification of Constatives and Directives, together with Searle's taxonomy of Commissives, Expressives and Declaratives (Fig. 1).

Based on the theory and the taxonomies, applications using computational methods became feasible on this field. Among the several works at the literature, Stolcke et al. [21] propose a probabilistic approach to dialogue act modeling for conversational speech that precludes the modern efforts for automatic extraction of speech acts. Tenschert and Lenz [22] state that, depending on the domain, the contents of a case and a representation of speech acts may vary. They also present definitions of the contents of a case, a representation for instances of speech acts, and speech act libraries to classify illocutionary speech acts forces.

Mavaddat [14] proposed an approach to facilitate the discovery of business process by analyzing emails and generating "Conversation for Action" diagrams, which can improve understanding of vague and unclear parts of business processes. However, the author does not elaborate on how to implement the conversation elements tagging. Also, Wang et al. [27] propose a framework to analyze discussion threads in online "Question and Answer" communities, relying on the identification of dialogue act patterns. After manually tagging thread messages, they applied process mining techniques to find frequent patterns (process fragments) that occur in helpful, unhelpful and solved threads.

2.2   Process Mining

Process mining is a discipline that aims to provide fact-based insights and to support process improvements. This discipline can be situated between computational intelligence and data mining on one hand, and process modeling and analysis on the other hand [23]. There are three types of process mining tasks: discovery, conformance and enhancement. As the interest of this work is on discovery, it is important to consider the following definition [1]: a process discovery algorithm is a function that maps an event log L onto a process model P such that model is representative for the behavior seen in the event log.

Real-world business processes may be supported or controlled by software systems. If these systems are capable of recording the execution history of their supported processes, it will be possible to apply process mining over their event logs. Event logs are the history of events over time, with specific attributes. Typical information stored in an event log is the case ID that groups all events that occurred in the same process instance. There is a unique identifier for an event occurred in a given timestamp, and the name of the executed activity is also presented. In addition, the resource information points to the user (or user role) who executed the activity, and additional information (e.g. "costs") provides more details about the circumstances in which the event was executed.

3   Methodology

This Section describes the methodology applied for the automated discovery of speech acts from business process message logs. We define concepts related to the control flow of process executions, as well as to the communication between process executors during the execution of process.

Our approach relies on some assumptions. First, messages related to a process instance correspond to the communicative interaction between actors of the process instance. Second, multiple speech acts may be discovered from each single message within a conversation [4]. Finally, a speech act is typically expressed as the complete sentence (although it may also occur in the form of an one-word sentence), usually with an "illocutionary force indicating device" such as a performative verb [19]. Based on these assumptions, concepts concerning the interactions are mapped to the concepts of the Process Mining field.

3.1   Mapping Control-Flow and Communication Perspectives

Since we assume no a-priori association between communication events and process execution events (which is the case in practice), this Section presents formal definitions in the control-flow perspective (event, trace, event log) and in the communication perspective (sentence, message, extended sentence, conversation message log), which are required for mapping a message log to an event log. This mapping may be employed on top of any message log with similar characteristics, such as chat message logs, email conversations and online forum threads.

Definition 1 (Event). An event e represents the occurrence of an activity observed during the execution of a process. An event is characterized by the tuple $e = (e_{id}, e_t, e_a, e_r)$, where e_{id} is the unique identifier of the occurrence of event e, e_t is the timestamp of the event occurrence assumed at its completion, e_a is the identifier of the activity, and e_r is the identifier of the event executor.

Definition 2 (Trace). Let T be the set of all logged executions of a single process P. A trace t (a.k.a. "process instance" or "case"), $t \in T$, is characterized by the tuple $t = (t_{id}, P, \sigma)$, where t_{id} is the unique identifier of the trace, P is the identification of the process and $\sigma = \{e_1, e_2, \ldots, e_n\}$ is a finite sequence of events e_i that occurred during the process instance execution. σ has at least one event ($|\sigma| > 0$).

Definition 3 (Event log). An event log L_E, characterized by the tuple $L_E = (T_E, P)$, contains all events from the set of traces T_E, $T_E \subseteq T$, that are related to the executions of one specific process P.

Definition 4 (Sentence). A sentence s consists of one or more phrases expressed in natural language speech, characterized by the tuple $s = (s_{id}, s_w, s_{ia})$, where s_{id} is the unique identifier of the sentence, s_w is the sequence of words and punctuation that forms the sentence and s_{ia} is a type of illocutionary act associated with the sentence, $s_{ia} \in \{$ *assertive, predictive, retrodictive, descriptive, ascriptive, informative, confirmative, concessive, retractive, assentive, dissentive, disputative, responsive, suggestive, suppositive, requestive, question, requirement, prohibitive, permissive, advisory, promise, offer, apologize, condole, congratulate, greet, thank, bid, accept, reject* $\}$.

In the context of an ICT troubleshooting process, three example messages are presented: $s_1 = (258, $ "I advise you to restore the system.", "suggestive"), $s_2 = (259, $ "Could you report me the situation after the system restoration?", "question") and $s_3 = (260, $ "I will perform the system restoration and will send the results to you.", "comissive").

Definition 5 (Message). A message m represents one communication from a sender to one or more receivers, possibly comprising several messages. It is characterized by a tuple $m = (m_{id}, m_s, m_r, m_t, S)$, where m_{id} is the unique identifier of the message, m_s is the single sender of the message, m_r is the set of receivers of the message, m_t is the timestamp of the message occurrence and S is the set of sentences $s \in S$ observed in the message.

Given the sentences defined above, the following messages are defined: $m_1 = (6241, $ "Alice", $\{$"Bob"$\}$, "2016-10-01 14:02:04", $\{s_1, s_2\}$) and $m_2 = (6242, $ "Bob", $\{$"Alice"$\}$, "2016-10-01 17:31:03", $\{s_3\}$).

Definition 6 (Extended Sentence). Since each sentence s is always comprised within exactly one message m, we further define an extended sentence s' as the sentence s enriched with characteristics of its message m. Formally, an extended sentence s' is characterized by the tuple $s' = (s_{id}, s_w, s_{ia}, m_{id}, m_s, m_r, m_t)$, where the first 3 elements correspond to the elements of the original sentence s, while the remaining 4 respectively represent the unique identifier of m, the sender of m, the set of receivers of m, and the timestamp of m.

The impact of this definition for process mining algorithms is that it leads to the assumption that sentences of the same message occurred at the same timestamp. Examples of extended sentences from the above defined sentences are: $s_1' = $ (258, "I advise you to restore the system.", "suggestive", 6241, "Alice", {"Bob"}, "2016-10-01 14:02:04"), $s_2' = $ (259, "Could you report me the situation after the system restoration?", "question", 6241, "Alice", {"Bob"}, "2016-10-01 14:02:04"), and $s_3' = $ (260, "I will perform the system restoration and will send the results to you.", "comissive", 6242, "Bob", "Alice", "2016-10-01 17:31:03").

Definition 7 (Conversation). A conversation c is characterized by the tuple $c = (c_{id}, S')$ where c_{id} is the unique identifier of the conversation, and S' is the set of all extended sentences $s' \in S'$ observed during a conversation. An example of a conversation is $c_1 = (987, S_1')$, where $S_1' = \{s_1', s_2', s_3'\}$.

Definition 8 (Message log). A message log L_M is characterized by the tuple $L_M = (S_M, C)$, where C is a set of conversations selected for analysis, and S_M is the union of the sets of extended sentences observed during each conversation c_i, $c_i \in C$. A message log in the ICT Troubleshooting scenario consists of the set of all conversations between clients and the technical staff about troubleshooting issues reported in the second semester of the previous year.

Given a set C of conversations $c_i = (id_i, S_i')$, $c_i \in C$, that occurred during the execution of traces of a process P to be analyzed. Let S' be the union of all sets of extended sentences of each c_i. Each extended sentence $s' = (s_{id}, s_w, s_{ia}, m_{id}, m_s, m_r, m_t)$, $s' \in S'$ is mapped to an event $e = (s_{id}, m_t, s_{ia}, m_r)$. Consequently, each conversation c_i is mapped to a trace $t = (id_i, P, S_i')$ and the message log $L_M = (S', C)$ is mapped to an event log $L_E = (S', P)$.

In the ICT troubleshooting scenario, the mapping will result in the events $e_1 = $ (258, "2016-10-01 14:02:04", "suggestive", "Alice"), $e_2 = $ (259, "2016-10-01 14:02:04", "question", "Alice") and $e_3 = $ (260, "2016-10-01 17:31:03", "comissive", "Bob") (respectively from the extended sentences s_1', s_2' and s_3'); trace $t_1 = $ (987, "ICT Troubleshooting", {s_1', s_2', s_3'}) from conversation c_1.

3.2   Extracting Speech Acts

An Information Extraction pipeline was defined for the extraction of speech acts, extending the proposal in [16]. The main differences from the original pipeline include the expansion of the gazetteer of performative verbs to be extracted, covering more categories (all Constatives and Directives classes from [3]) and more verbs (Commissives, Expressives and Declaratives categories from [24]).

We extended the rules of the original approach to extract the sentences from messages containing performative verbs, indicating speech acts of a specific type in our classification scheme. The classified speech acts are then used to extract KiP elements. More specifically, we assume that the class of the speech act performed by a process agent expresses one of the Mental Moments of the Agent taking the role of a "speaker" during a Communicative Interaction: a Constative speech act expresses a Belief, a Directive speech act expresses a Desire, a Commissive speech act expresses an Intention and an Expressive speech act expresses a Feeling. Additionally, a Declarative speech act expresses a Contingency event that triggers a KiA or, more specifically, a Decision. Finally, question speech acts were also identified in trivial cases where messages contained a question mark ("?"), which greatly improved the results in the case study.

Another distinct feature from the original approach relates to the overall objective of our method. We aim not only to classify speech acts from raw text data and identify the beliefs, desires and intentions of process participants, but also to be able to analyze the flow of interactions among them. Our proposed method structures a message flow related to a process instance into a series of speech acts, in order to express how a conversation between process participants happens during the instance execution. Taking it to the level of several instances present at the dataset, it reveals interesting patterns of speech acts being performed through interactions, related to the general process model as a whole (instead of an interaction related to an specific instance) that is able to be processed by ProM as interaction data related to an activity or process instance.

The pipeline has six distinct phases: (i) Pre-processing: First, all sentences are splitted and a word-level tokenizer is applied, transforming each message into a set of sentences, each containing a set of tokens (words and punctuation marks); (ii) POS-Tagging: The Hepple Tagger is applied to tag each token with its morphological classification (e.g. verbs, nouns, etc.); (iii) Lemmatization: All verbs are lemmatized and reduced to its infinite form; (iv) Feature Selection: A feature selection algorithm using a gazetteer list containing a list of performative verbs indicative of Speech Acts of each type is applied to each sentence, tagging each performative verb found according to the specific type of Speech Act (e.g. "believe" as a performative verb of an Assertive Act); (v) Speech Act extraction: Each sentence containing a performative verb is extracted as a Speech Act containing its performative verb and propositional content; (vi) Mímir Indexing: All extracted Speech Acts are loaded into the Mímir server, in order to be indexed and queried.

Thus, speech acts discovered from message logs are mapped to ordinary event logs following the definitions in Sect. 3.1, enabling the application of process mining techniques [1] to extract the flow of conversations that occurred during the execution of process instances.

4   Case Study

The methodology proposed for the case study is based on the concept of lag and lead measures [15]. Lag measures are the measures of success, the results that the company wants to achieve, often related to strategic goals of an organization. Lag measures focus on results achieved, so they tend to be less actionable since they look to the past. On the other hand, lead measures are predictive, meaning that if the lead measure changes, it should be possible to predict that a related lag measure will also change. Furthermore, a lead measure must be directly influenced by the process participants.

Since conversations mostly depend on process participants, process stakeholders can benefit from the analysis of speech acts to discover insights from the behavior of conversations, aiming to extract conversational patterns related to process performance that are monitored by traditional lag measures. To run the case study the following steps were planned and then employed: (1) Identification of existing lag measures that might be potentially influenced by communication aspects of the process; (2) Message log selection, preprocessing and cleansing; (3) Discovery of speech acts from the message log; (4) Map the message log to the XES [25] event log according to the definitions presented in Sect. 3; (5) Enrich the event log with performance data; (6) Run Inductive Visual Miner [11] and Multi-perspective Process Explorer [12] on ProM[1] to discover a Petri net of the flow of speech acts with data and performance perspectives; and, finally, (7) Inspect the results for evidences of possible speech acts influence on process performance.

4.1 Case Study Scenario

The case study was performed in a real life scenario of an Information and Communication Technology (ICT) outsourcing company, which has about a hundred contracts with diverse clients. One of the main services provided by the company is customer support, which intends to fulfill technical requests (e.g. e-mail configurations, backup and restore) or solve technical problems (e.g. system failures) that are reported by clients to the company's service desk.

The incident troubleshooting process was the object of our case study. When a client reports a new problem, this triggers the creation of an incident ticket in a process-aware system called OTRS[2] that supports company's operations. Within OTRS, incident tickets are registered, alternative solutions are considered, a solution approach is defined, executed, validated and then deployed. During the resolution of a ticket, messages are exchanged among process participants (both technical teams and client) and associated to a ticket in OTRS. These messages contain natural language texts in chronological order; from them, it is possible to retrieve speech acts uttered from process participants.

Processes of this nature essentially involve the application of technical skills, troubleshooting abilities, collaboration and information exchange between stakeholders, and ad-hoc decisions are frequently discussed and made, since most of the problems are situational. For all these reasons, this process is characterized as a KiP and should be managed as so.

4.2 Case Study Execution

In Step 1, we interviewed two company managers directly involved with this process, and asked them which were the most concerning performance issues they have to face for this process. They affirmed that the two main measures

---

[1] http://www.processmining.org.
[2] https://www.otrs.com/.

they report monthly to their clients were the volume of opened tickets and the total duration time of the troubleshooting sessions. Since the volume of opened tickets depends on the clients, this measure is not directly actionable by the technical teams (despite company's efforts to apply proactive monitoring of some ICT assets). The total duration of troubleshooting sessions, on the other hand, mostly depends on the complexity of the problems presented and on the solution strategy applied by the technical teams, which also involves communication with the clients and among different technical teams. Hence, we chose "total duration of troubleshooting sessions" as the lag measure to be analyzed from a communication perspective and the "presence of speech act" and "interplay of speech acts" as lead measures that might influence the lag measure.

In Step 2, the scope of analysis is the set of all tickets labeled as "incident" reported in the second semester of 2015, stored in the OTRS repository. A total of 5,714 tickets were gathered, comprising 25,380 messages exchanged in the system during the troubleshooting process.

In Step 3, we applied the GATE Developer tool [6] to operationalize our pipeline for extracting speech acts from messages. GATE is a platform to perform Text Mining and Natural Language Processing tasks from unstructured text. For storing, indexing and querying the annotated result data, we used GATE Mímir [9], a multi-paradigm information management index and repository which can be used to index and search over text, annotations, ontologies, and semantic meta-data. After the execution of our speech act discovery pipeline through GATE, the results were stored on the Mímir server. We defined and executed one query for each specific type of speech act according to the taxonomy of Bach and Harnish's [3] defined previously, so as to extract the discovered speech acts of each type as a comma separated text (CSV) file. The results contained a total of 50,800 extracted speech acts, distributed among 24 different types as illustrated in Table 1. Each CSV file contained all speech acts of the same type, as well as information on the corresponding messages from which each speech act was extracted.

In Step 4, each CSV file was enriched with complementary available information about incident tickets. At the case level, the following attributes were added to each speech act: ticket duration in hours, number of messages exchanged, ticket final priority (1 to 5, from lower to higher priority), customer anonymous identification and service id; at the event level, the message sender type (either "agent" or "customer") was included. The enriched CSV file was then imported to ProM and converted to a XES file using the plugin called "Convert CSV to XES", following our definitions for mapping message logs to event logs presented in Sect. 3. Due to the relevance of the speakers during a conversation, we also concatenated the sender type to the speech act type to name activities in the log (as in "*AssertiveSpeechAct|agent*", or "*InformativeSpeechAct|customer*"). This enabled the distinction between speech acts uttered by customers and by company agents during a conversation.

In Step 5, considering that the focus of our analysis is to search for possible associations from the flow of speech acts to the duration of a ticket, we split the

Table 1. Speech acts discovered using GATE pipeline and Mímir.

Type	#Total	%	Type	#Total	%
{DeclarativeSpeechAct}	4017	15.83	{CommissiveSpeechAct}	458	0.90
{AdvisorySpeechAct}	3505	6.91	{AssentiveSpeechAct}	396	0.78
{SuggestiveSpeechAct}	3447	6.79	{ConcessiveSpeechAct}	344	0.68
{InformativeSpeechAct}	3439	6.78	{AssertiveSpeechAct}	240	0.47
{RequestiveSpeechAct}	2521	4.97	{ExpressiveSpeechAct}	216	0.43
{QuestionSpeechAct}	1371	2.70	{SuppositiveSpeechAct}	181	0.36
{DescriptiveSpeechAct}	1168	2.30	{RetractiveSpeechAct}	142	0.28
{ConfirmativeSpeechAct}	1016	2.00	{DisputativeSpeechAct}	17	0.03
{ResponsiveSpeechAct}	910	1.79	{ProhibitiveSpeechAct}	8	0.02
{PermissiveSpeechAct}	809	1.59	{DissentiveSpeechAct}	7	0.01
{RetrodictiveSpeechAct}	664	1.31	{PredictiveSpeechAct}	6	0.01
{RequirementSpeechAct}	494	0.97	{AscriptiveSpeechAct}	4	0.01

event log in subsets of four duration ranges (in hours): 0.0 to 1.5 (very short), 1.5 to 16.0 (short), 16.0 to 72.0 (regular), more than 72.0 (long). These ranges uniformly distribute the total number of traces, and are according to company's interest for process performance analysis, thus enabling the identification of tickets violating the company service level agreement with customers (which states that incidents should take less than 72 h to be solved) and also mark interest intermediate durations. Therefore, the search for process improvement opportunities in such cases is very important for company managers.

4.3 Case Study Results and Discussion

In Step 6, the mainstream process (conversation) for each duration range was discovered, representing how most of the conversations happened for the cases solved within that duration range. To discover the mainstream conversation we applied Inductive Visual Miner, with the following parameters: 50% activities and 80% of paths. The resulting Petri Net represented the speech act behavior of each duration range. Then we applied Multi-perspective Process Explorer to replay each subset event log over its associated Petri Net. With Multi-perspective Process Explorer each of the four subset event logs were analyzed from the frequency of events, average time between events and data discovery perspectives. Figure 2 present the discovered models for each duration range with information about the frequency of events occurrence, and Table 2 shows the main characteristics of the discovered process models for each duration range. For the sake of space, discovered models images in high resolution and with complementary information are available in the following repository[3].

---

[3] https://bitbucket.org/pedrohr/speechactsprocessdiscovery/downloads/.

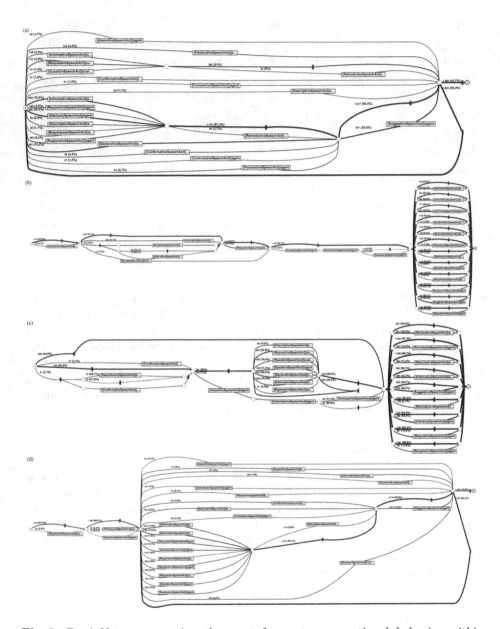

Fig. 2. Petri Nets representing the most frequent conversational behavior within (a) 0.0 to 1.5h (very short duration tickets), (b) 1.5 to 16h (short duration tickets), (c) 16 to 72h (regular duration tickets) and (d) more than 72h (long duration tickets).

Table 2. Main characteristics of discovered process models for each duration range.

Duration range (hours)	[0.0–1.5[	[1.5–16.0[	[16.0–72.0[	[72.0–Infinity[
% Trace frequency	24.48%	25.62%	26.22%	23.68%
# Traces	1,399	1,464	1,498	1,353
# Events	4,296	5,672	7,070	8,342
# Event classes	42	43	44	47
Avg. precision	47.50%	22.10%	27.90%	31.30%
Avg. fitness	97.10%	84.00%	78.40%	93.50%

In Step 7, we focus on the summary presented in Table 2, it is possible to note that the number of events grows as the duration ranges do the same; this is rather intuitive, meaning that with more time spent to solve a problem, more conversations events happened. Although the average precision of all models is low, fitness is very high, ranging from 78.40% (regular) to 97.10% (short). This points to a high variability scenario, where each mainstream model has enough flexibility to accept unobserved behavior. For each duration range, the following behavior was worth noting:

- **Very short duration**: incidents which take less than 1.5 h to be solved have few possible start events initiated by the customers, and some more events related to acts of communication among agents. This reflects a typical behavior when a customer directly calls the agent, the agent proactively starts the troubleshooting and solves the problem right after the first contact. This was made explicit by a multi-perspective rule in the bottom part of the model, indicating that tickets with only one interaction follow directly to the end of the process, without entering in the loop at the end of the process.
- **Short duration**: this subset presented the most frequent starting path, going directly to a XOR split-join where the agent plays all communication using 12 different speech acts, probably in the same conditions as in the very short duration range. However, the remaining paths before this XOR split-join present utterances from the customer, indicating what they mostly say when asking for technical support. The initial speech act types uttered by the customer are (in descending order of frequency): requestives, informatives, declaratives, descriptives and questions.
- **Regular duration**: in comparison to the previous shorter ranges, tickets solved in regular duration presented a higher number of initial participation from the customer, clearly identifiable by a XOR split-join in the middle portion of the process. Note the loop at the end of this join, maybe implying on more time spent on the customer asking questions to an agent. Also, the final actions applied by the agent to close the ticket are remarked by another XOR split-join with 9 different speech act types at the end of the process.
- **Long duration**: the conversational behavior during these cases resulted in a Flower-like model with a XOR split to 21 different speech act types, showing

the high variability of this range. Another remark on variability is the loop at the end of the process, where 80.2% of the cases return to another loop interaction, instead of moving to the end of the process. In practice, this model represents a set of undesired conversational patterns that could be further analyzed for improvement opportunities.

The process interaction models can be analyzed according to two dimensions: knowledge intensity (in the form of the "interaction loops" present in the models) and the most representative speech act types in the models. Beginning with the analysis of the interaction loops, they can be found in 3 out of 4 datasets. The very short duration dataset (0h–1.5h) has an interaction loop that repeats itself between 1–2 times in average, while the regular duration dataset (16h–72h) shows a new interaction loop, just before the parallel flow, that involves an interesting subset of Speech Acts; finally, the long duration dataset (72h–Infinity) has a more recurrent pattern in the form of an interaction loop that repeats itself 4–5 times. The only model without an interaction loop was the short duration model (1.5h–16h), composed of a trace with a final split-join XOR parallel flow. This dimension points out two important facts: First, that the main indicator that the incident will take more time than the usual is the "knowledge intensity" shown by the increase of speech acts exchanged in the form of a loop. Second, it shows that the incidents that are more "straightforward" to be solved (in the sense of less discussions between clients and agents occurring) typically last between 1.5 and 16 h. On the other hand, the high knowledge intensity present in the long duration model clearly raises the importance of monitoring conversational patterns during real-time ticket troubleshooting, so as to detect when a case is instantiating a number of interaction loops and will eventually violate the service level agreement. An improvement practice to be adopted in this case could be to raise the urgency of an intervention from an agent with more expertise, trying to shorten the time to close the incident ticket. The knowledge intensity (viewed as interaction loops) can act as a lead measure, and depending on the speech acts involved, they can proactively point to possible long duration incident cases.

The results also evidence frequent speech acts inside the conversational loops. We find two predominant patterns across all 4 datasets: (i) Suggestives + Advisories acts, and (ii) Declarative + Informative speech acts. The first pattern indicates an extended interaction at the end of the ticket, composed of final remarks from the agent (such as an advice or a suggestion to prevent the incident to happen again in the future). This suggests that human communication during incident troubleshooting not only comprises the resolution of problems, but also preventive advice from agents, which was previously unknown and may indicate a missing activity of the process that was made possible by analyzing the process from a communication perspective. The second pattern describes the interaction between the customer and the agent about the request status and action being taken to solve the incident; it also represent events such as increasing of priority of a ticket or informing that a system (or a peripheral) is back on-line.

5   Conclusions

In the context of Knowledge-intensive Processes, traditional control-flow and data perspectives may be not sufficient to explain the process behavior. This work presented an innovative approach, based on text and process mining tools, to provide an alternative view of the process flow in terms of the interplay of speech acts, representing the communication dimension of the process. The different types of speech acts and their relationship with process elements also bring a novel perspective to the KiPs and process mining, since subtle elements (such as beliefs, desires and intentions of the agents) are found to be somewhat related to goals, alternate flows of activities and activity instantiations during a KiP execution. In addition to count, data, time and condition measures used to evaluate process performance [7], the analysis of speech acts helps to explain how human communication may influence the process execution. The approach was assessed in a case study of a real company, resulting in innovative knowledge about the process that was not possible to be produced with state-of-the-art techniques available. At its current stage, the approach is limited to scenarios supported by software systems capable to register the conversations among participants during process execution. Although the proposal is general enough to address any scenario, generalization of the results of the method's application is limited to a single case study. Future work includes the usage of improved detection of speech acts applying logical operators and negations in sentences as well as a new case study at the scenario of online software development, focusing on the analysis of the interplay of speech acts during the process.

References

1. van der Aalst, W.M.P.: Process Mining - Discovery, Conformance and Enhancement of Business Processes. Springer, Heidelberg (2011). http://dx.doi.org/10.1007/978-3-642-19345-3
2. Austin, J.L.: How to Do Things with Words. Oxford University Press, Oxford (1975)
3. Bach, K., Harnish, R.: Linguistic Communication and Speech Acts. MIT Press, Cambridge (1979)
4. de Carvalho, V.R., Cohen, W.W.: On the collective classification of email "speech acts". In: SIGIR 2005: Proceedings of the 28th Annual International ACM SIGIR Conference on Research and Development in Information Retrieval, Salvador, Brazil, 15–19 August 2005, pp. 345–352 (2005). http://doi.acm.org/10.1145/1076034.1076094
5. Cohen, W.W., Carvalho, V.R., Mitchell, T.M.: Learning to classify email into speech acts. In: Proceedings of the 2004 Conference on Empirical Methods in Natural Language Processing, EMNLP 2004, Barcelona, Spain, pp. 309–316 (2004)
6. Cunningham, H., Maynard, D., Bontcheva, K.: Text Processing with Gate. Gateway Press CA, Murphys (2011)
7. del-Río-Ortega, A., Resinas, M., Ruiz-Cortés, A.: Defining process performance indicators: an ontological approach. In: Meersman, R., Dillon, T., Herrero, P. (eds.) OTM 2010. LNCS, vol. 6426, pp. 555–572. Springer, Heidelberg (2010). doi:10.1007/978-3-642-16934-2_41

8. Di Ciccio, C., Marrella, A., Russo, A.: Knowledge-intensive processes: An overview of contemporary approaches. In: Proceedings of the 1st International Workshop on Knowledge-intensive Business Processes, KiBP@KR 2012, Rome, Italy, 15 June 2012, pp. 33–47 (2012)

9. Greenwood, M.A., Tablan, V., Maynard, D.: Gate mimir: Answering questions google can't. In: Proceedings of the 10th International Semantic Web Conference (ISWC2011), pp. 466–471 (2011)

10. Isik, Ö., Mertens, W., den Bergh, J.V.: Practices of knowledge intensive process management quantitative insights. Bus. Proc. Manag. J. **19**(3), 515–534 (2013). http://dx.doi.org/10.1108/14637151311319932

11. Leemans, S.J.J., Fahland, D., van der Aalst, W.M.P.: Process and deviation exploration with inductive visual miner. In: Proceedings of the BPM Demo Sessions 2014 Co-located with the 12th International Conference on Business Process Management (BPM 2014), The Netherlands, 10 September 2014, p. 46 (2014)

12. Mannhardt, F., de Leoni, M., Reijers, H.A.: The multi-perspective process explorer. In: Proceedings of the BPM Demo Session 2015 Co-located with the 13th International Conference on Business Process Management (BPM 2015), Innsbruck, Austria, 2 September 2015, pp. 130–134 (2015)

13. Mavaddat, M.: Business process discovery through conversation log analysis in pluralist and coercive problem contexts. Ph.D. thesis, University of the West of England (2013)

14. Mavaddat, M., Beeson, I., Green, S., Sa, J.: Facilitating business process discovery using email analysis. In: The First International Conference on Business Intelligence and Technology. Citeseer (2011)

15. McChesney, C., Covey, S., Huling, J.: The 4 Disciplines of Execution: Achieving Your Wildly Important Goals. Simon and Schuster, New York (2012)

16. Morales-Ramirez, I., Perini, A.: Discovering speech acts in online discussions: A tool-supported method. In: Joint Proceedings of the CAiSE 2014 Forum and CAiSE 2014 Doctoral Consortium co-located, Thessaloniki, Greece, 18–20 June 2014, pp. 137–144 (2014)

17. Rus, V., Graesser, A.C., Moldovan, C., Niraula, N.B.: Automatic discovery of speech act categories in educational games. In: Proceedings of the 5th International Conference on Educational Data Mining, Chania, Greece, 19–21 June 2012, pp. 25–32 (2012)

18. dos Santos França, J.B., Netto, J.M., do E. Santo Carvalho, J., Santoro, F.M., Baião, F.A., Pimentel, M.: KIPO: The knowledge-intensive process ontology. Softw. Syst. Model. **14**(3), 1127–1157 (2015)

19. Searle, J.R.: A Taxonomy of Illocutionary Acts. Linguistic Agency University of Trier (1976)

20. Searle, J.R., Vanderveken, D.: Foundations of Illocutionary Logic. CUP Archive (1985)

21. Stolcke, A., Ries, K., Coccaro, N., Shriberg, E., Bates, R.A., Jurafsky, D., Taylor, P., Martin, R., Ess-Dykema, C.V., Meteer, M.: Dialogue act modeling for automatic tagging and recognition of conversational speech. CoRR cs.CL/0006023 (2000)

22. Tenschert, J., Lenz, R.: Towards speech-act-based adaptive case management. In: 20th IEEE International Enterprise Distributed Object Computing Workshop, EDOC Workshops 2016, Vienna, Austria, 5–9 September 2016, pp. 1–8 (2016). http://dx.doi.org/10.1109/EDOCW.2016.7584393

23. Van Der Aalst, W., Adriansyah, A., De Medeiros, A.K.A., Arcieri, F., Baier, T., Blickle, T., Bose, J.C., van den Brand, P., Brandtjen, R., Buijs, J., et al.: Process mining manifesto. In: International Conference on Business Process Management, pp. 169–194. Springer, Heidelberg (2011)
24. Vanderveeken, D.: Meaning and Speech Acts: Principles of Language Use. Cambridge University Press, Cambridge (1990)
25. Verbeek, H.M.W., Buijs, J.C.A.M., Dongen, B.F., Aalst, W.M.P.: XES, XESame, and ProM 6. In: Soffer, P., Proper, E. (eds.) CAiSE Forum 2010. LNBIP, vol. 72, pp. 60–75. Springer, Heidelberg (2011). doi:10.1007/978-3-642-17722-4_5
26. Vosoughi, S., Roy, D.: Tweet acts: A speech act classifier for twitter. In: Proceedings of the Tenth International Conference on Web and Social Media, Cologne, Germany, 17–20 May 2016, pp. 711–715 (2016)
27. Wang, G.A., Wang, H.J., Li, J., Abrahams, A.S., Fan, W.: An analytical framework for understanding knowledge-sharing processes in online Q&A communities. ACM Trans. Manage. Inf. Syst. 5(4), 18:1–18:31 (2015)
28. Zhang, R., Li, W., Gao, D., You, O.: Automatic twitter topic summarization with speech acts. IEEE Trans. Audio Speech Lang. Process. 21(3), 649–658 (2013). http://dx.doi.org/10.1109/TASL.2012.2229984

Process Mining 2

TESSERACT: Time-Drifts in Event Streams Using Series of Evolving Rolling Averages of Completion Times

Florian Richter$^{(\boxtimes)}$ and Thomas Seidl

Ludwig-Maximilians-Universität München, Munich, Germany
{richter,seidl}@dbs.ifi.lmu.de

Abstract. Business processes are dynamic and change due to diverse factors. While existing approaches aim to detect drifts in the process structure, TESSERACT looks for temporal drifts in activity interim times. This orthogonal view on the process extends the traditional data cube of events - case id, activities and timestamps - by a fourth dimension and improves the operational support by a visualization of temporal drifts in real-time.

Insights about temporal deviations lead to an augmented awareness of imminent failures or improved service times. The detection of related structural concept drifts can be improved by early warning, as operation times of critical parts often increase before they catastrophically fail.

Keywords: Process mining · Event streams · Temporal drift detection · Operational support

1 Introduction

Since years, information systems assist in collecting data from daily business operations and have superseded manual logging. However, mid to large-scale enterprises produce increasing amounts of data and the need for superior analysis tools has arisen. Process mining refers to this task by collecting many analysis techniques for discovery and evaluation of processes. Besides the static event log files, the online process mining - often referred to as operational support - is applied to streams of process data. It often supports analysts with the latest view over the scenario. This allows managers to react to circumstances faster or to handle larger amounts of data. Process mining emerged as a discipline between data mining and process analysis, adapting many existing tools to mine the business data and to return human-readable models. The abstract data logs are transformed into high value information, which provide better knowledge for managers to reach more reasonable decisions.

In this work we focus on online mining of event streams. It is rather important for certain dynamic applications to detect anomalies early. Analyzing logs too late may result in outdated insights. Factors like stock markets, supply conditions, regulating laws, crisis effects or restructuring attempts within a company

© Springer International Publishing AG 2017
J. Carmona et al. (Eds.): BPM 2017, LNCS 10445, pp. 289–305, 2017.
DOI: 10.1007/978-3-319-65000-5_17

cause changes in the processes. Thus the need to identify sources of deviations and find structural dependencies in the process structure is a very timely problem. Solving this problem in a certain scenario can help to predict further cascading changes.

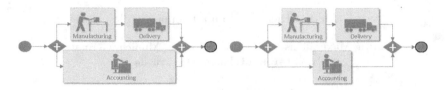

Fig. 1. Two models of the same production process can share the same structure, but can differ in their activity durations. These are indicated by the widths of the activity boxes. Note that the sojourn time of the left process is dominated by the accounting step. Some weeks later, a new norm simplified and accelerated the accounting. The resulting process at the right takes less time, but the process execution time is dominated by the upper branch now. Improving the accounting any further will not increase the process performance anymore.

Some methods for concept drift handling have been developed and discussed already [4–7,9,13–15]. We do not aim at competing with those methods by detecting deviations in the process structure itself. Our approach is orthogonal as our goal is to find temporal deviations, based on given timestamps. A process structure can remain unchanged while the operation times of process steps can change and vice versa. Therefore we do not propose a method to replace existing concept drift methods in this field. Instead we add another drift perspective by looking for temporal drifts.

As a motivation, we assume a set of local subsidiaries that might have to follow a certain process model as a common guideline. While all subsidiaries follow the same structural execution guidelines, small temporal deviations can cause a wide spectrum of the execution time for specific services. In case of an accelerated sub-operation, which still conforms to the guideline, traditional approaches would not detect the improvement. The company would lose the benefit after a change of the subsidiary staff without even recognizing the hidden asset. On the other hand, detecting an increase in delay can indicate a malfunction of certain hardware and an issue can be fixed before it gets a true problem. As a third important application, knowledge about service time deviations can improve existing sojourn time prediction approaches. Figure 1 contains a visual example showing the influence of changes of interim times.

The given examples lead to the introduction of the problem addressed in this work, the detection of deviations of activity durations in an event stream. This should help answer questions like: Can I rely on my process operation times for prediction? Are there periods when I should expect shorter or longer operation times? Can structural drifts be anticipated before they actually occur? Are there some recurring change patterns?

In this paper, we present our novel event-stream-based temporal drift detection method, called TESSERACT. The contributions are as follows:

1. Utilization of event interim times for process analysis
2. Adaptation of a scalable and noise adapting trend detection method from text mining
3. Anytime drift detection in events streams for operational support
4. Visualization for supervised domain expert interaction

The remainder of this paper is organized as follows: Sect. 2 lists some related work. In Sect. 3 we give some basic definitions needed the understanding of the approach. Section 4 describes our novel method called TESSERACT. In Sect. 5 we show our extensive experimental evaluation using synthetic datasets with known ground truth and two real-world datasets. Section 6 concludes this paper with some future directions.

2   Related Work

In online process mining or operational support business processes are analyzed during their runtime. Cases are considered even before they are closed, allowing a "pre mortem" analysis [20]. Our approach deals with online process mining to detect concept drifts in the time perspective. We will give a brief overview about related work in the areas regarding concept drift detection in processes and temporal process mining.

2.1   Drifts and Deviations

In the field of traditional data mining, concept drift is a very well established phenomenon. It was first mentioned that way in [17]. Concept Drift refers to changes in the resulting output, which is caused by a change in the input data. Over the years many different approaches have been developed to deal with concept drift in a variety of application fields. In [4,5], Bose et al. transferred the concept drift phenomenon to the process mining field. They performed statistical hypothesis testing over feature vectors to deal with change point detection and localization. The perspective of the drift was the structure of the underlying process model, which refers to the control-flow perspective. Carmona et al. used a geometrical approach with membership tests on abstract polyhedrons in [9] to declare a drift. Burattin et al. [6,7] and Hassani et al. [13] deal with concept drift in an evolving manner. There are further approaches with hypothesis testing on the event ordering in cases [14,15]. The mentioned works are dealing with concept drifts in the process structure. From now on we will refer to this perspective by structural drifts. To our best knowledge, there has been no work on temporal drifts in business data. But there has been research on drifts in time series. A survey on state-of-the-art concept drift methods in time-series is provided by Aminikhanghahi et al. [1].

2.2 Temporal Process Mining

This other major part of our approach relates to the time-perspective of business information. As most events in logs contain timestamps, some characteristics like service times and frequencies during certain seasons can be derived. Backus et al. considered factory cycle times in [2]. In [21], van der Aalst et al. enhanced a transition system of activities to enable the prediction of case remaining times. Polato et al. improved this further in [16] by building and using a regression model to estimate the following activities in a current uncompleted case. Bolt et al. [3] used query catalogs to estimate the remaining times. Although concept drift is mentioned in the literature, the approaches assume a stationary process. It is obvious that the prediction accuracy would decrease tremendously in case of a concept drift. Our novel approach aims at recognizing and pointing to changing activity completion times to update the remaining time predictors.

2.3 Event-Based Monitoring

In operational process mining, a stream of events is monitored to support the analysts with beneficial statistics and trends. The mining of process delays is a research focus in this area. Senderovich et al. [19] worked on the prediction of delays in service processes that are caused by queuing customers. Gal et al. [12] predict the travel time of buses considering their travel time which should be scheduled. [8] et al. developed a framework to monitor a business process and signal deviating behavior. In this context, TESSERACT can support these methods with information about temporal trends.

3 Preliminaries

Now we will give a more formal introduction to the terms in process mining. Traditional process mining deals with *event logs*. These are sets of cases, each labeled with a unique case identifier c. Every *case* consists of a sequence of events $c = \{e_1, e_2, \ldots, e_n\} \subseteq \mathbb{E}$, where $\mathbb{E}$ is the complete event space. It is assumed that each *event* contains at least the corresponding case identifier, an activity label to identify the related business action and the timestamp of the occurred action $e = (c, a, t) \in \mathbb{E}$. As a simplified notion, we define $c(e) = c, a(e) = a$ and $t(e) = t$ as the projected attributes of an event. The timestamps define a temporal order of events within a case. Using the timestamps only for ordering and projecting a case to its activity labels yields the *trace* of this case. Considering Fig. 2, which shows some actions happening in a short period of time, we can derive the stated process log below the timeline. Real-world datasets typically consists of a small activity space compared to the mostly large number of distinct cases.

We traverse from stationary processes to dynamic ones. Using the timestamps, we can define a stream of events to substitute the event log. Formally, we define the mapping $S : \mathbb{N} \to \mathbb{E}$ as an event stream with the following constraint: For any two events e_1, e_2, it must hold that $S^{-1}(e_1) < S^{-1}(e_2)$

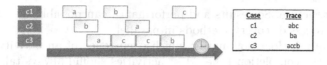

Fig. 2. Three cases are observed in parallel. They share the same activities a, b, c but the actions appear in different orders and number.

exactly if $t(e_1) < t(e_2)$, so the ordering of the timestamps must be conform to the stream ordering. This is more of theoretical importance, as a real application would assign timestamps based on the stream arrival time.

Activities usually have an execution time, which starts at some point in time and finishes at a later point. Sometimes, activities even occur in different life-cycle states like registered, started, completed, restarted or aborted. In this case, we can define the *activity completion time* as the time between the first and the last occurrence of the activity $t(a) = t((c, a : complete, t_2)) - t((c, a : start, t_1))$. Idle times between activities can also be defined by $t(a) = t((c, b : start, t_2)) - t((c, a : complete, t_1))$.

In many event logs, events are only treated as point-based observations. To deal with this limitation, we use the time between two events in the same case instead. If we know the process structure in advance, we can focus on specific consecutive activity pairs. Otherwise it is useful to observe every interim $t(e_2) - t(e_1)$ such that $e_1 = (c, a, t_1)$, $e_2 = (c, b, t_2)$ and $t_2 > t_1$. We are not able to split actual activity execution time and the following waiting or idle time, so we just define the complete interim between both activity timestamps as the completion time for a.

The process structure is slightly linked to the temporal structure. We will illustrate this with two examples. First, imagine a book reviewing process with three consecutive activities *getBook*, *readBook* and *writeReview*. A new book can be received on the beginning of the month and the review has to be finished at the end. The reading can happen anytime between both points in time. *getBook* and *writeReview* are not in a directly-follows relation, but they establish a very consistent time frame for the execution. By observing only the two interims (*getBook*, *readBook*) and (*readBook*, *writeReview*) the fluctuation makes it very difficult to detect slight changes in the global execution time of the process. So sometimes long-distant-follows relations can be more more relevant than directly-follows relations in this context.

Second, we consider the process with concurrency in Fig. 1 again. A valid observed trace could be (*manufacture*, *account*, *deliver*). We remember that the *account*-activity is concurrent to the remaining two activities. Due to the parallel execution of the accounting branch and the production branch, the relations (*manufacture*, *account*) and (*account*, *deliver*) are not relevant in the model and the important relation (*manufacture*, *deliver*) is overseen. This means we could ignore certain relation candidates if we had a good insight into the process structure. In the remaining paper we will focus on the worse case of not knowing the process structure. So results and performance can possibly be improved by using other process discovery methods in parallel.

Theoretically it does not matter which occurrence of an activity is chosen to determine the completion time. Most activities should always take the same amount of time. However, many factors cause an activity to take less or more time in different instances. For human-influenced activities, appointments can be set up vaguely or they can just be too late or too early. Traffic can cause deliveries to arrive unpredictably. And even automated production lines happen to add some noisy factors to the recognized timestamps. In this paper we assume that an activity has a certain average completion time. Most speedups or delays are clustered in the proximity of this average while only few outliers can be found at some distance to the expected time. Therefore, we want to focus on Gaussian distributed noise only as a reasonable distribution for completion time series. There are also other distributions, but this is planned for future work.

4   Algorithm

In the following, we will discuss the three parts of TESSERACT in detail. We have to deal with collecting the completion times in a stream environment, using the collected statistics to derive an indicator function and visualize the results.

4.1   Managing Cases and Completion Times

Our starting point is a stream of events. First we need to derive the completion times of activities. Without given activity life-cycles, we mine the interims between activity pairs within the same case. The used data structure has a high influence on the performance here. If we store naïvely all pairs of activities, we will probably overload our memory. We need a structure with fixed capacity to meet stream requirements, which has a fast performance for look-up, updating and deleting. In addition, it has to be capable of an aging mechanism, so we do not block the memory with deprecated cases.

In [11], Fan et al. developed a hashing data structure to maintain items very efficiently compared to the already efficient Bloom filters. We adapted and modified the method to deal with the large amount of concurrent active cases. Cuckoo hashing applies two different hash functions h_1, h_2 to an element, that has to be stored. All items are stored in an array of length n. The element is then kept in one of two positions $h_1(x), h_2(x) \in \{0 \ldots n - 1\}$, both defined by the two hash functions. If both positions are already occupied, the current element is swapped with one of the blocking items and the swap item has to be inserted elsewhere. As the swap item also has two possible storage positions, it can be inserted on its second position. This is repeated until an empty slot for the pending item has been found or until the swap maximum has been reached. The swap item is discarded then.

Using a traditional Cuckoo hash-table supports us with a good performance. Two problems arise, caused by properties of business processes. First, cases can be disrupted or canceled without any indication. The data structure needs an aging mechanism, which does not deteriorate the performance. Second, the pace of different traces can vary. Some cases are rather quick, others might have a low activity frequency. As we do not want to give the faster ones a higher influence, the data structure needs to find a good candidate to discard in a short time period.

Our modification uses short arrays of size m instead of single item positions. Then the data structure is a matrix of size $n \times m$. For a new observed event e, both hash functions determine the two candidate arrays at $h_1(c(e)), h_2(c(e))$. In case of free space and no existing event data for this case, we just store the event in one of both arrays. If there was already information for this particular case, we extract the old event data and update the information with e. The resulting information is a set of relations

$$R(e) = \{(a_{\text{previous}}, a(e), t(e) - t_{\text{previous}}) \mid a_{\text{previous}} \in c(e) \land t_{\text{previous}} < t(e)\}.$$

In case of a collision, both arrays are full and the current event case is not stored. Traversing over the complete table to delete old cases would be too time-consuming. Our strategy is to replace the new event case with the most obsolete case data we can find in both rows. Recursively we try to insert the swapped case data in its alternative position. This is repeated for finitely many times p_{max}. $(p_{\text{max}} + 1)m$ items are checked at most. Closing a case is as simple as inserting new information. Both hash functions define the only two possible locations of case related information. Two arrays have to be checked and the corresponding event data has to be removed by visiting at most $2m$ slots (Fig. 3).

Fig. 3. Processing of a new observation. Two hash functions are applied to the case identifier. If the case was already opened, it is guaranteed to be found in one of both rows. The current information is used to update the data in the table and the relation is built by both the old and the recent event data.

Regarding complexity so far, we have $\mathcal{O}((p_{\text{max}} + 1)m + \max |c|)$ for the maximal case length $\max |c|$. The first insertion part is constant because the parameters are set in advance. The second part is the update procedure. It can be assumed to be constant in practice as well. Cases have finite length in practical scenarios. Often we can assume an upper bound of 20 events per case. Otherwise we can force the size of the sets to be within a certain margin. Discarding the oldest events in a case, which are unlikely to be relevant for recent events, secures the constant complexity again for streaming purposes.

4.2 Rolling Averages

Now we want to assemble an indicator, that determines how far a new observation derives from the previous observations. Usually one is interested in the z-score, which states the distance between the observation and the mean value, measured as standard deviations:

$$z(x) = \frac{x - \text{mean}(x)}{\text{stddev}(x)}$$

Calculating the z-score for each observation needs knowledge about the mean and standard deviation of the complete time series. As we do not want to assume a stationary model, we need a mechanism to update the statistics incrementally. In the text mining area, Schubert et al. [18] presented the SigniTrend approach, which uses such an incremental z-score. The statistics are exponentially weighted moving averages and variances, namely EWMA and EWMVAR as rolling mean $\hat{\mu}$ and rolling variance $\hat{\sigma}^2$:

$$\hat{\mu}_0 = x_0, \qquad\qquad \hat{\mu}_n = (1 - \lambda)\hat{\mu}_{n-1} + \lambda x_n$$
$$\hat{\sigma}_0^2 = x_0, \qquad\qquad \hat{\sigma}_n^2 = (1 - \lambda)(\hat{\sigma}_{n-1}^2 + \lambda(x_n - \hat{\mu}_n)^2)$$

The tradeoff between a smoothed and history-considering or a very up-to-date but noise function can be adjusted with the decay factor $\lambda \in (0, 1)$. The moving z-score is then computed as

$$\hat{z}(x_i) = \frac{x_i - \hat{\mu}_i}{\sqrt{\hat{\sigma}_i^2}}$$

The expectation value of $\hat{z}(x)$ is zero. Its variance is more complex and dominated by λ while very robust towards noise in x. Figure 4 shows the results of our experiments regarding the standard deviation when influenced by different decay factors and degrees of noise. According to our experiment, we calculated a linear approximation for the functional influence of λ and define $c(\lambda) = 1 - \frac{3}{4}\lambda$ with a coefficient of determination of 0.95. Dividing the z-score by $c(\lambda)$ yields a standard

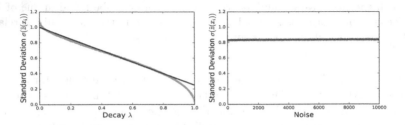

Fig. 4. The standard deviation of the moving z-score, dependent on the decay factor (left) and a noise level as standard deviation of the input between 0 and 10000 (right). For the noise experiment, a decay factor of 0.2 is chosen. $\sigma(\hat{z}(x)) = c(\lambda) = 1 - (0.2 * 0.75) = 0.85$, which explaines the offset on the right-hand side.

deviation of 1 then. It is interesting to notice that the noise has almost no influence on the z-score statistical measures. This property is crucial for this method as we can use the same values on activity relations with different noise levels without further adjustments. Even time series with changing noise levels can be handled as the z-score will just adapt to the deviated noise.

Algorithm 1. Tesseract Value

Input: S_r: relation stream, λ, θ: decay parameters
 1: Initialize HashMap for relation data R
 2: **loop**
 3:     $r = (a, b, \delta_t) \leftarrow S_r$.observeStream()
 4:     data $\leftarrow R$.getData($a \rightarrow b$)
 5:     **if** data is empty **then**
 6:         data.EWMA $\leftarrow \delta_t$
 7:         data.EWMVAR $\leftarrow \delta_t$
 8:         data.out $\leftarrow \delta_t$
 9:     **else**
10:         diff $\leftarrow \delta_t -$ data.EWMA
11:         incr $\leftarrow \lambda *$ diff
12:         data.EWMA $\leftarrow$ data.EWMA + incr
13:         data.EWMVAR $\leftarrow (1 - \lambda)($data.EWMVAR + diff $*$ incr$)$
14:         $\hat{z} \leftarrow \frac{x - \text{data.EWMA}}{\sqrt{\text{data.EWMVAR}} * c(\lambda)}$
15:         data.out $\leftarrow (1 - \theta) *$ data.out $+ \theta * \hat{z}$
16:         output data.out as Tesseract value
17:     **end if**
18: **end loop**

Control charts are a common tool in statistical process control. If a process is "under control", the significant time series oscillates around the center line within a certain range. We can use $\hat{z}(x) \cdot \frac{1}{c(\lambda)}$ as an indicator function in a control chart, which is expected to stay almost equal to zero for stationary values. The function filters a large ratio of noise, but short peaks are possible and cause false alerts. Additional smoothing, for example with another EWMA-filter, is adequate but will slow down detection times.

TESSERACT as presented so far is flexible in its case management and detection methods. The Cuckoo table is easily interchangeable with another data structure capable of storing case data. The moving z-score can also be exchanged with a different approach. For example, an adaptive window approach is possible, but handling different windows depending on the various relations is challenging. A second issue is the different pace and frequency of relations, so a window approach needs more effort for user-defined parameters. This is the reason we preferred the presented approach.

4.3   Implementation and Visualization

We implemented TESSERACT in the ProM framework. A screenshot is given in Fig. 5. On the left-hand side, the user can specify the pair of activities to monitor on the right. In the main frame, we give the observed time intervals between both activities. Below, the smoothed incremental z-score is plotted. We provide

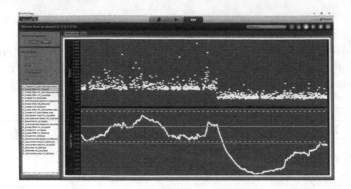

Fig. 5. A screenshot of TESSERACT while observing the loan application process [22]. A temporal drift in *O_Create_Offer* → *O_Created* can be seen, which halved the time between both activities from about 1477 ms to roughly 670 ms around July 12th. The drift can be noticed in Fig. 11 as the most intense red block in the diagram.

a zero line (green) and two alert thresholds (red). The two threshold lines are positioned at 3σ and -3σ. The white Tesseract value triggers a temporal drift, if a minimum number of events shift the value out of the in-control limits.

The last step is to use the results in a graphical visualization. TESSERACT produces weighted interval data, so our first choice is to use a Gantt-chart for its representation. As you can see in Fig. 6, we use a red/blue diverging colormap. A strong blue marks a decrease in speed, a red indicates acceleration. The intensities are given by the integrals of the Tesseract values, which exceed the threshold lines.

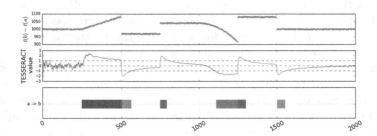

Fig. 6. The generation of the final Gantt-Chart. TESSERACT uses the durations or activity interim times to derive the Tesseract value. The values exceeding the thresholds are used to generate horizontal Gantt-bars.

5   Experiments

We performed some experiments on our novel temporal drift detection method. We tested our adaptation of the Cuckoo Table and the performance regarding collisions and load-factor using a real world dataset. Then we tested the detection delay of drifts with a large synthetic dataset. Thirdly, we performed complete runs on real world datasets. The datasets are the Road Traffic Fine Management Process [10] and the BPI 2017 challenge dataset [22].

5.1 Collision Performance of Cuckoo Hashing

Our variant of a Cuckoo Table is a matrix containing n positions with m buckets each. Every observed case event causes a look-up at two positions and if it is a new case, it is swapped with the most obsolete case information, which is then maybe stored at another position. Having many events in a stream can cause collisions, which leads to p attempts to reassign oldest items to another position. Theoretically, this can end in an infinite cycle, especially for almost full tables. We set a limiter $p_{max} = 10$ for our experiments and tested for several table sizes $m \times n$ the average number of reassignment steps, the average load factor, the total number of collisions and the total number of discards. The results can be seen in Fig. 7.

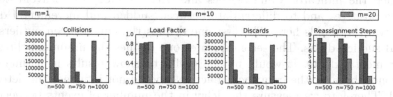

Fig. 7. The plots show the detection delay in case of incremental drifts. For several levels of noise and different drift intensities as gaps, the delay is given as the number of events between drift start and drift alert. The decay factor is set to 0.05 and the smoothing factor is 0.005. For each parameter setting, 10k drifts are simulated.

It is not very surprising that collisions are more unlikely for larger storage capacities. The load factor is high, but it is important to notice, that it only reaches a factor of 0.8. The usage of short arrays seems to be a good technique here. Although the load factor is not at maximum capacity, there are still a lot of discards, which means loosing information. On the other hand, reassigning elements to other locations takes more time. The m buckets at each position helps reducing these reassignment steps and discards and saves processing time.

5.2 Detection Delay for Sudden Drifts

TESSERACT is supposed to detect temporal drifts based on data from an event stream to deliver early results. One wants to react to drifts as soon as possible to lessen the chance of bad consequences or to prepare for good business opportunities. Therefore we tested the delay of the detection on synthetic data to ensure an accurate starting time for a drift as the ground-truth. We estimated the noise and the degree of deviation as the most important sources of delay. So we generated several event streams containing about 10k drifts each. The basis was a constant function for the time between two activities. For every drift, we increase or decrease the constant function by a uniformly distributed random value between -1000 and 1000 to simulate a sudden drift. Between drifts we include a constant phase so that the trigger function can normalize again and we are able to map each drift alert to a distinct drift. Finally, we add a normally distributed noise with a mean of 0 and a variable standard deviation as the noise level to the signal. We assume a normal distribution to be valid for most applications. Especially business data contains other distributions as we discovered in a real world dataset, but for this paper we followed the intuition that a particular action takes mostly a constant amount of time. A meeting can be scheduled to take one hour for example. In many cases, it will take approximately the scheduled amount but in rare cases, this action can require a lot less or more time. To simulate incremental drifts, we used a similar strategy by adding incrementally a value between -1000 and 1000 to the time series for a brief interval per drift.

We measured the time between the start point of the drift in the time series and the point in time when our method triggers a temporal drift. This is the case when the indicator value reaches and exceeds 1 or -1. To avoid detecting false positives and to differentiate between sudden and incremental drifts, we use two thresholds here. The first one defines the cut between outliers and declared sudden drifts. The second greater one defines the minimum threshold for incremental drifts. It would not be useful to show results in a time scale like seconds, because the frequency of actions should not influence the detection method. Therefore we measure the delay in the number of events between the first deviating event and the alert.

The results can be seen in Fig. 8 for the sudden drift experiments and in Fig. 9 for the incremental drift experiments. The sudden drifts are more influenced by noise than the incremental drifts. This is not very surprising as the indicator function starts to normalize again as soon as the drift is over. The noise forces the method to adapt to be robust, which degrades the ability to detect the

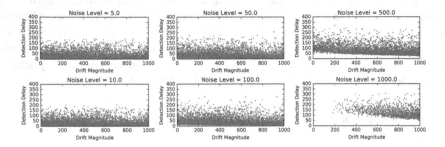

Fig. 8. The plots show the detection delay in case of sudden drifts. For several levels of noise and different drift intensities as slopes, the delay is given as the number of events between drift start and drift alert. The decay factor is set to 0.005 and the smoothing factor is 0.005. For each parameter setting, 10k drifts are simulated.

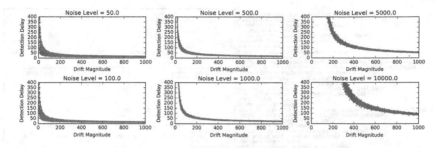

Fig. 9. The plots show the detection delay in case of incremental drifts. For several levels of noise and different drift intensities as gaps, the delay is given as the number of events between drift start and drift alert. The decay factor is set to 0.05 and the smoothing factor is 0.005. For each parameter setting, 10k drifts are simulated.

sudden drifts slightly. The incremental drifts consist of multiple increasing time values and can be differentiated better from the noise with this method. The figure shows that our method is capable of detecting incremental drifts in a very noisy signal.

While the incremental drift detection is rather dependent on the intensity of the drift, the sudden drifts are almost completely invariant to it. This can be explained by the fact that incremental drifts are technically a series of sudden drifts. The incremental aspect of the moving z-score detects longer periods of growing values more easily. Thus the noise affects sudden drift detection independently of the drift intensity. Although noise causes a growing delay for increasing levels of fluctuation, our method is still able to detect drifts but it might take more events.

5.3   Real World Application

Finally we want to prove the usefulness of TESSERACT by applying it to a real-world dataset. To the best of our knowledge, there is no similar approach to compare our method with regarding performance measures. Thus we try to find temporal drifts in case of a structural stationary process. The publicly available Road Traffic Fine Management Process [10] contains about 150k cases regarding processing of traffic fines. All traces follow the same guidelines. In this data, we should not find any drifts due to the stationary model and in compliance with state-of-the-art structural concept drift tools in ProM [4]. The results can be seen in Fig. 10. There seems to be a significant temporal recurring drift between the activities *Insert Fine Notification* and *Add penalty*. Every spring, the interim time between both activities is decreased and in the second half of the year increased again. We took a closer look at the corresponding data and this relation uses almost exactly 60 days. Only for two nearly 60-day-periods it takes 1 h more or 1 h less respectively. The explanation is rather trivial: This is an effect of the daylight saving time, causing artificial drifts in a very constant time series. On the one hand, it would be a good idea to avoid seasonal time differences in

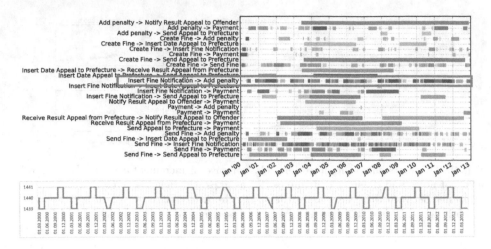

Fig. 10. The application of TESSERACT to the Traffic Loan Dataset [10]. Although the process is static, we find a significant recurring drift between in insert fine notifications and add penalty. The time series below shows the extracted interim times, shifting between 1439 and 1441 h.

event logs and refer to a standardized time. On the other hand, this showed that TESSERACT successfully detected a drift which is clearly present in the data.

For the second application on a real-world dataset we used the recently published loan application process [22] for the BPI 2017 challenge. We used both datasets, but due to space constraints we will discuss the compact offers-dataset here only. The resulting Gantt-chart is represented in Fig. 11. A very significant drift is observed in *O_Create_Offer* → *O_Created* around July 10th. According to the short duration and the bright color, it has to be a sudden and intensive acceleration. In the screenshot in Fig. 5 the response of the TESSERACT value is noticeable. Looking on the time between both activities before and after July 10th, there is a major difference between the mean interim time. While there is almost a constant time period for both intervals, the mean of the first interval is around 1477 ms, the second one is about 670 ms, which is more or less half of the former duration. This time series is quite noisy as we can see in the screenshot, but TESSERACT was able to detect the drift. Besides, by analyzing the Grantt-chart visually, we can identify recurring drifts before the *O_Accepted* activity, indicating that there is a monthly delay of some sort. Nevertheless there seems to be some attempts to accelerate the acceptance around June 2016 and November 2016. More insights are possible using the complete application dataset.

5.4 Limitations

Our approach is capable of detecting temporal drifts in an event stream. On the contrary, our approach does not take structural drifts into account. This is

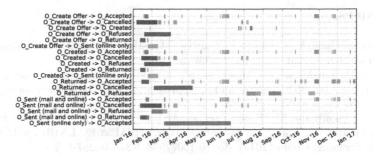

Fig. 11. The application of TESSERACT to the loan application process [22] is presented. There is a delay in the acceptance of offers on the 20th of each month. There are some improvements for the acceptance speed around June 2016 and November 2016. The most significant temporal drift is between July 10th-14th. The time to create an offer has halved, see also Fig. 5.

a limitation as structural and temporal drifts are related sometimes. A careful analysis should take all perspectives to create results. Including background knowledge about the ground model and changes in the structure is expected to increase the confidence of detected drifts and decrease of memory usage. We monitor many interim times which are less important and can be ignored. Due to a lack of information we store them to achieve a complete analysis. A general problem of the exponential weighted averages is a high sensitivity at the beginning. Few almost constant starting values with low standard deviation will lead to a high divergence peak for the first slightly different data value. A naïve approach defines a training period. However, in process mining different activities have deviating frequencies. While some activity pairs normalize after minutes, others occur only once per day and need weeks to yield reliable results about their drift behavior.

6   Conclusion

In this work we presented a novel algorithm TESSERACT, which detects concept drifts in the temporal dimension of events. It extracts the completion times of activities or inter-activity times and uses exponentially moving statistics to derive a drift indicator. Both parts are modular and can be exchanged with other techniques. Using an intuitive visualization, the results can be delivered to domain experts for further reasoning. We showed that TESSERACT works well on streams of real-world event data and is able to enhance the discovery of drifts by adding insights about temporal behavior. It is able to detect sudden and incremental drifts, but the results have shown that recurring drifts are detectable only by visual inspection so far. An automated approach for those higher-order patterns would be beneficial to enhance the knowledge about a process once more. We see a promising direction in embedding the drift information into other discovery algorithms to investigate whether our view on the completion times can improve the discovery process by analyzing the distribution patterns of time intervals.

References

1. Aminikhanghahi, S., Cook, D.J.: A survey of methods for time series change point detection. Knowl. Inf. Syst. **51**(2), 339–367 (2017)
2. Backus, P., Janakiram, M., Mowzoon, S., Runger, C., Bhargava, A.: Factory cycletime prediction with a data-mining approach. IEEE Trans. Semicond. Manuf. **19**(2), 252–258 (2006)
3. Bolt, A., Sepúlveda, M.: Process remaining time prediction using query catalogs. In: Lohmann, N., Song, M., Wohed, P. (eds.) BPM 2013. LNBIP, vol. 171, pp. 54–65. Springer, Cham (2014). doi:10.1007/978-3-319-06257-0_5
4. Bose, R.P., van der Aalst, W.M.P., Zliobaite, I., Pechenizkiy, M.: Dealing with concept drifts in process mining. IEEE Trans. Neural Netw. Learn. Syst. **25**(1), 154–171 (2014)
5. Bose, R.P.J.C., Aalst, W.M.P., Žliobaitė, I., Pechenizkiy, M.: Handling concept drift in process mining. In: Mouratidis, H., Rolland, C. (eds.) CAiSE 2011. LNCS, vol. 6741, pp. 391–405. Springer, Heidelberg (2011). doi:10.1007/978-3-642-21640-4_30
6. Burattin, A., Sperduti, A., van der Aalst, W.M.P.: Heuristics miners for streaming event data (2012). arXiv:1212.6383
7. Burattin, A., Sperduti, A., van der Aalst, W.M.P.: Control-flow discovery from event streams. In: Congress on Evolutionary Computation (IEEE WCCI CEC) (2014)
8. Cabanillas, C., Di Ciccio, C., Mendling, J., Baumgrass, A.: Predictive task monitoring for business processes. In: International Conference on Business Process Management, pp. 424–432. Springer (2014)
9. Carmona, J., Gavaldà, R.: Online techniques for dealing with concept drift in process mining. In: Hollmén, J., Klawonn, F., Tucker, A. (eds.) IDA 2012. LNCS, vol. 7619, pp. 90–102. Springer, Heidelberg (2012). doi:10.1007/978-3-642-34156-4_10
10. de Leoni, M., Mannhardt, F.: Road traffic fine management process (2015)
11. Fan, B., Andersen, D.G., Kaminsky, M., Mitzenmacher, M.D.: Cuckoo filter: Practically better than bloom. In: Proceedings of the 10th ACM International on Conference on emerging Networking Experiments and Technologies, pp. 75–88. ACM (2014)
12. Gal, A., Mandelbaum, A., Schnitzler, F., Senderovich, A., Weidlich, M.: On predicting traveling times in scheduled transportation. In: Proceedings of the 2nd International Conference on Mining Urban Data, Vol. 1392, pp. 88–89. CEUR-WS. org (2015)
13. Hassani, M., Siccha, S., Richter, F., Seidl, T.: Efficient process discovery from event streams using sequential pattern mining. In: IEEE Symposium on Computational Intelligence and Data Mining (CIDM), pp. 1366–1373 (2015)
14. Maaradji, A., Dumas, M., Rosa, M., Ostovar, A.: Fast and accurate business process drift detection. In: Motahari-Nezhad, H.R., Recker, J., Weidlich, M. (eds.) BPM 2015. LNCS, vol. 9253, pp. 406–422. Springer, Cham (2015). doi:10.1007/978-3-319-23063-4_27
15. Kuma, M.V.M., Thomas, L., Annappa, B: Capturing the sudden concept drift in process mining. In: BPM Workshops, pp. 132–143 (2015)
16. Polato, M., Sperduti, A., Burattin, A., de Leoni, M.: Data-aware remaining time prediction of business process instances. In: 2014 International Joint Conference on Neural Networks (IJCNN), pp. 816–823. IEEE (2014)

17. Schlimmer, J.C., Granger, R.H.: Beyond incremental processing: Tracking concept drift. In: National Conference AI, pp. 502–507 (1986)
18. Schubert, E., Weiler, M., Kriegel, H.-P.: Signitrend: scalable detection of emerging topics in textual streams by hashed significance thresholds. In: Proceedings of the 20th ACM SIGKDD International Conference on Knowledge Discovery and Data Mining, pp. 871–880. ACM (2014)
19. Senderovich, A., Weidlich, M., Gal, A., Mandelbaum, A.: Queue mining – Predicting delays in service processes. In: Jarke, M., Mylopoulos, J., Quix, C., Rolland, C., Manolopoulos, Y., Mouratidis, H., Horkoff, J. (eds.) CAiSE 2014. LNCS, vol. 8484, pp. 42–57. Springer, Cham (2014). doi:10.1007/978-3-319-07881-6_4
20. van der Aalst, W.: Process Mining: Data science in action. Springer, Heidelberg (2016)
21. Van der Aalst, W.M.P., Schonenberg, M.H., Song, M.: Time prediction based on process mining. Inf. Syst. **36**(2), 450–475 (2011)
22. van Dongen, B.F.: Bpi challenge 2017 - offer log (2017)

Intra and Inter-case Features in Predictive Process Monitoring: A Tale of Two Dimensions

Arik Senderovich[1(✉)], Chiara Di Francescomarino[2], Chiara Ghidini[2],
Kerwin Jorbina[3], and Fabrizio Maria Maggi[3]

[1] Technion - Israel Institute of Technology, Haifa, Israel
sariks@technion.ac.il
[2] Fondazione Bruno Kessler, Trento, Italy
{dfmchiara,ghidini}@fbk.eu
[3] University of Tartu, Tartu, Estonia
{kerwin.jorbina,f.m.maggi}@ut.ee

Abstract. Predictive process monitoring is concerned with predicting measures of interest for a running case (e.g., a business outcome or the remaining time) based on historical event logs. Most of the current predictive process monitoring approaches only consider intra-case information that comes from the case whose measures of interest one wishes to predict. However, in many systems, the outcome of a running case depends on the interplay of all cases that are being executed concurrently. For example, in many situations, running cases compete over scarce resources. In this paper, following standard predictive process monitoring approaches, we employ supervised machine learning for prediction. In particular, we present a method for feature encoding of process cases that relies on a bi-dimensional state space representation: the first dimension corresponds to intra-case dependencies, while the second dimension reflects inter-case dependencies to represent shared information among running cases. The inter-case encoding derives features based on the notion of *case types* that can be used to partition the event log into clusters of cases that share common characteristics. To demonstrate the usefulness and applicability of the method, we evaluated it against two real-life datasets coming from an Israeli emergency department process, and an open dataset of a manufacturing process.

Keywords: Predictive Process Monitoring · Inter-case Features · Bi-dimensional Feature Encoding

1 Introduction

Business processes are supported by process-aware information systems (PAIS) that record execution data into event logs [1]. Predictive process monitoring is concerned with analyzing these event logs to predict measures of interest (MOI) at runtime [2,14]. Key measures include the business outcome of a process execution (a.k.a. case), the remaining time and the remaining activities up to the

© Springer International Publishing AG 2017
J. Carmona et al. (Eds.): BPM 2017, LNCS 10445, pp. 306–323, 2017.
DOI: 10.1007/978-3-319-65000-5_18

completion of an ongoing case. State-of-the art approaches for solving predictive monitoring problems employ supervised machine learning to make predictions [13,20]. In particular, event logs are encoded into feature-outcome pairs to build training sets required to run machine learning algorithms [10].

In previous works, predictive process monitoring methods assumed that the MOIs of an ongoing case only depend on intra-case information, e.g., on the execution history of that specific case. This assumption results in encodings that include past events, inter-event durations, and other case attributes [13]. However, the intra-case assumption does not hold in many real-life scenarios. For example, in situations where cases share limited resources, the remaining time of a case heavily depends on other cases that are running at the same time [17].

In this paper, we propose a method for feature encoding in predictive monitoring that relies on a bi-dimensional state space. The first dimension expresses intra-case dependencies, while the second dimension represents inter-case dependencies capturing the interplay of all running cases. However, encoding features into the inter-case dimension is challenging, since (1) there is a large variance in the levels of dependencies for different business processes, and (2) encoding multiple features for a large number of simultaneously running cases may lead to feature space explosion. To overcome these two limitations, we propose an *inter-case encoding* general method that is grounded in the partitioning of recorded (and running) cases into *case types*. Case types represent a flexible instrument to cluster together cases being in the same *state* since the characteristics that define a case state can be different for different scenarios. In addition, to avoid feature space explosion, inter-case features are encoded using a *derivation* function that enables for their compact representation.

We evaluate the proposed method against two real-life datasets coming from an Israeli emergency department (ED), and a manufacturing process. The experiments show that our solution yields a 27% improvement when predicting remaining times of ED patients, and a 52% improvement when predicting the remaining time for the manufacturing process.

The remainder of the paper is structured as follows. We start by motivating our contribution using scenarios taken from a real-life hospital setting that describe different levels of inter-case dependencies (Sect. 2). In Sect. 3, we provide essential background concepts, before stating the main problem that we solve in this paper (Sect. 4). Section 5 presents the core contribution of the paper by specifying the details of our method. An empirical evaluation of the method is presented in Sect. 7, while Sects. 8 and 9 close the paper with related work and conclusive remarks.

2   Motivating Scenarios

When considering predictive process monitoring problems in real-life settings, one can identify several scenarios that reflect different levels of inter-case dependencies. Below, we report four realistic scenarios from an ED. The examples are targeting the remaining time prediction as the main MOI.

Scenario 1: Urgent Patients. An urgent patient who arrives into the ED requires first aid, and thus receives high priority. Hence, the patient does not compete over shared resources, as she gets immediate help. Here, the dependency between the urgent case and all other cases is negligible, and the remaining time of that case will depend on the clinical history of the patient. This is a case where intra-case features are most predictive for the MOI.

Scenario 2: Homogeneous Patients. Consider the distribution of food among patients in the ED. Here, assuming that no patient type has priority over the others, service times are independent and identically distributed for all patients, and the distribution order is random. Hence, we expect that the remaining time for a specific case would depend only on the total number of patients waiting to receive food.

Scenario 3: Heterogeneous Patient Types with Priorities. Consider the radiology unit, where patients compete over several types of machines (e.g., XRAY, CT, and MRI). Patients are prioritized according to their diagnosis, age, and recent events. For example, patients who recently went through an Oncology consultation receive the highest priority class. Patients with orthopedic trauma who were recently triaged (seen by a nurse) will receive the lowest priority class. In this setting, it is important to discriminate case types, and consider the number of patients in each priority class [17].

Scenario 4: Heterogeneous Patient Types with History-Dependent Priorities. Consider the surgery unit, where patients compete over surgeons and operating rooms. Patients are prioritized according to their diagnosis, age, and their clinical history. For example, patients with a severe diagnosis and who already had other similar suspected diagnosis or other severe diseases in the past, receive the highest priority class. Patients who never had health problems and with a minor trauma will receive the lowest priority class. In this setting, a more fine-grained case typing than in Scenario 3 must be considered. In particular, one needs to take into account the history of the patients when creating priorities.

To summarize, we observe that different processes may require different feature encodings to capture inter-case dependencies. We shall return to these scenarios in Sect. 5, where we present our solution to inter-case feature encoding.

3 Preliminaries

In this section, we provide the required preliminaries for our work, namely the definition of event logs, the formulation of the predictive process monitoring problem, and the description of the supervised machine learning framework.

3.1 The Event Log

In what follows, we define an event log, our representation of process execution recordings. Let $\mathcal{E}$ be the universe of all events. An event log L is a K-sized set of event sequences (or cases), $L = \{\sigma_i : i = 1, \ldots, K\}$ with $\sigma_i = (e_i^1, \ldots, e_i^{n_i}) \in \mathcal{E}^*$

being the ith case of length n_i. We denote by $\mathcal{L}$ the universe of event logs. Without loss of generality, we assume that L comprises complete cases only. However, for running case prediction, we are interested in prefixes of complete cases. Hence, we define the prefix function, $\phi : \mathcal{E}^* \times \mathbf{N}^+ \rightarrow \mathcal{E}^*$, which returns a prefix of size n, given a sequence, namely $\phi(\sigma_i, n) = (e_i^1, \ldots, e_i^n) : n \leq n_i$. The function returns σ_i in case $n > n_i$. We define the extended event log L^* to be the event log that contains all prefixes of L, i.e., $L^* = \{\phi(\sigma_i, n) : \sigma_i \in L, n \leq n_i\}$.

Further, we assume that every event in a sequence is associated with data in the form of attribute-value (AV) pairs. Formally, we define the AV function α to be $\alpha : \mathcal{E} \rightarrow \mathcal{A}_1 \times \cdots \times \mathcal{A}_p$ with $\mathcal{A}_j, j = 1, \ldots, p$ being p attribute domains. We assume that $\exists j : \mathcal{A}_j$ is the universe of timestamps (e.g., UNIX time recordings). We denote the domain of timestamps by $\mathbb{TS}$. Moreover, we assume that, for $j = 1, \ldots, p$, $\mathcal{A}_j$ contains the *unknown* value $\perp$. Lastly, each prefix of σ_i is associated with a label (e.g., the remaining time for a running case) such that $y(\phi(\sigma_i, n)) \in \mathcal{Y}$ with $\mathcal{Y}$ being the domain of labels. Note that the labels can be dynamic and change as the case progresses (e.g., the remaining time for a running case changes as the case progresses).

σ_1 = (consultation,{33, radiotherapy, 10:30AM},50), (ultrasound, {33, nursing ward, 10:55AM},25),...

...

σ_K = (order blood,{56, general lab, 12:30PM},120), ..., (payment, {56, financial dept., 2:30PM},0)

Fig. 1. Example of an event log.

As a running example, consider the event log presented in Fig. 1 pertaining to a medical treatment process. Each case relates to a different patient and the corresponding sequence of events indicates the activities executed for a medical treatment of that patient. In the example, *consultation* is the first event of sequence σ_1. Its AV function maps the *consultation* event to "{33, radiotherapy, 10:30AM}" corresponding to the data associated to attributes *age*, *department*, and *timestamp*. The *remaining time* (in minutes) is the label that is present for every event (with 50 min being the remaining time after the first event in the first case, and 25 min being the remaining time after the second event). Note that the value of *age* is static, i.e., it does not change for all the events in a case, while the values of *department*, *timestamp*, and *remaining time* are dynamic, i.e., different for every event.

3.2 Predictive Process Monitoring

Having defined our data model, we turn to formulate the problem of predictive process monitoring, following the notation in [13, 14].

Definition 1 (PPM). *Given a (possibly running) case σ_x, the predictive process monitoring problem (PPM) is to find a function $f : \mathcal{E}^* \rightarrow \mathcal{Y}$ that accurately maps σ_x to the corresponding label $y(\sigma_x)$.*

Note that differently from previous works on predictive process monitoring, here we do not necessarily observe a single outcome per case. In contrast, every prefix of a running case can have a corresponding label in $\mathcal{Y}$. Since historical observations of cases (with all their prefixes and dynamic labels) are provided in L^*, it is natural to employ supervised learning for solving PPM. Hence, we complete the section by defining the supervised learning framework.

3.3 Supervised Learning

Below, we introduce the basic definition of a supervised learning task based on [18], which we later utilize to solve the PPM problem.

Let $\mathcal{X}$ be some vector-feature space, and let $\mathcal{Y}$ be the outcome space (we abuse notation on purpose as we consider the same domain as for sequence labels). Further, we denote by r some nonnegative real-valued risk function $r : \mathcal{Y} \times \mathcal{Y} \to \mathbb{R}^{0+}$ that given a predicted outcome and the real label, returns the risk value (here, high risk corresponds to low accuracy and vice versa). For example, r can be the squared error risk function. The supervised learning task (SLT) is defined as follows.

Definition 2 (Supervised Learning Task (SLT)). *Given the tuple $\langle \mathcal{S}, F, r \rangle$ with*

- *$\mathcal{S} \subseteq \mathcal{X} \times \mathcal{Y}$ being the training set of independent and identically distributed samples from the distribution of $\mathcal{X} \times \mathcal{Y}$,*
- *$F \subseteq \mathcal{Y}^{\mathcal{X}}$ being the class of learning functions, and,*
- *r being the risk function that we wish to minimize,*

the supervised learning task (SLT) is to find a function $\hat{f} \in F$ s.t. for a test observation $(x_0, y_0) \in \mathcal{X} \times \mathcal{Y}$, the excepted risk is minimized:

$$\hat{f} = \arg \min_{f \in F} \mathbb{E}[r(\hat{f}(x_0), y_0)]. \tag{1}$$

Note that, in real-life situations, we may observe only x_0, while y_0 is unobserved. Hence, the true risk can be assessed only in hindsight. In practice, to obtain $\hat{f}$ that solves the SLT, machine learning methods minimize the empirical risk with respect to the training pairs, i.e., for $\mathcal{S} = \{(x_i, y_i), i = 1, \ldots, K\}$, we select $\hat{f}$ such that:

$$\hat{f} = \arg \min_{f \in F} \frac{1}{K} \sum_{i=1}^{K} r(\hat{f}(x_i), y_i). \tag{2}$$

This approach can lead to over-fitting, which can in turn be avoided by proper selection of F, the class of learners. In this paper, we avoid over-fitting by selecting well-established supervised learning methods that balance over- and under-fitting.

4   Problem Setup

In this section, we present the problem that we solve in this paper, namely the *Sequence-To-feature Encoding Problem* (STEP). The problem arises when we aim at casting the predictive monitoring problem (PPM) into a supervised learning task (SLT). The casting appears straightforward. As solution to the PPM, f, one may consider using $\hat{f}$ that results from setting: (i) the training set S to be L^* (all prefixes of historical cases); (ii) F to be some class of functions (e.g., the set of linear functions); and (iii) r to be some risk function (e.g., the squared error).

However, the training data in the SLT setting, S, is assumed to be a set of independent and identically distributed (i.id.) observations of feature-outcome pairs, (x_i, y_i). In contrast, the training data that stems from L^* contains a set of highly dependent prefix-outcome pairs, $(z_i, y(z_i)), z_i = \phi(\sigma_i, n)$ with $i = 1, \ldots, K, n = 1, \ldots, n_i$: any two prefixes of the same case are highly correlated as they represent the same process execution (intra-case dependencies), and every two prefixes that run in the process at the same time potentially share limited resources (inter-case dependencies). Furthermore, the SLT solution function $\hat{f}$ maps a newly observed feature value to a label, while in the PPM problem, the new value is a (possibly running) case. This leads to the need for transforming sequences into features.

Problem 1 (Sequence-To-feature Encoding Problem (STEP)). *Let L^* be an extended event log that contains all prefixes of the sequences in L. Solving the STEP problem is to find a function $g : \mathcal{E}^* \times \mathcal{Y} \times 2^L \to \mathcal{X} \times \mathcal{Y}$ such that the result of its operation, $\{g(\sigma_i, y(\sigma_i), L^*))\} \subset \mathcal{X} \times \mathcal{Y}$, is an i.id. sample of feature-outcome from $\mathcal{X} \times \mathcal{Y}$.*

Clearly, a STEP solution includes a joint choice of g and $\mathcal{X}$. The STEP has been solved in [8,13,14,20] by various intra-case feature encodings. The main contribution of this paper is a novel solution to STEP that leverages inter-case information, while bounding the feature space size.

5   Bi-dimensional STEP Solution

In this section, we show our solution to STEP by means of the construction of a bi-dimensional STEP function g_β that maps every prefix in L^* and its corresponding label into $\mathcal{X} \times \mathcal{Y}$ with $\mathcal{X}$ being a bi-dimensional feature space $\mathcal{X}_1 \times \mathcal{X}_2$. The first component of the feature space, $\mathcal{X}_1$, captures intra-case dependencies, while the second component, $\mathcal{X}_2$, represents inter-case dependencies. The first dimension is defined using existing techniques from [7,13], while the contribution of this paper is in the definition of $\mathcal{X}_2$, and in the construction of g_β. In the remainder, we specify the basic requirements that a STEP solution needs to satisfy. Then, guided by these requirements, we provide a method to encode the intra- and inter-case dimensions.

5.1   STEP Requirements

Denote $\mathcal{S}^\beta$ the set of feature-outcome pairs such that

$$\mathcal{S}^\beta = \{(x_i, y_i) = g_\beta(\sigma_i, y(\sigma_i), L^*) \mid \sigma_i \in L^*\}. \tag{3}$$

We are looking for a STEP solution such that the following requirements hold:

R1: *Sufficiency.* Let σ_i and σ_j be two different prefixes coming from L^*. We require that $(x_i, y_i) = g_\beta(\sigma_i, y(\sigma_i), L^*)$ is independent of $(x_j, y_j) = g_\beta(\sigma_j, y(\sigma_j), L^*)$, i.e., the resulting label y_i depends only on x_i and not on any other (x_j, y_j) in $\mathcal{S}^\beta$.

R2: *Accuracy.* When applying the SLT framework with a class of predefined functions F to the training set that results from g_β, $\mathcal{S}^\beta \subset \mathcal{X} \times \mathcal{Y}$, we require that g_β provides the minimal empirical risk with respect to any other function g that solves the STEP, i.e., for $(x_i, y_i) = g(\sigma_i, y(\sigma_i), L^*)$ we require:

$$(\hat{f}, g_\beta) = \arg\min_{f,g} \frac{1}{K} \sum_{i=1}^{K} r(f(x_i), y_i). \tag{4}$$

R3: *Compactness.* We require that the dimension of the feature space $|\mathcal{X}|$ will not exceed the minimal required representation for R1 and R2 to hold.

Requirement R1 assures that g_β provides sufficient information such that for $(x_i, y_i) \in \mathcal{S}^\beta$, the label y_i will depend only on x_i. This implies that after the encoding, both inter-case and intra-case dependencies required to predict the label y_i are captured by the corresponding x_i. We use R1 to justify our decisions when constructing g_β.

Further, requirements R2 and R3 assure that $\mathcal{X}$ provides an accurate representation of the dependencies, and is compact. In order to demonstrate that requirements R2 and R3 hold for our solution g_β, we evaluate the method by using our STEP solution g_β to solve the PPM based on real-world data (Sect. 7). Below, we briefly outline the solution to the intra-case encoding, i.e., the construction of $\mathcal{X}_1$. Then, we focus on the main contribution of our work, namely the inter-case STEP encoding of sequences into $\mathcal{X}_2$.

5.2   Intra-case STEP Encoding

In this part, we present the construction of $\mathcal{X}_1$. Solving this aspect of the STEP is related to machine learning techniques for predicting labels in sequential data [7]. These methods transform training data that comprises correlated sequences into an independent representation of the same sample by adding relevant information to the observations. In particular, we select the *sliding window method* (Sect. 3 in [7]) to encode recent (up-to window size w) history of the prefix. We denote by $g_\beta^{(1)}$ the intra-case component of g_β. Then, for $\sigma_i = (e_i^1, \ldots, e_i^n)$, we get that

$$g_\beta^{(1)}(\sigma_i, y(\sigma_i), L^*) = ((e_i^{n-w+1}, \ldots, e_i^n), y(\sigma_i)), \tag{5}$$

assuming without loss of generality that $w \leq n$.

Note that one may consider the encoding of additional static and dynamic attribute-values as shown in [13]. For example, we set $w = 1$ and choose two AV functions as follows. Let $\alpha_1 : \mathcal{E} \to \mathcal{A}_1$ be the elapsed time from the arrival of the patient until every event (dynamic), and $\alpha_2 : \mathcal{E} \to \mathcal{A}_2$ be the age of the patient at every event (static if the age of the patient is the same at every event). Then, an encoding of σ_i that considers both AV functions and the last event (since $w = 1$) is:

$$g_\beta^{(1)}(\sigma_i, y(\sigma_i)) = (e_i^n, \alpha_1(e_i^n), \alpha_2(e_i^n), y(\sigma_i)). \tag{6}$$

Note that the value of static attributes is constant and hence mentioned only once in encodings involving multiple events. The sliding window approach assumes that the last w events (and their Attribute-Values) are sufficient to explain the MOI. Therefore, if this assumption holds, requirement R1 is satisfied.

5.3 Inter-case STEP Encoding

We now turn to introduce the inter-case STEP encoding. To give intuition, we use Fig. 2, which revisits the four scenarios from Sect. 2.

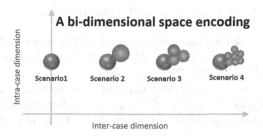

Fig. 2. A graphical representation of our proposed bi-dimensional encodings.

The vertical axis corresponds to the level of dependencies between the predicted label and intra-case features (e.g., recent history, elapsed time, and age), while the horizontal axis captures the level of dependencies for inter-case features (e.g., the number of acute patients in the ED). The blue circle corresponds to the target patient, whose MOI we wish to predict, while the red circles correspond to the patient types on whom the MOI may depend (*case types*). Each case type can be described in terms of intra-case features and the more case types we have, the more fine grained information is required (i.e., the more features are required for encoding this information). In this setting, Scenario 1 is placed on the intra-case axis, as there is no inter-case dimension. In Scenario 2, the intra-case component does not change with respect to Scenario 1, while the inter-case dimension increases in order to take into account all the other patients grouped in the same class (i.e., all of the same case type). Finally, for Scenarios 3 and 4, the inter-case component is further strengthened by means of a more fine-grained case partitioning into types.

The assumption that drives our method is that case types (e.g., urgency priorities) explain the inter-case dependencies between cases. If this assumption holds and the types are properly selected, requirement R1 is satisfied. We also assume that all events are timestamped with an AV function τ. The timestamp of a case $\sigma = (e^1, \ldots, e^n) \in L^*$ is $\tau(\sigma) = \tau(e^n) \in \mathbb{TS}$. Considering time when encoding inter-case dependencies is crucial, since we assume that these exist only between cases that run in the process at the same time.

Our inter-case STEP encoding relies on four basic concepts, namely *case types*, *discrimination*, *partition*, and (feature) *derivation*. Below, we go over these four concepts:

- *Case Types.* Denote $\mathcal{T}$ the set of m case types. Returning to our ED, a type of a patient can be set according to her severity grade.
- *Discrimination.* The discrimination function δ is used to set the features that distinguish between case types. In particular, the discrimination function $\delta(\sigma) \in \mathcal{T}$, maps a case σ into its type T. Note that AV functions can be used for discrimination: we may consider, for example, the age of a patient and the elapsed time from the arrival of the patient until the last event. The typed event log is

$$L_T^* = \{(\sigma, T) \subseteq \mathcal{E}^* \times \mathcal{T} \mid \sigma \in L^* \wedge \delta(\sigma) = T\}. \tag{7}$$

- *Partitioning.* In order to categorize cases into types, we define a function π that partitions a typed event log L_T^* into m event logs according to their types, namely $\pi(L_T^*) \in \mathcal{L}^m$.
- *Derivation.* A derivation function γ maps m event logs for some time t into the desired feature space $\mathcal{X}_2$, i.e., $\gamma : \mathcal{L}^m \times \mathbb{TS} \to \mathcal{X}_2$. For example, γ can produce the feature *number of patients of type $i = 1, \ldots, m$ in the ED at time t*.

To demonstrate the specification of the four concepts, consider the encoding of $\sigma_i = (e_i^1, \ldots, e_i^n) \in L^*$ in Scenario 2 (i.e., food distribution in the ED), where all patients are assumed to be homogeneous. Trivially, the case type set is $\mathcal{T} = \{Patient\}$, and the discrimination function results in a typed event log $L_T^* = \{(\sigma, Patient) \mid \sigma \in L^*\}$ as all patients are of the same type ('Patient'). The partitioning function returns L_T^* itself, while the derivation function γ can be set to the number of cases by type at time $\tau(\sigma_i)$:

$$\gamma(L_T^*, \tau(\sigma_i)) = |\{(\sigma, T) \in L_T^* \mid \tau(\sigma) \leq \tau(\sigma_i) \wedge \delta(\sigma) = T\}|. \tag{8}$$

In essence, once the quadruplet $(\mathcal{T}, \delta, \pi, \gamma)$ is specified, the inter-case component of g_β, which we denote by $g_\beta^{(2)}$, is constructed as follows:

$$g_\beta^{(2)}(\sigma_i, y(\sigma_i), L^*) = \gamma(\pi(\{\delta(\sigma) \mid \sigma \in L^*\}), \tau(\sigma_i)). \tag{9}$$

Let us consider the encoding of $\sigma_i = (e_i^1, \ldots, e_i^n) \in L^*$ in a less trivial scenario, Scenario 3, where patients are typed according to their last event (where the event represents the visited unit) into m urgency types, $\mathcal{T} = \{e_1, \ldots, e_m \mid e_i \in \mathcal{E}\}$

with $m = |\mathcal{E}|$ and e_i being the possible events. The discrimination function sets $\delta(\sigma) = e^n$, with e^n being the last event of σ. The partitioning is made according to the typed event log that results from δ, and the derivation function remains as in Eq. (8). This leads to m features: the number of patients of all possible urgencies at time $\tau(\sigma_i)$.

Each of the components δ, π and γ plays a different role in our method. The partitioning function π classifies cases according to the notion of similarity imposed by the discrimination function δ. To avoid feature space explosion, the derivation function allows an aggregation of the resulting typed event logs. Having fully defined the bi-dimensional STEP solution, we are now ready to test requirements R2 (accuracy) and R3 (complexity) against real-life event logs.

6   Implementation

In this section, we introduce the four instances of the STEP framework we have implemented. They take inspiration from the four scenarios presented in Sect. 2. Each of them aims at predicting the remaining time of an ongoing case. We believe indeed that the remaining time is among the most relevant MOIs when dealing with inter-case dependencies.

- LEVEL 0. Only the intra-case dimension is taken into account (as in *Scenario 1*) and it is instantiated as described in Sect. 5.2 with a window size of the maximal training trace length.
- LEVEL 1. As in *Scenario 2*, a coarse-grained level of inter-case dependencies is identified. There is only one case type, i.e., a unique priority class, and the derivation function is set to the number of cases running at time t.
- LEVEL 2. As in *Scenario 3*, a fine-grained level of inter-case dependencies is identified. There are m case types, each based on the last event carried out, the discrimination function is set to the last event in the case and the derivation function is set to the number of cases for each case type, i.e., with the same last event at time t.
- LEVEL 3. As in *Scenario 4*, an even more fine-grained level of inter-case dependencies is identified and the inter-case dimension is instantiated so that there are m case types, each defined according to the last w' events carried out (in particular, we set $w' = 3$). The discrimination function is set to the last w' events in the case and the derivation function is set to the number of cases for each case type, i.e., with the same last w' events at time t.

All feature encoding algorithms have been implemented in Python and are publicly available.[1] We use the implementations and default parameters of scikitlearn [15] for linear regression, cross-validated Lasso, and random forests. We further use the XGBoost package [5] for cross-validated XGBoost.

Finally, a web interface of our prediction system has been realized in the alpha version of a Predictive Process Analytics (PPA) tool.[2] In particular, the front end

---

[1] https://github.com/kerwinjorbina/predictive-process-webservice.

[2] The web application is available at http://obscure-springs-12588.herokuapp.com.

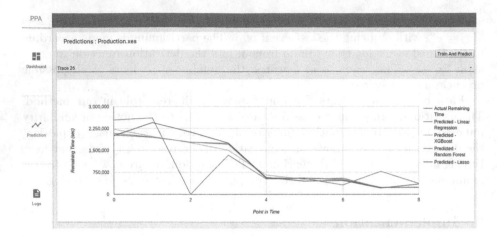

Fig. 3. Prediction results in the Predictive Process Analytics (PPA) interface.

application uses the template of Angular Material Dashboard[3] while the back end uses the Python Django Framework[4] to process the logs and provide predictions. Figure 3 reports a screenshot of the web interface showing the remaining time of a trace at specific points in time in a log.

7   Evaluation

In this section, we present the main findings we identified by evaluating our method against real-world event logs. In the remainder, we present our experimental setup and procedure, and conclude the section by providing a detailed overview of the results.

7.1   Experimental Setup and Procedure

Below, we describe the experimental setup in terms of the supervised learning methods and datasets that were used in the experiments. We then outline the overall experimental procedure and define our evaluation measures.

Machine Learning Methods. In our experiments, we relied on two families of learning algorithms: (1) linear models, including linear regression and Lasso; and (2) tree ensemble methods, including random forests and gradient tree boosting. The linear family was chosen as a weak regression baseline, to show that selecting a strong regression method is important for making the method applicable. Below, we briefly describe ensembles of regression trees.

Ensembles of Regression Trees. We selected ensembles of decision trees as learning algorithms due to their non-linearity (expressiveness) and their ability to

---

[3] https://angularjs.org/.
[4] https://www.djangoproject.com/.

automatically partition the feature space into interpretable regions. In particular, given the set of features, such as age, gender, queue lengths, and recent visit times, the resulting models end up with a clear feature segmentation. Furthermore, using an ensemble rather than a single regression tree typically leads to an improvement in accuracy [3]. We briefly describe the ensemble methods that we consider, namely random forests and XGBoost.

Random forest (RF) [4], is an ensemble built by learning each tree on a different bootstrap replica of the original training set. A bootstrap replica is obtained by randomly drawing (with replacement) original samples and copying them into the replica. Each tree is learned by starting with a single node and greedily extending the tree. Leaves are split using the test (features and values) that maximizes the reduction in quadratic error. The RF algorithm is known not to over-fit as the number of trees increases. We present the results for RF with 50 trees since increasing the number of trees beyond 50 up to 1000 resulted in an improvement of less than 5% in our accuracy measures.

The second ensemble algorithm, eXtreme Gradient Boosting (XGBoost) [5], is the most recent version of the gradient tree boosting algorithm first introduced by Friedman [11]. Unlike RF, instead of weighting the samples, gradient boosting modifies the target variable value for learning each tree.

Datasets. Our experiments used two real-world datasets, namely DS1 and DS2.

DS1. The data comes from the Electronic Health Record of an Israeli ED that has a maximal capacity of approximately 100 patients per day. Every patient that enters the ED receives a bar-code that gets tracked at the start and end of every medical procedure. Patient treatment events include 'Reception_End', 'Triage_Start', 'Treatment_Start', etc. The events are manually recorded by scanning the patient's bar-code. The dataset ranges between March 2014 and May 2015, with approximately 42 000 patient visits. The median length-of-stay in the ED is 290 minutes. Lastly, the data comprises approximately 350 000 events coming from the 42 000 cases.

DS2. This log published in the IEEE TF on Process Mining site[5] contains a comma separated value dataset in a production process. This provides data on activities, resources, machines and type of items that are manufactured. The dataset occurred from January 2012 to March 2012, with 225 cases. Listed in this log are 55 types of activities, 49 human resources and 31 machines. The median case duration is almost 14 days. The dataset contains approximately 7 200 events.

Experimental Procedure. We exploit the training-validation-test paradigm [10] to evaluate our method and randomly partition the two datasets into 60% training data, 20% validation data, and 20% test data. Note that we operate our method to encode inter- and intra-case features prior to dataset partitioning. The encoded data is used as an input to the supervised learning algorithms. While random forests are robust against over-fitting, an increased number of trees raises the computational complexity. Hence, we evaluated the number of

---

[5] http://data.4tu.nl/repository/uuid:68726926-5ac5-4fab-b873-ee76ea412399.

required trees by adding trees and stopping when the improvement is small. The Lasso model was cross-validated to select the regularization parameter. We used Lasso only for Level 3 encoding, as in all other cases linear regression dominated against the validation set. Our experiments were conducted on an 8 core Intel Xeon CPU E5-2660 v4 @ 2.00 GHz each core with 32 GB memory, running on Linux Centos 7.3 OS.

We measure the accuracy of predictions with two empirical loss functions: RMSE and MAE. Root Mean Squared Error (RMSE) is grounded in the squared difference between the actual time and the predicted value. Let y_l^* be the actual value of y_l, the time of interest for a log entry of the test set $l \in L_{test}$. With $\hat{y}_l$ being the predicted value, the RMSE is defined as:

$$RMSE = \sqrt{\sum_{l \in L_{test}} [\hat{y}_l - y_l^*]^2}.$$

The RMSE quantifies the error in the time units of the original measurements (in our case seconds, which are converted to minutes and hours in the experiment reporting, for convenience).

The RMSE is sensitive to outliers [10]. Therefore, we also consider the absolute error, which is known to be more robust [10]. In particular, the Mean Absolute Error (MAE) is defined as:

$$MAE = \sum_{l \in L_{test}} |\hat{y}_l - y_l^*|.$$

In the next section, we present these errors for the two datasets and different levels of inter-case feature encoding. We compare them with a purely static baseline, i.e., which takes into account only intra-case features. Furthermore, we plot the RMSE errors for different predicted remaining times.

7.2   Results

Table 1 summarizes the results of our experiments. The columns of the table (*Linear*, *RF*, and *XGB*) correspond to the learning methods: linear model (linear regression or Lasso), random forests, and XGBoost, respectively. The rows correspond to the combination of dataset (*DS1* or *DS2*), accuracy measure (*RMSE* or *MAE*), and feature encoding level (LEVEL0, LEVEL1, LEVEL2 and LEVEL3 introduced at the beginning of Sect. 6). The RMSE and MAE are presented in minutes for the ED, and in hours for the production process. In the table, the boldfaced values are the best combination of machine learning method and feature encoding function.

Observing the results, we notice that for both datasets, for all types of encodings and for both error metrics, the best results are obtained by using XGBoost, followed by Random Forests and Linear models. This implies that the remaining times in both datasets are better described by the gradient boosting model. Moreover, by looking at the different levels of encoding, we observe an increase of accuracy for both DS1 and DS2, as more inter-case information is added. This can be explained by the hypothesis that there are strong inter-case dependencies for both datasets. An interesting phenomenon is that case interplay is well-captured by non-linear machine learning techniques (i.e., RF and XGBoost), but not by linear models. This can be explained by a non-linear relation between the remaining time of a running case, and the features coming from the bi-dimensional representation. Overall, we can observe that by taking into account inter-case dependencies, the proposed method increases accuracy of remaining times prediction by 27% in *DS1*, and by 52% in *DS2* with respect to LEVEL0 (intra-case only).

Next, we set to investigate the influence of the predicted value (the remaining time) on the error. Specifically, we chose to focus on the best machine learning method (XGBoost), running on our second dataset (DS2). Figure 4 presents the RMSE as function of the remaining times in the test set. Note that we do not provide the same plot with the MAE, as both present similar trends. Observing the plot, we conclude the following results: (1) as expected, for all levels, accuracy is worse for longer remaining times (e.g., $> 1\,000$ hours); (2) few (or none) inter-case features cause that the prediction accuracy increases as we get more information about the running case, although with some oscillations; (3) more inter-case features guarantee that the accuracy stabilizes faster and remains stable also for short remaining times. For instance, LEVEL3 has almost constantly lower RMSE values compared to the other encodings. This result confirms and strengthens the aggregated results shown in Table 1. Therefore, our method meets requirement R2.

Finally, we report that the training phase in our experiments took from few minutes (for LEVEL0) to about 1 h (for LEVEL3) for *DS1* and from few seconds to 5 min for *DS2*. From this, we can conclude that our method also meets requirement R3.

Table 1. Prediction accuracy.

DS	Encoding	Measure	Linear	RF	XGB
DS1	LEVEL0	*RMSE*	213	206	176
		MAE	142	135	113
	LEVEL1	*RMSE*	212	187	151
		MAE	142	123	95
	LEVEL2	*RMSE*	212	177	138
		MAE	142	116	85
	LEVEL3	*RMSE*	211	169	**130**
		MAE	142	105	**80**
DS2	LEVEL0	*RMSE*	256	195	189
		MAE	136	72	67
	LEVEL1	*RMSE*	253	158	142
		MAE	137	58	52
	LEVEL2	*RMSE*	221	121	97
		MAE	137	50	90
	LEVEL3	*RMSE*	216	109	**91**
		MAE	135	42	**33**

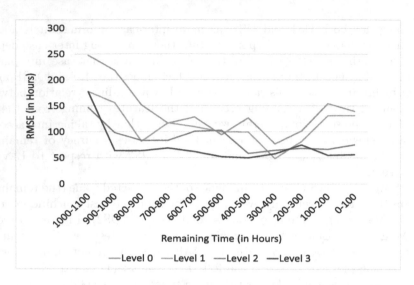

Fig. 4. XGBoost RMSE errors for different predicted remaining times in DS2.

8   Related Work

Predictive process monitoring (PPM) was studied extensively over the past. The methods for solving PPM can be divided into two categories, namely *model-based* and *sequence-to-feature encoding* (STEP). Below, we go over the related work in these two categories.

Model-Based Solutions to Predictive Monitoring. In [2], an annotated transition system is mined from an event log. The model is then used to predict remaining times based on remaining times of historical prefixes. The two main limitations of this approach are: (1) the curse of dimensionality since the transition system 'explodes' in size as the number of process variants grows, and (2) generalization, since the use of historical remaining times without additional features has a poor predictive power. The curse of dimensionality can be controlled by abstracting the transition system state space representation (e.g., by considering bag of activities instead of sequences). To overcome the second limitation, Senderovich et al. [17] proposed the use of a featured transition system, which includes additional contextual attributes such as the queue length, and the current workload. The featured transition system is then used to transform the data into a feature-outcome representation, and non-linear regression methods are applied. In [16], the authors broaden the variety of machine learning techniques that can be applied to the featured transitions system. The model-based solutions balance the trade-off between dimensionality (too many features) and accuracy (too few features) by reducing the model size first. In encoding-based approaches, this trade-off is struck by using the generalization power of machine learning methods to create a compact number of accurate predictors.

STEP solutions to Predictive Monitoring. The seminal work by van Dongen et al. [8] employs non-parametric regression to estimate remaining times. The work lays the ground for all STEP solutions to follow, in that it shows an encoding of an event log into feature-outcome pairs. However, the work is limited to control-flow encoding. The work of Folino et al. [9] proposes to add context features (e.g., resource availability) when predicting processing times. In [14], these approaches are extended and the problem of predictive process monitoring is formalized and generalized. Additional related work suggested a general framework for correlating traces in an event log for prediction of MOIs [12]. However, the authors did not propose a feature space reduction mechanism, and therefore the framework can lead to feature explosion.

Continuing the line of work on encoding sequences into features, Leontjeva et al. [13] introduce the so-called *complex sequence encoding*, which utilizes Hidden-Markov Models to extend control-flow features. Here, separate HMMs are trained for each possible label. Then, the likelihood of a trace prefix to belong to each of these two models is measured. The difference in likelihoods is expressed in terms of odds-ratios, which are then used as features to train the classifier. Further, in [19], a Deep Learning approach that utilizes feature encoding was proposed. The main shortcoming of the STEP-based approaches we outlined above is the assumption that the MOIs depend only on the information payload that accompanies the case for which the prediction is made. However, inter-case dependencies that stem from the interplay between running cases (e.g., competition over scarce resources), may influence the MOIs [21]. In [6], the authors develop a method that is tailored to risk prediction in business process executions, while considering the inter-case dependencies. However, their method does not accommodate for temporal predictions, such as the remaining time of a running case. In our work, we propose a general STEP solution that captures inter-case dependencies for (dynamic) MOIs of an arbitrary type.

9   Conclusion

This paper proposes an adaptive method for feature encoding in predictive process monitoring. The method provides possible feature encodings within a bi-dimensional space characterized by both intra- and inter-case features that can be tailored for specific datasets. The proposed inter-case encodings rely on the concept of partitioning an event log and the running cases according to a discrimination function that differentiates between case types. A derivation function is applied to the resulting partition to limit the feature space size, thus preventing the curse of dimensionality. The proposed encodings have been evaluated against two real-life datasets coming from an emergency department and a manufacturing process. Our experiments demonstrated a 27% improvement for the emergency department event log, and up-to 52% improvement for the production log, when predicting remaining times of running cases.

In the future, we plan to investigate other instances of the proposed method by exploring additional case types, as well as new discrimination, partitioning and derivation functions. Finally, we aim at exploring other dimensions that can affect the future of a running case and their interplay with the intra- and the inter-case ones. One of them is the dimension related to contextual variables, i.e., exogenous variables that are independent of cases but can influence their future such as weather conditions or strikes.

References

1. van der Aalst, W.M.: Process Mining: Data Science in Action. Springer, Heidelberg (2016)
2. van der Aalst, W.M., Schonenberg, M.H., Song, M.: Time prediction based on process mining. Inf. Syst. **36**(2), 450–475 (2011)
3. Breiman, L.: Bagging predictors. Mach. Learn. **24**(2), 123–140 (1996). http://dx.doi.org/10.1023/A:1018054314350
4. Breiman, L.: Random forests. Mach. Learn. **45**(1), 5–32 (2001)
5. Chen, T., Guestrin, C.: XGboost: A scalable tree boosting system. In: Proceedings of the 22nd ACM SIGKDD International Conference on Knowledge Discovery and Data Mining, pp. 785–794. ACM (2016)
6. Conforti, R., de Leoni, M., La Rosa, M., van der Aalst, W.M.P., ter Hofstede, A.H.M.: A recommendation system for predicting risks across multiple business process instances. Decis. Support Syst. **69**, 1–19 (2015)
7. Dietterich, T.G.: Machine learning for sequential data: a review. In: Caelli, T., Amin, A., Duin, R.P.W., Ridder, D., Kamel, M. (eds.) SSPR /SPR 2002. LNCS, vol. 2396, pp. 15–30. Springer, Heidelberg (2002). doi:10.1007/3-540-70659-3_2
8. van Dongen, B.F., Crooy, R.A., van der Aalst, W.M.P.: Cycle time prediction: when will this case finally be finished? In: Meersman, R., Tari, Z. (eds.) OTM 2008. LNCS, vol. 5331, pp. 319–336. Springer, Heidelberg (2008). doi:10.1007/978-3-540-88871-0_22
9. Folino, F., Guarascio, M., Pontieri, L.: Discovering context-aware models for predicting business process performances. In: Meersman, R., Panetto, H., Dillon, T., Rinderle-Ma, S., Dadam, P., Zhou, X., Pearson, S., Ferscha, A., Bergamaschi, S., Cruz, I.F. (eds.) OTM 2012. LNCS, vol. 7565, pp. 287–304. Springer, Heidelberg (2012). doi:10.1007/978-3-642-33606-5_18
10. Friedman, J., Hastie, T., Tibshirani, R.: The Elements of Statistical Learning. Springer Series in Statistics, vol. 1. Springer, Berlin (2001)
11. Friedman, J.H.: Greedy function approximation: a gradient boosting machine. Ann. Stat. **29**(5), 1189–1232 (2001)
12. de Leoni, M., van der Aalst, W.M., Dees, M.: A general process mining framework for correlating, predicting and clustering dynamic behavior based on event logs. Inf. Syst. **56**, 235–257 (2016)
13. Leontjeva, A., Conforti, R., Di Francescomarino, C., Dumas, M., Maggi, F.M.: Complex symbolic sequence encodings for predictive monitoring of business processes. In: Motahari-Nezhad, H.R., Recker, J., Weidlich, M. (eds.) BPM 2015. LNCS, vol. 9253, pp. 297–313. Springer, Cham (2015). doi:10.1007/978-3-319-23063-4_21
14. Maggi, F.M., Di Francescomarino, C., Dumas, M., Ghidini, C.: Predictive monitoring of business processes. In: Jarke, M., Mylopoulos, J., Quix, C., Rolland, C., Manolopoulos, Y., Mouratidis, H., Horkoff, J. (eds.) CAiSE 2014. LNCS, vol. 8484, pp. 457–472. Springer, Cham (2014). doi:10.1007/978-3-319-07881-6_31

15. Pedregosa, F., Varoquaux, G., Gramfort, A., Michel, V., Thirion, B., Grisel, O., Blondel, M., Prettenhofer, P., Weiss, R., Dubourg, V., Vanderplas, J., Passos, A., Cournapeau, D., Brucher, M., Perrot, M., Duchesnay, E.: Scikit-learn: Machine learning in python. J. Mach. Learn. Res. **12**, 2825–2830 (2011)
16. Polato, M., Sperduti, A., Burattin, A., de Leoni, M.: Time and activity sequence prediction of business process instances. CoRR abs/1602.07566 (2016)
17. Senderovich, A., Weidlich, M., Gal, A., Mandelbaum, A.: Queue mining for delay prediction in multi-class service processes. Inf. Syst. **53**, 278–295 (2015)
18. Shalev-Shwartz, S., Ben-David, S.: Understanding Machine Learning: From Theory to Algorithms. Cambridge University Press, New York (2014)
19. Tax, N., Verenich, I., La Rosa, M., Dumas, M.: Predictive business process monitoring with LSTM neural networks. In: Dubois, E., Pohl, K. (eds.) CAiSE 2017. LNCS, vol. 10253, pp. 477–492. Springer, Cham (2017). doi:10.1007/978-3-319-59536-8_30
20. Teinemaa, I., Dumas, M., Maggi, F.M., Di Francescomarino, C.: Predictive business process monitoring with structured and unstructured data. In: La Rosa, M., Loos, P., Pastor, O. (eds.) BPM 2016. LNCS, vol. 9850, pp. 401–417. Springer, Cham (2016). doi:10.1007/978-3-319-45348-4_23
21. Verenich, I.: A general framework for predictive business process monitoring. In: Proceedings of CAiSE 2016 (2016)

Discovering Infrequent Behavioral Patterns in Process Models

David Chapela-Campa[✉], Manuel Mucientes, and Manuel Lama

Centro Singular de Investigación en Tecnoloxías da Información (CiTIUS),
Universidade de Santiago de Compostela, Santiago de Compostela, Spain
{david.chapela,manuel.mucientes,manuel.lama}@usc.es

Abstract. Process mining has focused, among others, on the discovery of frequent behavior with the aim to understand what is mainly happening in a process. Little work has been done involving uncommon behavior, and mostly centered on the detection of anomalies or deviations. But infrequent behavior can be also important for the management of a process, as it can reveal, for instance, an uncommon wrong realization of a part of the process. In this paper, we present WoMine-i, a novel algorithm to retrieve infrequent behavioral patterns from a process model. Our approach searches in a process model extracting structures with sequences, selections, parallels and loops, which are infrequently executed in the logs. This proposal has been validated with a set of synthetic and real process models, and compared with state of the art techniques. Experiments show that WoMine-i can find all types of patterns, extracting information that cannot be mined with the state of the art techniques.

Keywords: Infrequent patterns · Process mining · Process discovery

1 Introduction

One of the aims of process mining during the past years has been, among others, the study of frequent behavior in order to focus on the more common parts of a process during the different tasks of process mining —discovery, monitoring, and enhancement. Under this scope, several algorithms have been proposed to discover process models covering the most common behavior [1–3], and to search frequent structures either directly in the logs [4,5] or extracting the structures from the process models [6]. The search of infrequent cases —deviations or anomalous traces— has been also used during the discovery of a process model, removing them to reduce the complexity of the model without a high decrease in the fitness [7,8]. Nevertheless, the discovery of infrequent behavior can be also interesting in order to monitor and enhance a process, and discarding it might not be a proper solution.

There are scenarios where an infrequent subprocess in the model can hint a wrong behavior which must be examined. For instance, in insurance companies,

© Springer International Publishing AG 2017
J. Carmona et al. (Eds.): BPM 2017, LNCS 10445, pp. 324–340, 2017.
DOI: 10.1007/978-3-319-65000-5_19

infrequent behavior can be used to recognize fraudulent claims [9]. It can be also useful to detect intrusions in networks [10], or failures in software behavior [11]. Additionally, in well-structured processes, the behavior supported by a model is designed and expected to be executed. A substructure of the model with a low frequency of execution can hint a path in the process that must be reinforced in order to increase its frequency or, conversely, where the assigned resources could be restructured to optimize the process.

There are few approaches related with the search of infrequent behavior, most of them focused on the detection of uncommon anomalous traces in process logs [7,8,12,13]. Nevertheless, these techniques focus on the identification of infrequent traces considering the whole trace as a unit. Further knowledge can be obtained searching for infrequent patterns, as they allow to focus on infrequent subprocesses, instead of discovering infrequent full traces —an infrequent trace can contain frequent behavior, hindering the study of infrequent behavior.

In this paper we present WoMine-i, a novel algorithm to detect infrequent behavioral patterns from a process model, measuring their frequency with the instances of the log. The main novelty of our approach is that it can detect infrequent substructures of the process model, i.e., behavioral patterns, with all type of structures —sequences, selections, parallels and loops. The ability to work with these structures prevents WoMine-i of interpreting the traces as sequences of events. Furthermore, the extracted information allows to focus on infrequent subprocesses, and not to analyze infrequent full traces. The algorithm has been qualitatively compared using various synthetic process models with all related techniques, showing our algorithm finds the correct infrequent patterns and estimates precisely its frequency while related techniques do not. Experiments have also been run with real logs of two Business Process Intelligence Challenges, 2012 and 2013.

2   Related Work

There are no approaches in the literature focused directly on the extraction of infrequent substructures in a process. There are a few techniques, like Heat Maps, that can be used to detect infrequent behavior in a process model, although that is not their main objective. Heat Maps provide a simple way to highlight the frequent structures of a process model considering the individual frequency of each arc. If the arcs with a frequency higher than a threshold are removed, the remaining structures are formed by infrequent arcs. The drawback of this approach is that the frequency of each arc is measured individually. An infrequent pattern can be composed by arcs that are individually frequent and, therefore, they will not be part of the result of this technique. Figure 1 shows an example of a process model represented by a C-net, and the result of the Heap Maps technique over this model (Fig. 1a). As can be seen, there are arcs —e.g. $(E \rightarrow G)$— executed in all the traces of the log which are part of an infrequent pattern (Fig. 1b).

In [7], a state automaton with each state representing an activity of the log is built. A valuated arc between two states is added when one of them is followed by the other one in the log. Its value increases as this relation appears in the log.

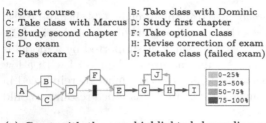

Id	Trace	
0	ABDFEGHI	(×16)
1	ABDEGHI	(×3)
2	ABDFEGHJGHI	(×6)
3	ABDFEGHJGHJGHI	(×4)
4	ABDEGHJGHI	(×8)
5	ABDEGHJGHJGHJGHI	(×11)
6	ACDFEGHI	(×19)
7	ACDEGHI	(×2)
8	ACDFEGHJGHI	(×3)
9	ACDFEGHJGHJGHJGHJGHI	(×8)
10	ACDEGHJGHI	(×11)
11	ACDEGHJGHJGHI	(×9)

(a) C-net with the arcs highlighted depending on its individual frequency (Heat Maps).

(b) Infrequent pattern with a frequency of 5%.

(c) 100-trace log.

Fig. 1. An example of a process model with an infrequent pattern that cannot be discovered through related techniques.

Afterwards, the infrequent arcs are used to filter infrequent traces. The drawback of this technique is the same as the Heat Maps approach, because the frequency is measured individually. Furthermore, the automaton interprets the log as a sequence, without parallels nor other dependencies.

The technique used by Lu et al. in [13] also performs a filtering of traces using the infrequent parts of a process model. In this case, models are built by merging the behavior in a subset of traces. The drawback of this approach is also its inability to measure the frequency of a structure as a whole, analyzing instead the number of individual executions of an arc.

Bezerra et al. search in [12] for infrequent or anomalous traces in the log analyzing the whole trace. They present three approaches to filter infrequent traces depending on their frequency and conformance. The drawback of this technique is that it takes into account the whole trace, and instead of an infrequent pattern, this approximation returns a set of traces. This makes impossible, without further analysis, to know which parts of the traces are infrequent. Figure 1c shows an example: this log contains four instances with a frequency under 5% (1, 3, 7, 8). Two of them (1, 7) contain the pattern from Fig. 1b, but they also contain frequent patterns as A-B-D, A-C-D or G-H-J-G-H-I. Trace-clustering techniques [14] could also be used to obtain traces containing infrequent behavior, but the problem would be the same.

Finally, techniques searching for frequent structures could be adapted to search infrequent behavior, inverting the main search. The drawback of this alternative lies in the way these algorithms measure the frequency of the patterns. For instance, the approach of Tax et al. [5] performs an alignment-based method to detect if the pattern is executed in a trace. When an activity from the trace does not appear in the pattern, the method performs a *move on log* without a penalty because this activity might belong to a parallel branch in the model. This method gets a frequency of 40% for the pattern in Fig. 1b (traces 0, 1, 6, 7) which is far from the real value (5%). The shortcoming of this method is that it analyses the frequency based in the order of the activities in the log, not in the real path that is being executed in the model. Similarly, pattern-based search techniques that only use the log to extract frequent behavior would present the same drawback. Furthermore, the search space of these approaches would be extremely large, as they do not use the model to build the patterns.

As far as we know there is not algorithms to retrieve infrequent patterns from a process model. The algorithm presented in this paper, WoMine-i, is able to retrieve infrequent subgraphs ensuring the low frequency of the entire structure. This allows to focus on infrequent subprocesses and to abstract from the traces containing them, simplifying the analysis of the process.

3 Preliminaries

In this paper, we will represent the examples with place/transition Petri nets [15] due to its comprehensibility. Nevertheless, our algorithm represents the process with a Causal net (Definition 1).

Definition 1 (Causal net [16]). A Causal net (C-net) is a tuple $C = (A, a_i, a_o, D, I, O)$ where:

- A is a finite set of activities;
- $a_i \in A$ is the start activity;
- $a_o \in A$ is the end activity;
- $D \subseteq A \times A$ is the dependency relation, item $AS = \{X \subseteq \mathcal{P}(A) \mid X = \{\emptyset\} \vee \emptyset \notin X\}$;[1]
- $I \in A \to AS$ defines the set of possible input bindings per activity;
- $O \in A \to AS$ defines the set of possible output bindings per activity,

such that:

- $D = \{(a_1, a_2) \in A \times A \mid a_1 \in \bigcup_{as \in I(a_2)} as\}$;
- $D = \{(a_1, a_2) \in A \times A \mid a_2 \in \bigcup_{as \in I(a_1)} as\}$;
- $\{a_i\} = \{a \in A \mid I(a) = \{\emptyset\}\}$;
- $\{a_o\} = \{a \in A \mid O(a) = \{\emptyset\}\}$;
- all activities in the graph (A, D) are on a path from a_i to a_o.

Definition 2 (Trace). Let A be the set of activities of a process model, and ε an event —the execution of an activity $\alpha \in A$. A trace is a list (sequence) $\tau = \varepsilon_1, ..., \varepsilon_n$ of events ε_i occurring at a time index i relative to the other events in τ. Each trace corresponds to an execution of the process, i.e., a process instance.

Definition 3 (Log). An event log $L = [\tau_1, ..., \tau_m]$ is a multiset of traces τ_i. In this simple definition, events only specify the name of the activity, but usually, event logs store more information as timestamps, resources, etc.

---

[1] $\mathcal{P}(A) = \{A' \mid A' \subseteq A\}$ is the powerset of A. Hence, elements of AS are *sets of sets* of activities.

Definition 4 (Pattern). Let $C = (A, a_i, a_o, D, I, O)$ be a C-net representing a process model M. A connected subgraph represented by the C-net $P = (A', A'_i, A'_o, D', I', O')$, where $A'_i \subseteq A'$ and $A'_o \subseteq A'$ represent respectively the start and end activities, is a pattern of M if and only if:

- $A' \subseteq A$;
- $D' \subseteq D$;
- for any $\alpha \in A'$: $I'(\alpha) \subseteq I(\alpha), O'(\alpha) \subseteq O(\alpha)$

A *pattern* (Definition 4) is a subgraph of the process model that represents the behavior of a part of the process. For each activity α in the pattern, its inputs, $I'(\alpha)$, must be a subset of $I(\alpha)$; and the outputs, $O'(\alpha)$, must be also a subset of $O(\alpha)$. This ensures that a pattern has not a partial parallel connection. Figure 2 shows some examples of valid and invalid patterns.

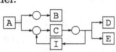

(a) Petri net, with parallels, selections, and a loop.

(b) Valid pattern with a selection and a parallel.

(c) Invalid pattern. J has incomplete input combinations —$\{\{H\}\} \not\subseteq \{\{H,D\}, \{H,E\}\}$.

(d) Valid pattern with a parallel, a selection and a loop.

Fig. 2. Examples of a process model, valid and invalid patterns.

Definition 5 (Simple pattern). A pattern $P = (A', A'_i, A'_o, D', I', O')$ is a simple pattern if and only if, for all activities $\alpha \in A'$:

- $[\exists!\varPhi \in I'(\alpha): \varPhi \not\subseteq R_\alpha^+] \vee [\forall\varPhi \in I'(\alpha): \varPhi \subseteq R^+(\alpha)]$;
- $[\exists!\varTheta \in O'(\alpha): \varTheta \not\subseteq R_\alpha^-] \vee [\forall\varTheta \in O'(\alpha): \varTheta \subseteq R^-(\alpha)]$

Being R_α^+ the set of successors[2] of an activity α, and R_α^- the set of predecessors[3] of an activity α.

The simple patterns (Definition 5) are those patterns whose behavior can be entirely executed in at least one trace. If the inputs or outputs of an activity have a selection, it must be able to execute each path in the same trace —at most, one of the paths is not a loop. For this, the inputs of each activity α must have all activities reachable from α except, at most, the activities of one path. The outputs present the same constraint, but in this case they must reach α, not be reachable from α. Figure 3 shows two valid simple patterns and an invalid one.

---

[2] The successors of an activity α are the activities with a path from α to them, e.g., the successors of B in Fig. 2a are F, G, H and J.

[3] The predecessors of an activity α are the activities with a path from them to α, e.g., the predecessors of C in Fig. 2a are C, I and A.

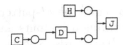

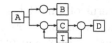

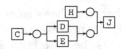

(a) Valid simple pattern. The pattern is executed in the instance [C D H J].

(b) Valid simple pattern with a loop. The pattern is executed in the instance [A B C I C D].

(c) Invalid simple pattern. D and E cannot be in the same instance of the pattern.

Fig. 3. Examples of valid and invalid simple patterns of the process model shown in Fig. 2a.

Definition 6 (Minimal pattern, M-pattern). Each activity of the process model belongs to, at least, one minimal pattern. The M-pattern of an activity α corresponds to the closure of α, i.e., the structure that is going to be executed when α is executed. An exception is made with parallel structures: if α has a parallel in its inputs or outputs, there must be an M-pattern containing each parallel path.

Given a C-net $C = (A, a_i, a_o, D, I, O)$ representing a process model M and an activity $\alpha' \in A$, a pattern $P = (A', A'_i, A'_o, D', I', O')$ is a Minimal Pattern of α' if and only if is a maximum simple pattern containing α' and fulfilling the following rules:

- if $|I(\alpha')| > 1$ then $[I'(\alpha') = \emptyset] \vee [|I'(\alpha')| = 1, \Phi \in I'(\alpha'): |\Phi| > 1]$;
- if $|O(\alpha')| > 1$ then $[O'(\alpha') = \emptyset] \vee [|O'(\alpha')| = 1, \Theta \in O'(\alpha'): |\Theta| > 1]$;
- $\forall \alpha \in R^+_{\alpha'}$: if $|O(\alpha)| \neq 1$ then $O'(\alpha) = \emptyset$;
- $\forall \alpha \in R^-_{\alpha'}$: if $|I(\alpha)| \neq 1$ then $I'(\alpha) = \emptyset$;
- $\forall \alpha \in A', \alpha \neq \alpha', \alpha \notin (R^+_{\alpha'} \bigcup R^-_{\alpha'})$: if $|I(\alpha)| \neq 1$ then $I'(\alpha) = \emptyset$, and if $|O(\alpha)| \neq 1$ then $O'(\alpha) = \emptyset$

In WoMine-i each activity α' is associated, at least, to an M-pattern. The M-patterns of an activity α' are obtained through an expansion process that starts in α' and continues through its inputs and outputs fulfilling the following rules: (i) the process will not expand through the inputs of α' with size 1 and being part of a selection; (ii) the same stands for the outputs of α'; (iii) for all the successors of α' the expansion stops if the outputs are formed by a selection; (iv) the same stands for the inputs of the predecessors of α'; (v) finally, the process does not expand either through the inputs or outputs of the activities not fitting the previous constraints if those are formed by an XOR structure in the model.

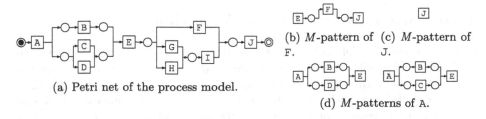

(a) Petri net of the process model.

(b) M-pattern of F.

(c) M-pattern of J.

(d) M-patterns of A.

Fig. 4. A process model and three examples of M-patterns.

Figure 4 shows some M-patterns of a model. Figure 4b shows the M-pattern of F: the process starts in F and expands the M-pattern through F inputs and outputs, because both are formed by only one path. The backwards expansion stops in E because its inputs are part of a selection. Figure 4c depicts the M-pattern of J. It is formed only by itself, because its inputs are part of a selection and its outputs are empty. Finally, Fig. 4d presents the two M-patterns of A. As A is an AND-split with a selection, two M-patterns are created, each one related to one of the possible paths.

Definition 7 (Candidate arcs). Let $C = (A, a_i, a_o, D, I, O)$ be a causal net representing a process model M. An arc $\langle \alpha_i \to \alpha_j \rangle \colon \alpha_i, \alpha_j \in A$ is part of the $A^<$ set, i.e., a candidate arc, if and only if:

- $O(\alpha_i) = \{\Theta \in AS \mid \Theta = \{\alpha_j\} \vee \alpha_j \notin \Theta\}$
- $I(\alpha_j) = \{\Phi \in AS \mid \Phi = \{\alpha_i\} \vee \alpha_i \notin \Phi\}$

The set of candidate arcs, or $A^<$, is a subset of the arcs in the model which are not part of an AND structure. For instance, all arcs of Fig. 4a, but those starting in A or ending in E, are included in the $A^<$ set.

Definition 8 (Compliance). Given a trace $\tau \in L$ and a simple pattern SP belonging to the process model, the trace is compliant with SP, denoted as $SP \vdash \tau$, when the replay of the trace in the process model contains the replay of the pattern, i.e., all the arcs and activities of SP are executed in a correct order, and each activity fires the execution of its output activities in the pattern.

Definition 9 (Frequency of pattern and simple pattern). Let L be the set of traces of the process log. The frequency of a simple pattern SP is the number of traces compliant with SP divided by the size of the log:

$$freq(SP) = \frac{|\{\tau \in L \colon SP \vdash \tau\}|}{|L|} \tag{1}$$

And the frequency of a pattern P is the maximum frequency of the simple patterns it represents:

$$freq(P) = \max_{\forall SP \in P} freq(SP) \tag{2}$$

Definition 10 (Infrequent Pattern). Given a frequency threshold $\sigma \in \mathbb{R} \colon 0 < \sigma \leq 1$, a pattern P is an infrequent pattern if and only if $freq(P) < \sigma$.

4   Infrequent Pattern Mining Algorithm

Given a process model and a set of instances, i.e., traces, the objective is to extract the subgraphs of the process model that are executed in a percentage of the traces under a threshold. A naive approach might be a brute-force algorithm, checking the frequency of every existent subgraph inside the process model, and retrieving the infrequent ones. The computational cost of this approach makes it a non-viable option. The algorithm presented in this paper performs an a priori search[4] starting with the minimal patterns (Definition 6) of the model. In this search, there is an expansion stage done in two ways: (i) adding M-patterns not contained in the current pattern, and (ii) adding arcs of the $A^<$ set (Definition 7). This expansion is followed by a pruning strategy that verifies the upward-closure property of support —also known as monotonicity [17]. This property ensures that if a pattern is infrequent, all patterns containing it will be infrequent and, thus, it is no necessary to continue expanding it —the minimum pattern itself expresses all the infrequent behavior containing it.

This pruning presents an exception in order to simplify the results: if a pattern is infrequent and maintains the value of its frequency with the expansion, it is not removed from the expansion stage —it means the pattern is being expanded with a selection branch with less frequency (cf. Definition 9). In this way, WoMine-i returns the largest patterns expressing the minimum infrequent behavior.

Figure 5 shows an example of a part of the expansion process, assuming a threshold under 40%. The example starts with the M-pattern of C and shows three expansions of the first iteration: the M-pattern of A, one of the M-patterns of I and one of the M-patterns of J. Each of the patterns obtained in the first iteration is again expanded in the second iteration with the M-patterns of J, an M-pattern of I, and the arc $\langle I \rightarrow C \rangle$.

The pseudocode in Algorithm 1 shows the main structure of the search made by the algorithm. First, the candidate arcs and the minimal patterns are initialized (Algorithm 1:2). These M-patterns will be the used to start the iterative process. Then, using the algorithm described in Sect. 5, the infrequent patterns are included in the final set (Algorithm 1:5).

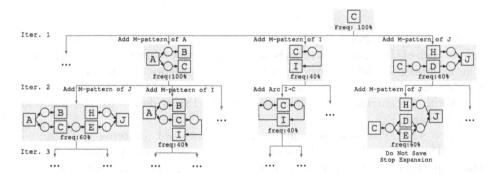

Fig. 5. Example of a part of the expansion process starting with the M-pattern of C. The example shows only three branches of expansion and two iterations. Some of the expansions have been omitted for the sake of clarity.

---

[4] An a priori search uses the previous —a priori— knowledge. It reduces the search space by pruning the exploration of the paths that will not finish in a valuable result.

Algorithm 1. Main structure of WoMine-i.

Input: A process model W, a set $T = \{T_1, T_2, \ldots, T_n\}$ of traces of W and a threshold tr.
Output: A set of maximum infrequent patterns of W w.r.t. T.

```
1   Algorithm infrequentSearch(W, T, tr)
2       M ← {m | m ∈ W, m is an M-pattern } // Def. 6
3       A< ← {a | a ∈ W, a is a Candidate Arc } // Def. 7
4       currentPatt ← M
5       infreqPatt ← {m | m ∈ M, m is infrequent w.r.t. T} // using Alg. 2
6       while currentPatt ≠ ∅ do
7           candPatt ← ∅
8           forall p ∈ currentPatt do
9               candPatt ← candPatt ∪ addArcs(p)
10              complementaryM ← {m | m ∈ M, m ∉ p}
11              forall m ∈ complementaryM do
12              |   candPatt ← candPatt ∪ addMPattern(p, m)
13              end
14          end
15          currentPatt ← filterCandidatePatterns(candPatt, infreqPatt)
16      end
17      Delete the redundant patterns of infreqPatt
18      return infreqPatt
19  Function filterCandidatePatterns(candPatt, infreqPatt)
20      currentPatt ← ∅
21      forall p ∈ candPatt do
22          measure current frequency of p // using Alg. 2
23          if p has no previous frequency || p's frequency has not increased then
24              if p is frequent then
25              |   currentPatt ← currentPatt ∪ p
26              else if p is infrequent then
27                  if p has no previous frequency || p was frequent || p's frequency has
                       maintained then
28                  |   currentPatt ← currentPatt ∪ p
29                  |   infreqPatt ← infreqPatt ∪ p
30                  end
31              end
32          end
33      end
34      return currentPatt
```

Afterwards, the iterative part starts (Algorithm 1:6). In this stage, an expansion of each of the current patterns is done, followed by a filtering of the patterns. The expansion by adding arcs from the $A^<$ set (Algorithm 1:9) is done with the function addArcs. The other expansion, the addition of M-patterns that are not in the current pattern (Algorithm 1:10–13), is done with the function addMPattern.

Once the expansion is completed, the obtained patterns are filtered (Algorithm 1:15) to distinguish the promising from the unpromising ones. Firstly, the frequency of the new pattern is measured, comparing it with the frequency of the pattern before the expansion (Algorithm 1:22). If this expansion has caused its frequency to grow, the pattern is discarded, otherwise the analysis continues (Algorithm 1:23). Then, if the pattern is frequent, it is saved for the next iteration (Algorithm 1:25) —because any frequent pattern can become infrequent by expanding it. And otherwise, if the pattern is infrequent, it is saved in the results as infrequent one (Algorithm 1:28). But, this is only done if (i) it is the first iteration and the pattern has no previous frequency, (ii) the pattern was frequent before the expansion, i.e., it has become infrequent or (*iii*) the frequency has maintained, i.e., the pattern was already infrequent and the expansion has not changed its frequency (Algorithm 1:27).

Finally, once the iterative process finishes, a simplification is made to delete the patterns which provide redundant information (Algorithm 1:17). This redundancy is because there are patterns in the k-th iteration which are expanded and thus are subpatterns of those in the $k + 1$-th iteration. A naive approach to reduce the redundancy generated by the expansion might be to remove the patterns from iteration k-th that are expanded in iteration $k + 1$-th but, with the existence of loops, there is no assurance that the behavior of a pattern is contained in all its superpatterns.

The simplification process consists in the deletion of the patterns that are contained into others, but whose difference is not a loop. For this, each pattern is compared with its previous patterns in the expansion. If the arcs and activities of a pattern are contained into the other, and the difference between them does not contain a complete closed loop, one of the two patterns must be deleted. The subpattern is deleted if its frequency is higher or equal to the frequency of the pattern under analysis. Otherwise the pattern under analysis is deleted.

5   Measuring the Frequency of a Pattern

In each step of the iterative process, WoMine-i reduces the search space by pruning the infrequent patterns (Algorithm 1:15). For this, an algorithm to check the frequency of a pattern is needed (Algorithm 2). Following Definitions 9 and 10, the algorithm generates the simple patterns of a pattern and checks the frequency of each one (Algorithm 2:2–6). After calculating the frequency of the simple patterns, the function checks if this is considered infrequent w.r.t. the threshold (Algorithm 2:12). The frequency of a simple pattern is measured in function getPatternFrequency by parsing all the traces and checking how many of them are compliant with it (Algorithm 2:15–19). Finally, to check if a trace is compliant with a simple pattern, function isTraceCompliant is executed: it goes over the activities in the trace (Algorithm 2:22), replaying its execution in the model, and retrieving the activities that have fired the current one (Algorithm 2:23–24). The simulation (simulateExecutionInPattern) consists in a replay of the trace, checking if the pattern is executed correctly (Algorithm 2:25).

With the current activity —the fired one— and the activities that have fired it —the firing activities, retrieved by the simulation—, the executed activities and arcs are saved, in order to analyze and to detect if the execution of the pattern is being disrupted before it is completed. Figure 6 shows an example of this process. The algorithm starts (#0) with the empty sets of *executed arcs* and *last executed activities*. The first step (#1) executes A. There are no firing activities because A is the initial activity of the process model. As A is also one of the initial activities of the pattern, it is saved correctly in the *last executed activities* set.

The following activity (#2) in the trace is B. As there is only one firing activity (A), a single arc is executed ($\langle A \rightarrow B \rangle$). The arc is added to the *executed*

Algorithm 2. Check if a pattern is infrequent.

Input: A set $T = \{T_1, T_2, \ldots, T_n\}$ of traces, a pattern *pattern* to measure its frequency
w.r.t. T and a threshold to establish the bound of frequency.
Output: A Boolean value indicating if the pattern is infrequent or not.

```
 1  Algorithm isInfrequentPattern(pattern, T, threshold)
 2  |   simplePatterns ← generate the simple patterns of pattern
 3  |   frequencies ← ∅
 4  |   forall simplePattern ∈ simplePatterns do
 5  |   |   frequencies ← frequencies ∪ getPatternFrequency(simplePattern, T)
 6  |   end
 7  |   maxFreq ← 0
 8  |   if frequencies.length > 0 then
 9  |   |   maxFreq ← maximum of frequencies
10  |   end
11  |   realFreq ← maxFreq/T.length
12  |   return realFreq < threshold
13  Function getPatternFrequency(pattern, T)
14  |   executed ← 0
15  |   forall trace ∈ T do
16  |   |   if isTraceCompliant(pattern, trace) then
17  |   |   |   executed ← executed + 1
18  |   |   end
19  |   end
20  |   return executed
21  Function isTraceCompliant(pattern, trace)
22  |   forall activity ∈ trace do
23  |   |   Replay activity in the process model
24  |   |   sources ← get the activities that fired the execution of activity
25  |   |   simulateExecutionInPattern(sources, activity, pattern)
26  |   |   if pattern has been successfully executed then
27  |   |   |   return true
28  |   |   end
29  |   end
30  |   return false
```

---

arcs set, and the activity B to the *last executed activities* set. The A activity is not deleted because the set of outputs is formed by {B, C}, and C is still pending.

The next step, activity E (#3), has the same behavior. There is only one firing activity, i.e., one executed arc. The arc is in the pattern and its source activity is in the *last executed activities* set. Hence, the *executed arcs* set is updated and B replaced by E in the *last executed activities* set. After this process, the following activity is C (#4). Its execution has the same behavior as the execution of B, but with the deletion of A from the *last executed activities*, because the set of outputs {B, C} has been fired.

Finally (#5), F has two firing activities and, thus, two arcs are executed. In both cases, the source activity of the arcs —C and E— is in the *last executed activities* set, and the arc is in the pattern. Thus, a simple addition of F to the *last executed activities* set is done when the last of its branches is executed.

At the end of each step, the algorithm checks if the pattern has been correctly executed (Algorithm 2:26), i.e., all its arcs have been correctly executed and the *last executed activities* set corresponds with the end activities of the pattern (A_o). Unlike the other steps, this testing has a positive result when F is executed. Thus, the trace is compliant with the pattern.

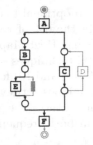

Trace: A B E C F			
Initial activities: {A}			
End activities: {F}			
#	executed activities	executed arcs	last executed activities
0	-	∅	∅
1	A	∅	A
2	B	$\langle A \to B \rangle$	A, B
3	E	$\langle A \to B \rangle, \langle B \to E \rangle$	A, E
4	C	$\langle A \to B \rangle, \langle B \to E \rangle, \langle A \to C \rangle$	E, C
5	F	$\langle A \to B \rangle, \langle B \to E \rangle, \langle A \to C \rangle, \langle C \to F \rangle, \langle E \to F \rangle$	F

(a) Petri net of a process model with a pattern highlighted in black (the un-named activity is an invisible activity).

(b) Check of the execution of a trace for the pattern highlighted in Fig. 6a: '#' is the step of the algorithm; *'executed activity'* is the activity currently executed; *'executed arcs'* is the set with the arcs belonging to the pattern which execution was correctly saved; *'Last executed activities'* is the set of activities which have not fired an entire set of their outputs.

Fig. 6. An example that shows how the algorithm checks if a trace is compliant with a pattern of the process model.

The process of saving the executed arcs and activities has to be restarted when the executed arc is disrupting the execution of the pattern. For instance, in step #5, if the arc $\langle C \to D \rangle$ was executed, this would cause this saving process to go back by removing the arcs and activities of the failed path and to continue with the trace to check if the execution of the pattern is resumed later. This analysis is able to detect the correct execution of a pattern in 1-safe Petri nets[5].

6 Experimentation

In this section we evaluate the performance of WoMine-i. First (Sect. 6.1), we qualitatively compare WoMine-i with the related techniques and, then (Sect. 6.2), we test WoMine-i on four logs from two Business Process Intelligence Challenges. These experiments have been executed in a laptop with an Intel i7-3612QM (2.1 GHz) processor and 8 GB of RAM (1600 MHz)[6].

6.1 Qualitative Comparison Between WoMine-I and the State of the Art Approaches

We present a qualitative comparison between WoMine-i and related techniques through a set of illustrative synthetic models. We have classified the related techniques into three groups: (*i*) individual frequency-based, (*ii*) pattern extraction-based and (*iii*) trace-based.

---

[5] A Petri net is 1-safe when there can be only one mark in a place at the same time.
[6] The algorithm and datasets can be downloaded from http://tec.citius.usc.es/processmining/womine/.

The first process model (Fig. 7) presents several selections, an optional task and a loop. WoMine-i finds the pattern in Fig. 7b appearing in the 6% of the traces. On the contrary, individual frequency-based techniques —e.g. Heat Maps (Fig. 7a)— detect parts of the pattern as frequent. Pattern-based techniques — as Local Process Models— get a frequency of 48% (traces 0, 2, 4, 6, 8 and 10) for the pattern —the correct value is 6%. Finally, trace-based techniques retrieve full traces, being necessary a post analysis to extract infrequent patterns —e.g. traces 0, 2, 5, 7 and 11 have a frequency under 6% but contain both frequent and infrequent behavior.

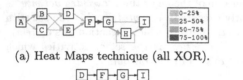

(a) Heat Maps technique (all XOR).

(b) Infrequent pattern (frequency of 6%).

Id	Trace		Id	Trace	
0	ABDFGI	(×4)	6	ACDFGHI	(×12)
1	ABEFGI	(×17)	7	ACEFGHI	(×5)
2	ACDFGI	(×2)	8	ABDFGHHI	(×11)
3	ACEFGI	(×15)	9	ABEFGHHHI	(×7)
4	ABDFGHI	(×10)	10	ACDFGHHHI	(×9)
5	ABEFGHI	(×4)	11	ACEFGHHI	(×4)

(c) 100-trace log.

Fig. 7. Process model, infrequent pattern and event log of a process.

The second example presents a more complex model with loops, parallels and selections (Fig. 8). The approach presented in this paper discovers, with a 5% of frequency, an infrequent pattern denoting as uncommon the execution of the C–E parallel structure after the loop of G–H. As can be seen, based in the individual frequency of the arcs is impossible to extract this infrequent behavior. Local Process Models can extract this pattern successfully but the obtained frequency is not reliable. Also, the search space is larger because they do not rely on the process model. Trace-based techniques present the same problem as in the previous example but, as traces are longer, the post analysis becomes more difficult.

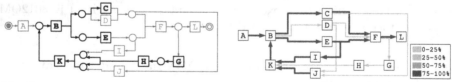

(a) Infrequent pattern detected by WoMine-i (frequency 5%).

(b) Heat Maps technique applied to the process model.

Fig. 8. Results of WoMine-i and Heat Maps for a process model composed by a sequence with a selection, and two loops.

6.2   Infrequent Patterns for the BPI Challenges

The objective of this section is twofold: on the one hand, to test WoMine-i on complex real logs from the Business Process Intelligence Challenge (BPIC)[7,8,9] demonstrating the ability to retrieve all type of structures and, on the other hand, to analyze the influence of the model in the retrieved patterns. We used 4 BPIC logs, and we mined the process models with two different discovery algorithms, ProDiGen (PDG) [3] and the Inductive Miner (IM) [2].

A series of experiments have been run for these logs and models with different thresholds. Table 1 shows the structural characteristics of the mined infrequent patterns for a threshold of 5%. As explained in Sect. 5, the algorithm needs to replay the trace in the model to retrieve the executed arcs. This process is independent of the threshold —it only depends on the traces (log) and on the model. Thus, the runtime is divided in two parts to distinguish this preprocessing time and the time spent by the algorithm. As can be seen, WoMine-i is able to retrieve infrequent patterns with all type of structures. Regarding the runtime, the preprocessing time is short, being 208 ms the longest time. The time spent by the algorithm is longer, and depends on the model and patterns extracted. Log *2012_o* shows a difference in the runtime due to the number of patterns extracted —a model with more uncommon structures will return as infrequent this behavior, increasing the runtime. Nevertheless, the other runtimes are under 20 s (*2012_a*), and 7 s (*2013*).

Besides, we have compared the number of patterns discovered for the PDG and the IM models. As can be seen, except for one log (*2012_a*), the algorithm retrieves more patterns with the PDG model than with the IM one. This is due to the structure of the models: the higher number of relations in the IM model allows to embrace more infrequent behavior with few small patterns, while with the PDG model is necessary to build larger patterns —smaller patterns with IM. Results with *2012_a* show a case where the infrequent patterns represent behavior not recorded in the log, but allowed by the models —frequency 0.

Figure 9 shows an example of a pattern extracted by WoMine-i from the PDG model of the *BPIC 2012_o* log, which corresponds to a Dutch Financial Institute. The extracted pattern appears in the 2% of the traces, and models the selection of a procedure, followed by the creation and shipment of it, and ended by sending it back and canceling the procedure, but with a return to the selection, instead of a finalization of the instance. This behavior might be from a illegal execution where the procedure is restarted after a cancellation, while the normal execution should be the ending of it. Trace-based approaches extract complete traces —the traces of the log have up to 35 activities—, hindering the identification of the pattern. On the other hand, pattern-based approaches might

---

[7] BPIC 2012 - 10.4121/uuid:3926db30-f712-4394-aebc-75976070e91f. This dataset has been split into two logs: 2012_a contains the events related with the state of an application process, while 2012_o has the events related with the state of an offer belonging to an application process.

[8] BPIC 2013 clo - 10.4121/uuid:c2c3b154-ab26-4b31-a0e8-8f2350ddac11.

[9] BPIC 2013 op - 10.4121/uuid:3537c19d-6c64-4b1d-815d-915ab0e479da.

Table 1. Behavioral structure of the infrequent patterns extracted for a threshold of 5% from the process models of the BPICs. It shows the results for two process models (ProDiGen and Inductive Miner) on each log.

			Threshold: 5%								
			Runtime (secs)		#patt	frequency	#activities	#sequences	#choices	#parallels	#loops
			pre	alg							
PDG	*2012	a	0.208	16.439	1	0±0	11.00±0.00	2.00±0.00	4.00±0.00	0±0	0±0
		o	0.202	343.106	21	0.02±0.01	6.67±0.91	0.52±0.51	3.33±1.62	0±0	0.38±0.50
	2013	clo	0.056	0.700	3	0.03±0.02	2.00±1.73	0±0	0±0	0.33±0.58	0.33±0.58
		op	0.036	4.806	12	0.02±0.02	6.00±0.43	0±0	1.58±1.08	1.50±0.52	1.50±1.09
IM	2012	a	0.208	16.766	2	0±0	10±0	1.50±0.71	1.50±2.12	0±0	0±0
		o	0.202	79.027	4	0.02±0.02	5.25±2.87	0±0	2.25±3.30	1.75±1.26	1.25±1.89
	2013	clo	0.056	6.630	1	0.01±0.00	1.00±0.00	0±0	0±0	0±0	0±0
		op	0.036	0.474	5	0.02±0.01	3.00±2.74	0±0	0±0	0.40±0.55	0.60±0.55

consider the infrequent pattern as executed although other activities, that are not part of the pattern, are executed before the end of the it.

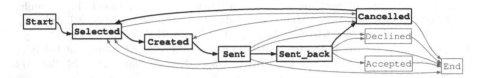

Fig. 9. Infrequent pattern (2%) retrieved from the BPIC 2012_o. All relations are selections (XOR)

7   Conclusion and Future Work

We have presented WoMine-i, an algorithm designed to search infrequent behavioral patterns in an already discovered process model, being able to discover patterns with the most common control structures, including loops. This structures allow to discover, for instance, subprocesses executed less than the expected, or uncommon wrong behavior. We have compared WoMine-i with other proposals, showing that our approach discovers uncommon behavior that other techniques are not able to detect. Moreover, we have also tested our algorithm with complex real logs from the BPICs. Results show the importance of the infrequent patterns to analyze and optimize the process model.

Acknowledgments. This research was supported by the Spanish Ministry of Economy and Competitiveness (grant TIN2014-56633-C3-1-R) and the Galician Ministry of Education, Culture and Universities (grants GRC2014/030 and accreditation 2016-2019, ED431G/08). These grants are co-funded by the European Regional Development Fund (ERDF/FEDER program).

References

1. Weijters, A., van Der Aalst, W.M., De Medeiros, A.A.: Process mining with the heuristics miner-algorithm. Technische Universiteit Eindhoven, Technical report WP 166 1–34 (2006)
2. Leemans, S.J.J., Fahland, D., Aalst, W.M.P.: Discovering block-structured process models from event logs - a constructive approach. In: Colom, J.-M., Desel, J. (eds.) PETRI NETS 2013. LNCS, vol. 7927, pp. 311–329. Springer, Heidelberg (2013). doi:10.1007/978-3-642-38697-8_17
3. Vázquez-Barreiros, B., Mucientes, M., Lama, M.: Prodigen: mining complete, precise and minimal structure process models with a genetic algorithm. Inf. Sci. **294**, 315–333 (2015)
4. Han, J., Pei, J., Mortazavi-Asl, B., Pinto, H., Chen, Q., Dayal, U., Hsu, M.: Prefixspan: mining sequential patterns efficiently by prefix-projected pattern growth. In: Proceedings of the 17th International Conference on Data Engineering, pp. 215–224 (2001)

5. Tax, N., Sidorova, N., Haakma, R., van der Aalst, W.M.: Mining local process models. J. Innov. Digit. Ecosyst. **3**, 183–196 (2016)
6. Greco, G., Guzzo, A., Manco, G., Pontieri, L., Saccà, D.: Mining constrained graphs: the case of workflow systems. In: Boulicaut, J.-F., Raedt, L., Mannila, H. (eds.) Constraint-Based Mining and Inductive Databases. LNCS, vol. 3848, pp. 155–171. Springer, Heidelberg (2006). doi:10.1007/11615576_8
7. Conforti, R., La Rosa, M., ter Hofstede, A.H.: Filtering out infrequent behavior from business process event logs. IEEE Trans. Knowl. Data Eng. **29**, 300–314 (2016)
8. Ghionna, L., Greco, G., Guzzo, A., Pontieri, L.: Outlier detection techniques for process mining applications. In: An, A., Matwin, S., Raś, Z.W., Ślęzak, D. (eds.) ISMIS 2008. LNCS, vol. 4994, pp. 150–159. Springer, Heidelberg (2008). doi:10.1007/978-3-540-68123-6_17
9. Yang, W.S., Hwang, S.Y.: A process-mining framework for the detection of health-care fraud and abuse. Expert Syst. Appl. **31**(1), 56–68 (2006)
10. Münz, G., Li, S., Carle, G.: Traffic anomaly detection using k-means clustering. In: GI/ITG Workshop MMBnet (2007)
11. Lo, D., Cheng, H., Han, J., Khoo, S.C., Sun, C.: Classification of software behaviors for failure detection: a discriminative pattern mining approach. In: Proceedings of the 15th ACM SIGKDD International Conference on Knowledge Discovery and Data Mining, pp. 557–566. ACM (2009)
12. Bezerra, F., Wainer, J.: Algorithms for anomaly detection of traces in logs of process aware information systems. Inf. Syst. **38**(1), 33–44 (2013)
13. Lu, X., Fahland, D., Biggelaar, F.J.H.M., Aalst, W.M.P.: Detecting deviating behaviors without models. In: Reichert, M., Reijers, H.A. (eds.) BPM 2015. LNBIP, vol. 256, pp. 126–139. Springer, Cham (2016). doi:10.1007/978-3-319-42887-1_11
14. De Weerdt, J., vanden Broucke, S., Vanthienen, J., Baesens, B.: Active trace clustering for improved process discovery. IEEE Trans. Knowl. Data Eng. **25**(12), 2708–2720 (2013)
15. Desel, J., Reisig, W.: Place/transition petri nets. In: Reisig, W., Rozenberg, G. (eds.) ACPN 1996. LNCS, vol. 1491, pp. 122–173. Springer, Heidelberg (1998). doi:10.1007/3-540-65306-6_15
16. Aalst, W., Adriansyah, A., Dongen, B.: Causal nets: a modeling language tailored towards process discovery. In: Katoen, J.-P., König, B. (eds.) CONCUR 2011. LNCS, vol. 6901, pp. 28–42. Springer, Heidelberg (2011). doi:10.1007/978-3-642-23217-6_3
17. Leung, C.K.S.: Monotone Constraints. In: Liu, L., Özsu, M.T. (eds.) Encyclopedia of Database Systems, pp. 1769–1769. Springer, Boston (2009)

Author Index

Printed in the United States
By Bookmasters